Technologies for Electrical Power Conversion, Efficiency, and Distribution:
Methods and Processes

Mihail Hristov Antchev
Technical University of Sofia, Bulgaria

ENGINEERING SCIENCE REFERENCE
Hershey · New York

Director of Editorial Content: Kristin Klinger
Senior Managing Editor: Jamie Snavely
Assistant Managing Editor: Michael Brehm
Publishing Assistant: Sean Woznicki
Typesetter: Carole Coulson, Kurt Smith, Jamie Snavely
Cover Design: Lisa Tosheff
Printed at: Yurchak Printing Inc.

Published in the United States of America by
Engineering Science Reference (an imprint of IGI Global)
701 E. Chocolate Avenue
Hershey PA 17033
Tel: 717-533-8845
Fax: 717-533-8661
E-mail: cust@igi-global.com
Web site: http://www.igi-global.com/reference

Library of Congress Cataloging-in-Publication Data

Antchev, Mihail Hristov, 1955-
Technologies for electrical power conversion, efficiency, and distribution : methods and processes / by Mihail Hristov Antchev.
p. cm.
Summary: "This book presents an overall description of electrical energy conversion technologies and required power electronic converters"--Provided by publisher.
Includes bibliographical references and index.
ISBN 978-1-61520-647-6 (hardcover) -- ISBN 978-1-61520-648-3 (ebook) 1. Electric power. 2. Energy conversion. 3. Electric power production. I. Title.
TK1001.A586 2010
621.31--dc22
2009036154

British Cataloguing in Publication Data
A Cataloguing in Publication record for this book is available from the British Library.

Table of Contents

Section 2
Electronic Energy Converters

Section 3
Applications of Electronic Energy Converters

Foreword

During recent years the attention of many of the scientists and researchers is paid to processes of generation, conversion and storage of energy and to not only technical problems, but also global problems concerning the population of the Earth connected to these processes. A trend to increase the share of the converted electrical energy of total energy balance in the world is shown. In this book, the author reviews the processes of conversion electrical energy by the use of electronic switches. The author presents different power electronic converters realizing conversion of AC or DC energy into AC or DC energy. Principles of implementation of their control systems are also listed. Furthermore, the description of practically approved methods and schemas is followed by a presentation of known newer solutions which are with a perspective to enhance their future implementation. Special attention is paid to the topics connected with the increase of power efficiency of power electronic converters, as well as to methods and schemas used. Applications of electronic converters and systems in industry, transportation and everyday life are analyzed. A special part of the book is dedicated to uninterruptible power supplies.

Nature and structure of the material presented permit this book to be useful both in education in electrical and electronic engineering and in solving practical problems by specialists in these fields.

The problems with generation, transmission and conversion of electrical energy have not only technical aspects but also economical and humanitarian aspects. The character of Power Electronics as a quickly and successfully developing scientific field is interdisciplinary. A problem in this field to be solved successfully often requires combine efforts of different specialists. To this effect, the style and way of the presentation make the book accessible to a wider reader's circle that is interesting in the questions of electronically conversion of electrical energy and the methods to increase the power efficiency during this conversion.

Professor Stefan Tabakov, PhD

Stefan Tabakov *has BS and MS degrees in electronic engineering. He received his PhD degree in power electronics from the Moscow Power University, Russia. During his length of service, he was the department chief of the Power Electronics Department and he was also the dean of the Faculty of Electronic Engineering and Technologies. Currently, he is a professor with the Power Electronics Department, the president of the Electronics, Electrical Engineering and Communication Union with Federation of Scientific - Technical Unions, Bulgaria, and the chief editor of transaction Electrical Engineering and Electronics. Professor Tabakov has more than 130 publications in the field of power electronics, 10 patents, more than 20 books and manuals on electronics. He has worked upon over 40 contracts – both research and implanted contracts.*

Preface

Today, basic energy sources are fossil fuels – coal, oil, and natural gas. Coal was the most important energy source in 18th century, while coal and natural gas emerged at the end of 19th century. These energy sources provide 90% of energy in the world. They are formed as a result of plants and animals rotting for thousands of years. Their basic feature is their quicker depletion compared to their production. According to different studies, supplies of oil and natural gas will be enough for the following 40-60 years, while the supplies of coal for the next 200 years. Burning these solid and liquid fuels causes pollution and other harmful effects on ambient air. A continuous increase of average temperatures in the world is a statistical fact and it is called global warming. A main agent for global warming is enhanced concentration of carbon dioxide in the air. This concentration causes an effect of slimming particular parts of the the Earth's ozone layer. Results of this slimming are a rise in sea level, people suffering from temperature stress, gradual extinction of different kinds of animals, crop destruction, spread of tropical diseases, etc. Another harmful effect of the usage of the abovementioned fuels is acid rains. They cause fatal destruction of woods, flora and fauna in water pools, as well as gradual destruction of buildings and century-old cultural monuments. As a result of burning the before-mentioned energy sources, air quality in big cities all around the Earth has worsened. After the mid-20th century, nuclear fuel was considered a potential energy source. At the beginning, it was thought that it would entirely replace all other fuels. Gradually, these forecasts have become less optimistic because the supply of uranium is limited and there have also been accidents at nuclear plants worldwide. One of the main problems in using nuclear energy is the problem with a deposit of nuclear scraps.

Since the population of the Earth increases, there are more people who need energy. Also, of course, the demand of energy continues to increase (IEA, 2006). Recently, special attention has been paid to the use of renewable energy sources – sun, air, water, geothermic energy, bio mass. The increase in using renewable source energy is expected to decrease the hothouse gases and cause an improvement in air quality. For the time being, the main emphasis has been placed upon the positive effects of renewable sources. There is a possibility that they can also have negative effects but until this stage of usage, it is impossible to assess them.

The development of science and technology is not only an integral part of advancing human civilization but is also, in a great degree, a formative part of its progress. The discovery of electrical energy started a new era of the human progress. A need to convert energy of different sources into electrical energy has arisen. Every day we witness more and more significant achievements where natural resources are used, for example in spacecraft. Recently, there has been rapid development of information and communication technologies connected to the possibility of processing a large quantity of data with high speed, and of data transfer around the Earth and into space. Furthermore, we observe development in the field of material science a result of new treatment technologies. Moreover, achievements in using the contemporary computer, as well as communication and electronic technologies in medicine are extremely significant. Operation of the technical means and equipment applied in the above mentioned fields is impossible without using electrical energy at current stage of human development. Electrical energy usage for different needs increases and, at the same time, traditional energy carriers are replaced (Bertoldi, 2007). An example is a car with hybrid motion.

Individuals are accustomed to methods and processes and even to the most ordinary aspects of electrical energy that they consider them to be something that has always existed. Sometimes, in those moments when a common way of their lives changes as a result of absence of electrical energy, they have to question this assumption. Fortunately, these moments are, for the greater portion of the Earth's population, very rare and they are quickly forgotten. Individuals in different countries have connected their professional lives with the processes of generation, distribution, conversion and consumption of electrical energy for variety of needs. Knowledge in the field of electrical energy has been passed from generation to generation of specialists. This handing over process will also continue in the future. Recently, there has been increased activity in scientific research connected to new energy sources and to the increase of power efficiency. Different research programs concerning new energy sources are financed at international and national levels. The importance of solving tasks is proved by the fact that funds come not only from medium and large enterprises but also from different governments. In many countries, special groups, such as committees, agencies, etc, are founded to operate in the field of power efficiency.

The primary equipment connected to human activity in industry and everyday life operates using electrical energy from the electrical supply network. Most consumers are sensitive to changes of power quality. Its deviations cause undesired results such as fault operation. The consequences are unpredictable and relevant not only to economic resources and time but sometimes also to humans. Such critical consumers are everywhere around us – at technological process control, banks, hospitals, airports, telecommunication apparatus, in warcraft, during obtaining oil or using atom energy, in space, etc. To secure equipment's flawless operation, different solutions have been applied and are applied. One of the most popular solutions is the use of uninterruptible power systems.

The processes of generation of electrical energy, its conversion and distribution, and its storage and rational use again as electrical or other types of energy of a certain quality and sometimes as an uninterrupted source, are a question of continuous study and improvement. A major share in these processes is taken by Power Electronics, whose objects are power electronic converters. The converters are complicated technical products and they are basically a combination of power schema, based on power semiconductor switches, and a control system. Generally, at the input of the power electronic converters AC or DC power, generated in different ways, is applied. Moreover, the power electronic converters have to give at their outputs AC or DC power in accordance with a load type. Proceedings with different orientations in the field of electrical energy conversion are available at the market. Most of them are strictly specialized and they scrutinize topics connected with the converters' operation.

Under the formed trends of enlarging the usage of energy as electrical energy, the problems to increase power efficiency have appeared to be of great importance. Usually, these problems are topics of articles published at specialized proceedings and conference papers. They present a particular solution of a problem.

The purpose of this book is to present to readers an overall description of electrical energy conversion technologies and required power electronic converters. During the description, special terms and specific methods, which require a preliminary background, are used. But the main attention is paid to indicators of these converters that determine power efficiency. From this point of view during every discussion, contemporary achievements and emerging trends to increase power efficiency of the power electronic converters are presented. The operation of their control systems is of great importance to proprer operations of converters. Chosen control algorithm and used element base for converter implementation mainly determine its whole operational reliability. The control systems have different special features depending on the electrical energy converter type, source and load. Thus, it is necessary to allocate space to discuss them during illustration of the particular converters. After the reader has become acquainted with technologies for electrical energy conversion and equipment needed, it could be interesting for him or her to discover where and how they are implemented in different fields, for example in industry, transportation, home appliances, etc., also, to find out the possibilities of these converters to convert energy of the renewable energy sources and what are the trends for the conversion.

Developing processes of globalization and free exchange of information could be considered not only as processes between scientists and specialists in separated fields, working in different parts of the World, but also as processes requiring the information of strictly specific fields to be made accessible to people working in other fields without material similarity among them. Usually, the deep knowledge in the field of energy conversion requires specific preparation which can be obtained at the existing specialized technical universities. Very often, graduates from different levels of engineering study successfully communicate among themselves, but it appears as if they are isolated from people working at humanitarian or economic fields. Therefore, this book is written in a simple and easy way that will make it understandable for specialists in fields other than n engineering. This book will contribute to enhancing the possibilities of communication about the problems of power efficiency at energy conversion among a variety of specialists. Perhaps new ideas, whose solutions may be found in the future, can arise. I have reached these conclusions as a result of service for many years as a lecturer and researcher in an extremely interesting field – Power Electronics. To be able to make essential achievements in this field, mankind should incorporate the efforts of a variety of specialists because of Power Electronics' interdisciplinary character. Guided by these thoughts and ideas, I decided to present the information in the book in the following basic sections:

Chapter 1 ***"Energy and Energy Efficiency"*** identifies the present state and trends of the use of electrical energy in emerging and developing and advanced countries. The trends are determined by the economic activity of the emerging and developing countries and by geopolitical plans of the advanced countries and also by those countries which have basic energy resources. History of the recent past is presented. The forecast for development of the World Energy Market divided by different energy source is included. There is a discussion on the use of these sources per basic sectors - transportation, industry, commercial and residential sectors. Possible ways to increase energy efficiency during the processes of producing, transportation and using energy are described.

Chapter 2 ***"Storage and Usage of Energy"*** reviews ways and technical means to store energy as electrochemical, electromagnetic, electrostatic and mechanical energy. Storage of energy as different kinds is enforced because it is impossible to always consume energy at the moment of its generation. Also, in different occasions, the produced energy should be stored as a reserve one. The trends and forecasts show an increase of energy used as electrical energy. Therefore, energy conversion is required in accordance with a source – AC or DC electrical power, and with a consumer's type – AC or DC load.

Chapter 3 ***"Power Electronics and Its Role in Effective Conversion of Electrical Energy"*** treats the large share of Power Electronics into the energy conversion. Power Electronics' ability to increase power efficiency during the stage of transmission and use of electrical energy is discussed. Power electronic converters are basically a combination of power schema, based on power semiconductor switches, and a control system. The power electronic converters are complicated devices with a complex character; thus, when designing and implementing them, the knowledge and experience of experts in different fields of science and technique are required. Emphasis is placed the upon interdisciplinary nature of Power Electronics. The reader can find a description of a long process of implementation of a power electronic converter from a scientific idea to its batch production. Finally, I pay attention to the use of computers in simulating the operation of power electronic converters and the study of different physics processes. A review of available software for simulation and study is made.

Chapter 4 ***"AC/DC Conversion"*** scrutinizes schemas and operational principles of single-phase and three-phase uncontrolled rectifiers, single-phase and three-phase controlled rectifiers, as well as bidirectional converters when an AC source is available. Emphasis is placed upon the indicators of the converters in respect to the source – power factor and harmonic spectra of source current, also, upon different methods to increase the power efficiency in AC/DC conversion. Requirements and structural schemas to implement control systems of AC/DC converters are presented.

Chapter 5 ***"AC/AC Conversion"*** presents schemas and operational principles of single-phase and three-phase AC regulators. There is also a review of the indicators of the regulators that determine the electromagnetic compatibility to the source network – power factor and harmonic spectra of the consumed current. As this chapter concludes, methods and schemas to increase power efficiency in AC/AC conversion are presented. Requirements and structural schemas to implement control systems of AC/AC converters are also presented.

Chapter 6 ***"DC/DC Conversion"*** analyses schemas and specific operational characters of converters without galvanic isolation - buck, boost, buck-boost inverting, Cuck-converter, SEPIC-converter, followed by the analysis of galvanic isolated converters - forward, flyback, push-pull, half bridge and full bridge converters. Then, a review of schemas and operational principle of bidirectional DC/DC converter is made. Basic special characteristics, which determine converter efficiency coefficient, are defined. Methods to increase the efficiency coefficient and to improve electromagnetic compatibility in respect to the source and ambient area are shown. Moreover, Chapter 6 reviews schemas of DC/DC converters with a soft commutation, which uses a resonance operational principle. Requirements and structural schemas to implement control systems of DC/DC converters are presented.

Chapter 7 ***"DC/AC Conversion"*** is a description of schemas and operational principles of single-phase and three-phase converters, which have AC output when a DC input voltage is applied. These converters are known as inverters. Different control methods used to improve harmonic spectra of the output voltage are studied. A relationship among the basic indicators of the inverters and the chosen control method, as well as the methods to improve these indicators, is described. Requirements and structural schemas to implement control systems of DC/AC converters are presented.

Chapter 8 ***"Conversion of Electrical Energy in the Processes of Its Generation and Transmission"*** describes power electronic converters used in the processes of generating and transmitting electrical energy at a distance. Methods and schemas used in HVDC transmission, as well as methods to increase the power efficiency under this transmission, are presented. Furthermore, I describe the possible decisions for compensating the reactive power and improving current harmonic spectra in already implied equipment – static compensator of reactive power, active power filters and unified power quality conditioner.

Chapter 9 ***"Conversion of Electrical Power from Renewable Energy Sources"*** discusses the use of energy of renewable sources – sun and wind. Energy conversion requires different types of converters to be combined. The choice of the converters and their compatibility defines the power efficiency. Here, I show the special characteristics of the converters for renewable energy sources – photovoltaics, windgenerators, and also I show their basic characteristics. In addition, I also describe the principle of distributed generation of electrical energy and special operation characteristics of final output inverters connected to the source distribution network. Recently, together with these alternative sources for generation of electrical energy, generation of energy using fuel cells has become a question of a great interest. The final fact is valid not only for the automobile industry but also for other needs. This is the reason why the special features of the converters, which can be used with fuel cells, are listed. The probabilities to increase converter power efficiency are also listed. In the last section of this chapter, different structural schematics of converters of water energy applicable in small Hydro Power Plants are included.

Chapter 10 ***"Uninterruptible Power Supply Systems"*** analyses methods and schemas to insure uninterruptible supply for so-called critical consumers, which are crucial to the quality of AC source voltage. Possible disturbances of the quality are examined. Uninterruptible power supply systems (UPS) are a combination of an energy storage element and different converters for this energy. The converters convert the energy producing AC output voltage. Principles to implement dynamic and static UPS are included. I describe principles to implement passive standby, line interactive and double conversion UPS. Also, I analyze methods to implement supply systems with enhanced reliability using several AC voltage sources and UPS. The communication between UPS and other systems is also represented.

Chapter 11 ***"Other Applications of Converters and Systems of Converters"*** presents applications of converters and systems of converters in the following fields: industry, transportation, home appliances and communication technologies. Induction heating, welding, electrolysis, and air purifying are discussed from the industrial applications. Chapter 11 describes the converters applied for the needs of different vehicles. Contemporary energy efficient lighting also requires energy converters. Methods and schemas for lightning implementation are included in Chapter 11. Electrical energy converters applied in different home appliances are shown. Furthermore, electrical energy converters ensuring the operating reliability of different applications in communication technologies are discussed. Finally, special attention is paid to the requirements for electrical energy converters in medicine.

Chapter12 ***"State-of-the-Art Review on Power Electronics"*** aims at introducing several of the most recent achievements and research directions in power electronics field to the reader. Thus, readers who are already acquainted with the content of the book up to this chapter have an additional opportunity to see the basic trends in development in this interesting field. Information from scientific publications as well as from different materials published by companies known for their practical expansions are used for the conclusions made.

The Appendix includes harmonic analysis of some periodical functions applicable in power electronics. The analysis is made to facilitate the reader studying the different parts of the book. Also, probabilities of using several software programs to make analyses in power electronics are shown.

It is obvious from the contents of the chapters that the reader can get acquainted with many different features of electrical energy conversion after thoroughly reading the presented book. It is recommended to read this book according to the order of the presentation, chapter after chapter. Readers who have theoretical and practical backgrounds can find individual topics for them in the book by reading only selected parts of it. Readers who do not have engineering education should read Chapters 1, 2, 3, 8, 9, 10, 11 and 12. This book gives basic knowledge of electrical energy conversion and also emphasizes methods and means to increase power efficiency at that conversion. In this way, this book succeeds in its general aim – to be useful for a wide circle of readers working in different fields. I am pleased to bring it to the readers' attention and hope that its reading will content them.

Mihail Antchev

REFERENCES

International Energy Agency IEA (2006). *Key world energy statistics 2006*. Retrieved 2008 from http://www.iea.org

Bertoldi, P., & Atanasiu, B. (2007). *Electricity consumption and efficiency trends in the enlarged European Union - Status Report 2006.* Luxembourg: Office for Official Publications of the European Communities

Acknowledgment

The author would like to acknowledge the kind invitation received from IGI Global team, especially from Ms. Jan Travers, to write the published material. The publication of the book could not been satisfactorily completed without the hard work done by the team of IGI Global and extremely useful help of Ms. Julia Mosemann – Development Editor. With a great pleasure and gratitude, the author wants to underline the discussions of ideas and the work done during the preparation of the manuscript by PhD eng Mariya Petkova, Assistant Professor. The author also would like to acknowledge the recommendations made by his older colleague and foreword author PhD Stefan Tabakov, Professor

Moreover, I want to express my gratitude to all organizations, scientific institutes and companies, which gave their assents to include their materials in this book.

Special thanks to my family - Pepi, Hristo, Slavka and Stefka, for their moral support during the preparation of the manuscript.

I dedicate this book to my family and all my ex, present and future students.

Section 1
Energy, Conversion and Storage of Energy

Chapter 1
Energy and Energy Efficiency

ENERGY SOURCES

Originally, coal was the main source of energy. It remains so throughout the 18th century during the period of the rapid industry development. Later on, oil and naphtha began to be used as energy sources and their usage expanded especially in 19th century. A special feature of the above mentioned fossil fuels is their long creation period – requiring millennia. They are a result of rotting of different plant and animal kinds. In comparison to the period of their formation, the period of their utilization is far shorter. In accordance with a number of existing statistics about 2050 year it may be talked about a depletion of the liquid fossil fuels, also, the world coal supplies are considered to last within the next 200 years. Therefore, the development of nuclear power engineering is considered to be one of the alternatives to generate energy. Recently, the nuclear power energy generation has been denied in many countries because of the risks associated with its generation and because these risks have been confirmed by serious accidents throughout the World. The storage of worked nuclear waste is also a problem and risky. The renewable energy sources are another possibility to generate energy.

DOI: 10.4018/978-1-61520-647-6.ch001

Figure 1. Energy sources

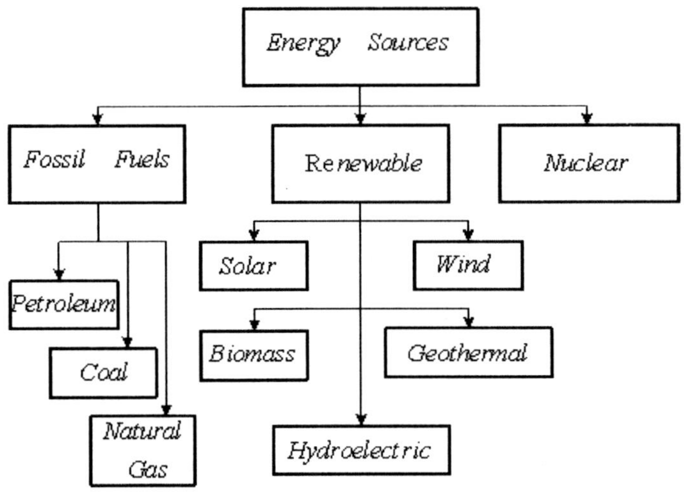

Different kinds of energy sources known at present are shown in Figure 1.

Significant achievements have been made in the 20th century. Many of them need additional supplies of energy. The needs are especially high in the developing countries.

The following questions are related to this point:

- How does the energy consumption change worldwide and what are the trends among branches and countries, and what are the future needs?
- What is the balance among the main energy sources, what are the forecasts for their development?
- How does the use of different energy sources reflect on the pollution of the environment?
- What are the possibilities to increase the energy efficiency of the different sources?
- Will the pollution of the environment continue and what should we expect for decreasing the harmful emissions?

The contemporary state and the forecasts for energy needs according to a specific kind of sources in the World are presented in Figure 2. It is evident that the necessity natural gas and other renewables keep on increasing. Detailed data for different countries can be found in (EIA, 2008).

Figure 3 shows the shares of the energy consumption per branches towards 2005 year (IEA, 2008).

It is seen that the basic consumption of energy covers industrial needs – 33% and it should be mentioned that the household consumption follows it closely – 29%. In the third place is energy consumed for transportation – 26%. Based on the data provided by (IEA, 2008) there is an increase in the energy consumption of 23% between years 1990 and 2005. The biggest increase of energy consumption is in the branches of transportation and services – a total of 37%.

The energy consumption varies from country to country. In the following study, the countries are separated in two groups –members of Organization for Economic Co-operation and Development (OECD), and non members of this organization (non-OECD).

There are several aggregate energy indicators that show energy use divided per measure of activity that drives energy demand used in assessment of the final energy use amongst the countries. The most commonly used indicators are total final energy consumption (TFC) per unit of gross domestic product (GDP) and energy per capita.

The ratio of TFC to GDP measures how much energy is needed to produce one unit of economic output. To be able to perform cross-country comparison, a common measure of GDP must be used. Two methods are used to convert GDP in national currency of a common unit of measure. The first is conversion at market exchange rate (MER) and the second is conversion at purchasing power parity (PPP). Results from both approaches for year 2005 are graphically presented in Figure 4 and it is seen that the two approaches produce different results for the level of TFC per GDP. Using GDP at PPP, aggregate final energy intensity in 2005 varies from 4.0MJ per USD in Mexico to 10.1MJ per USD in Russia. When using GDP at MER, TFC per GDP varies from 3.6MJ per USD in OECD Pacific to 39.8 MJ per USD in Russia. Using MER all the non-OECD countries presented in the analysis use more energy per unit of GDP than those in the OECD. However, these differences considerably narrow and sometimes completely disappear when based on GDP at PPP aggregate final energy intensity is calculated.

Figure 2. A consumption of energy of the different kinds (Source: Energy Information Administration Report #DOE/EIA – 0383(2008), Report date: June 2008. Used with permission)

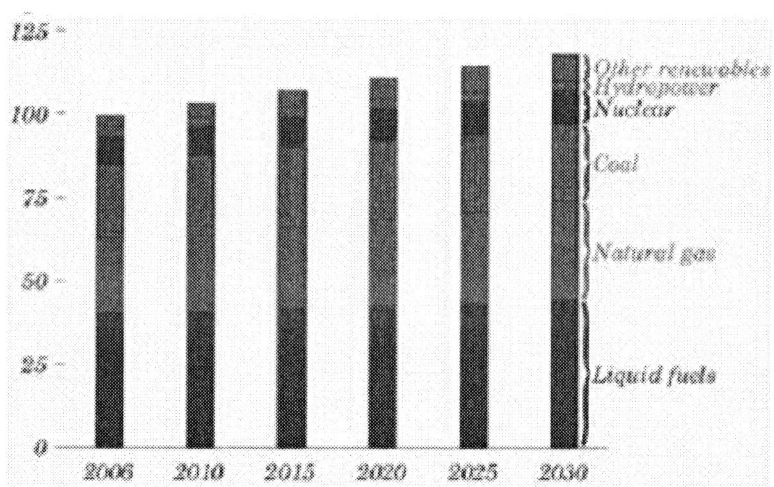

Figure 3. Shares of global final energy consumption and CO_2 emissions by sector, 2005 (Source: World-wide Trends in Energy Use and Efficiency: Key Insights from IEA Indicator Analysis ©OECD/IEA, 2008, Figure 2.1, page 17. Used with permission)

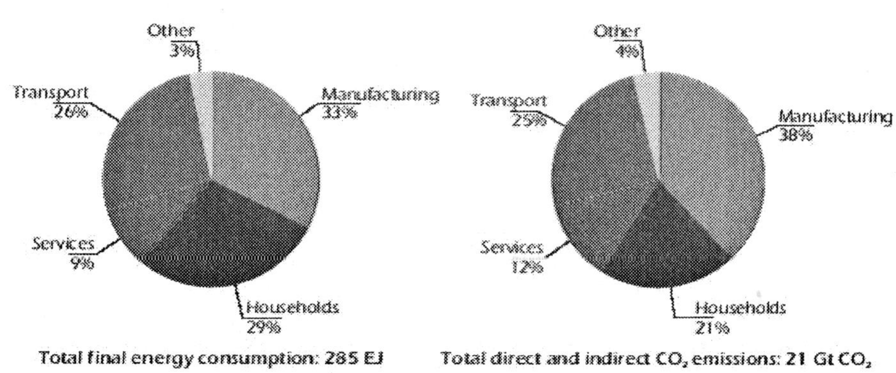

Figure 4. Total final energy consumption per unit of GDP (Source: Worldwide Trends in Energy Use and Efficiency: Key Insights from IEA Indicator Analysis ©OECD/IEA, 2008, Figure 2.5, page 21. Used with permission)

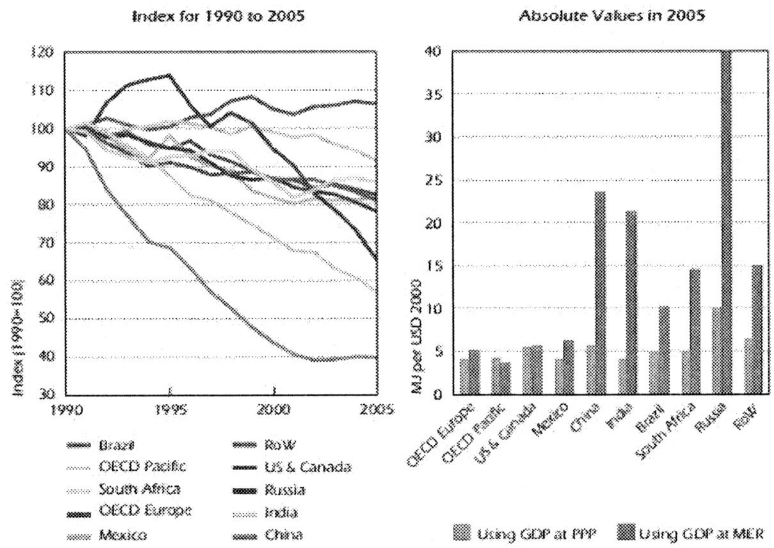

One of the reasons of the difference in variations of energy consumption levels per unit of economic output amongst countries is the variations in energy efficiency. The energy efficiency ratio is influenced by many non-energy factors such as climate, geography, travel distance, home size and manufacturing structure.

The results of the second indicator mentioned - final energy use per capita amongst the countries from 1990 to 2005 are presented in Figure 5.

Figure 5. Total final energy consumption per capita (Source: Worldwide Trends in Energy Use and Efficiency: Key Insights from IEA Indicator Analysis ©OECD/IEA, 2008, Figure 2.6, page 22. Used with permission)

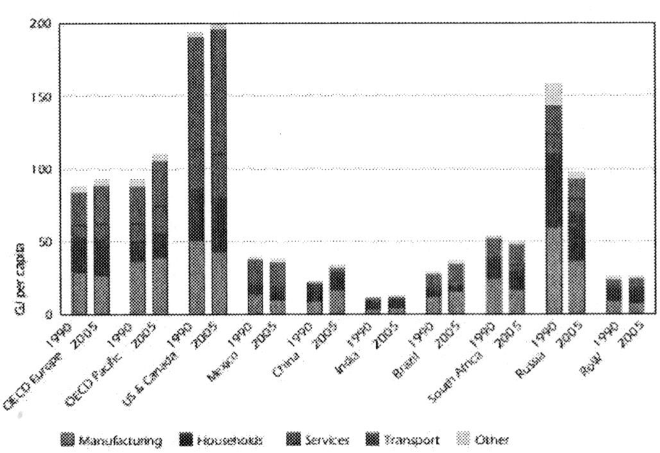

Energy use per capita increases with 6% for the OECD countries, and with 1% for the non-OECD countries (IEA, 2008). In average the energy used per capita in non-OECD countries is only 23% of that used in the OECD.

Burning fossil fuels though leads to a pollution of the environment. Continuous increase of average temperatures in the World is a statistical fact and it is called global warming. Because of the harmful emissions in the air, basically of carbon dioxide, the slimming effect of particular parts of the Earth atmosphere ozone layer has been discovered. Figure 3 displays the shares of the different branches of CO_2 emissions (IEA, 2008).

As it is seen, the industry and transportation represent the main part of CO_2 emissions. Their relative percentage of the air pollution is the same as the percentage of the consumed energy, while the pollution produced by the households share is less than its consumption. In this connection, it is interesting to analyze the types of the consumption for home use.

Figure 6 (IEA, 2008) shows, that there is a trend to stabilization of energy consumption for household heating at the expense of the house appliances increase. However, the energy consumption for heating purposes remains predominate towards 2005 year, too.

The contemporary distribution and forecasts for the emissions of carbon dioxide in the atmosphere by sectors and source types are presented in Figure 7 (EIA, 2008). Electrical energy is the most impressive one and a basic source for its generation the coal is predicted to remain. In the field of transportation as a basic source oil remains. Coal and the oil are exactly the reason for the prognosticated increase of the harmful emissions.

Other harmful result of the emissions produced is the more frequent presence of the so-called acid rains leading to a destruction of flora and fauna. Particular vegetable and animal kinds are gradually dying out. Under these circumstances, it is naturally for the human beings to search for other energy production possibilities.

Figure 6. Household energy use by end-use, IEA_{19}(Source: Worldwide Trends in Energy Use and Efficiency: Key insights from IEA Indicator Analysis ©OECD/IEA, 2008, Figure 4.3, page 46. Used with permission)

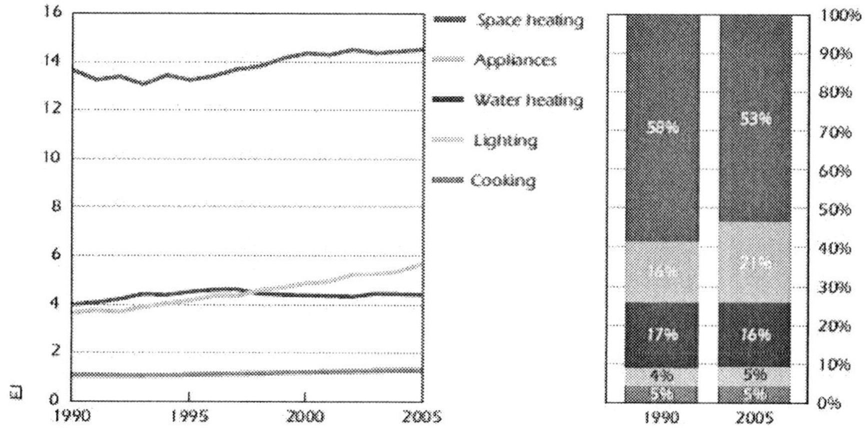

Figure 7. Distribution of the harmful emissions (Source: Energy Information Administration Report #DOE/EIA – 0383(2008), Report date: June 2008. Used with permission)

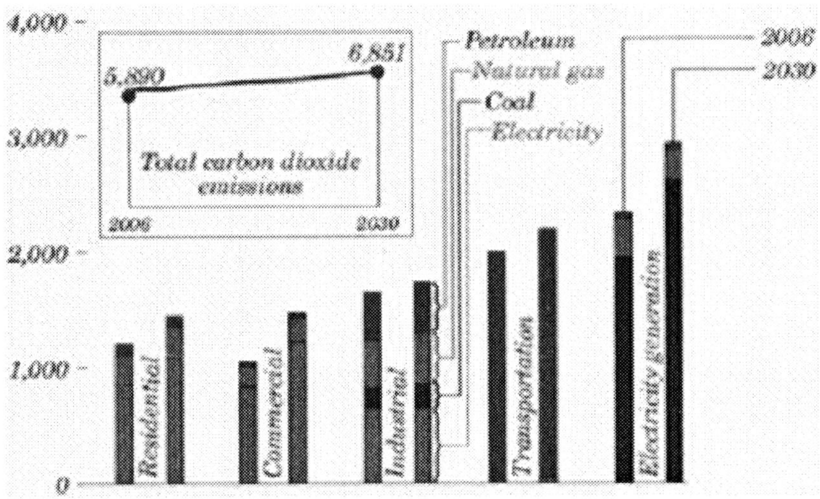

At the end of the 20th century, two clearly established facts of worldwide importance stood out:

- The world supplies of fossil fuels (coal, oil, naphtha, uranium) are limited
- The increasing rate of the industrial development connected with the "traditional" energy generation conceals risks for the environment

In this connection, the interest towards so-called renewable energy sources (RES) – sun, wind, water, biomass, geo thermal sources, has been quickened. As advantage of RES lack of harmful influence over the environment is mainly assessed, as well as the fact that they are considered to be inexhaustible. Of interest (National Energy Policy Development Group, 2001) are the researches for those sources use. A great attention is paid to RES implementation into detached residential buildings. It has to be mentioned, however, that their propagation is not still very wide and at this stage eventual harmful results of their use that might arise under other circumstances, can not be assessed. In Figure8 electricity from renewable energy sources is also added to the "classical" sources – coal, natural gas and liquid fuels.

In accordance with this prognosticated data, a significant increase of the role of the energy gained from the renewable sources and of electrical energy is expected in the forthcoming 20 years. The traditional fossil fuels however keep their utmost importance. Therefore the only way to decrease their harmful influence while the energy needs of the humanity increase is to increase their efficiency.

The next chart studies the question about the efficient use of fuels.

ENERGY EFFICIENCY AND CONTEMPORARY TRENDS

Graver attention on the topics about the energy efficiency has started to be paid to in 1970-ies. In the years after 1990, however, a delay in the implementation of the measurements for this regard is reported. Figure 9 depicts graphical relationships for the countries members of the International Energy Agency

Figure 8. Shares of the different energy sources into the total energy balance (Source: Energy Information Administration Report #DOE/EIA – 0383(2008), Report date: June 2008. Used with permission)

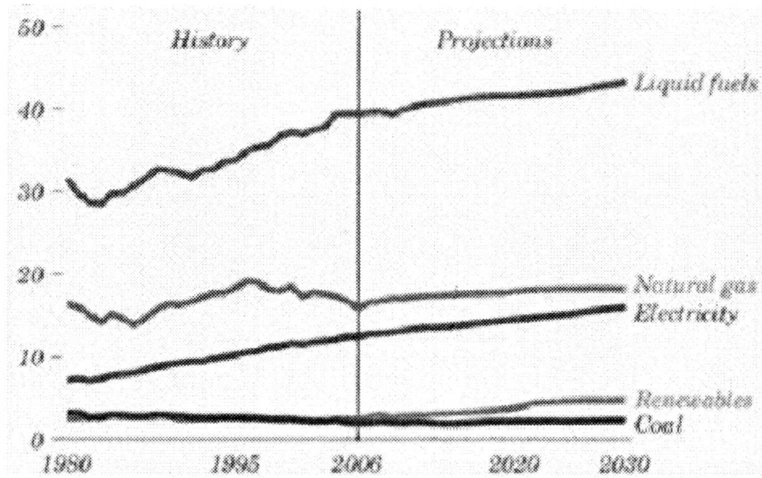

(IEA, 2008). Using the comparison of the actual and hypothetic states, and because of the measurements taken approximately 35 years ago, it appears that towards 2005 the energy savings in all branches should be about 60%. Positive results were found in the fields of energy efficiency, for example, in USA according to data of (National Energy Policy Development Group, 2001).

According to different forecasts, by 2040 year, the share of consumed electrical energy is expected to increase twice compared to its contemporary share. Therefore, the basic measurements have to follow the increase of energy efficiency in the cycle generation – transmission – distribution - usage of the electrical energy (Marlino, 2005). In this aspect, the data shown in Figure 10 displaying the

Figure 9. Long-term energy savings from improvements in energy efficiency, all sectors, IEA₁₁(Source: Worldwide Trends in Energy Use and Efficiency: Key Insights from IEA Indicator Analysis ©OECD/IEA, 2008, Figure 2.9, page 26. Used with permission)

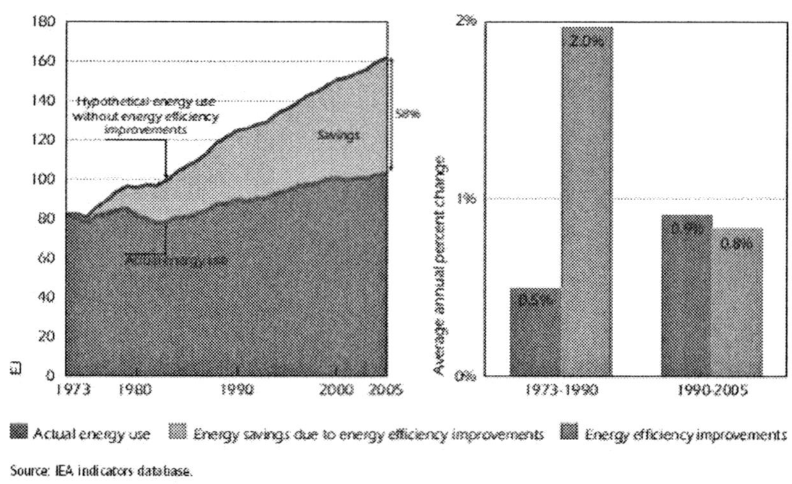

potential savings and decrease of emissions of CO_2 are of interest (IEA, 2008). The savings for the OECD countries and the non-OECD countries, as well as of world importance, are also presented by the type of the material – coal, oil, naphtha. The increase of coal production efficiency offers good possibilities as for OECD countries or non-OECD countries.

Data for the efficiency during the generation of electrical energy for different countries are presented in Figure 11 (IEA, 2008). In them, CHP states for combined heat and power.

And yet, the electrical power generation efficiency remains low – Figure 11 (IEA, 2008). Its average value in the World is 35-40%.

An estimation of possibilities for additional energy resulting from the energy efficiency increase is shown in Figure 12 (Infineon, 2008). It appears that during the processes of generation and transmission

Figure 10. Technical fuel and CO_2 savings potentials in 2005 from improving the efficiency of electricity production (Source: Worldwide Trends in Energy Use and Efficiency: Key Insights from IEA Indicator Analysis ©OECD/IEA, 2008, Figure 7.3, page 74. Used with permission)

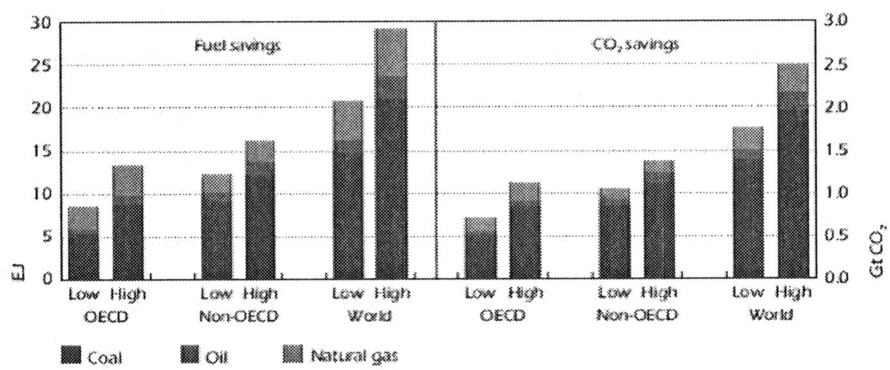

Figure 11. Efficiency of electricity production from fossil fuels in public electricity and CHP plants (Source: Worldwide Trends in Energy Use and Efficiency: Key Insights from IEA Indicator Analysis ©OECD/IEA, 2008, Figure 7.2, page 73. Used with permission)

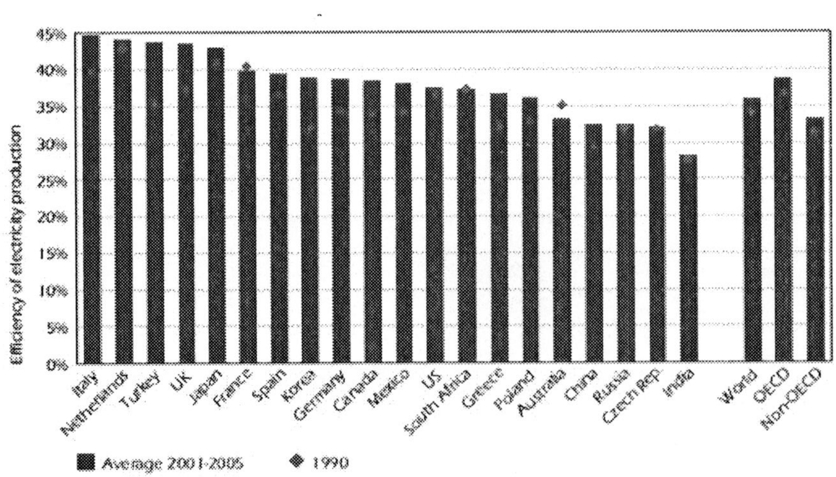

Figure 12. Possibilities to increase energy efficiency in electrical energy (Source: Saving Energy Through Innovation and Technology ©2008 Infineon Technologies North America Corp., Page 7, Figure 1.4 Used by permission of World Energy Council, London, www.worldenergy.org Used with permission)

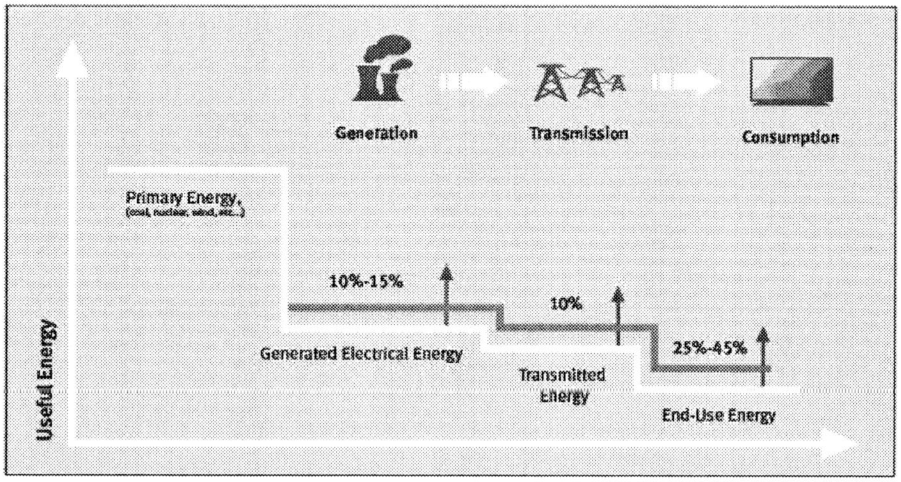

the additional share is between 10 and 15%. The achievements of additional energy efficiency increase are left with the end-user – up to 45%.

The end-user is exactly the party through which the electrical energy is converted into an appropriate type to be consumed. Therefore the main attention in this book is focused on the technologies for electrical energy conversion for its use as AC or DC energies, as well as on the power electronic converters designed for this purpose and on the methods of increasing the energy efficiency under the abovementioned conversions. The role of Power Electronics is especially significant for the distributed generation of electrical energy. The implementation of the power electronic converters for this use is also subject to discussion in a separate chapter of the book.

REFERENCES

Energy information Administration EIA, (2008). *Annual energy outlook 2008 with projections to 2030.* USA/Washington. Retrieved 2008 from http://www.eia.doe.gov

Infineon, (2008). *Saving energy through innovation and technology.* Retrieved 2008 from http://www.techonline.com

International Energy Agency IEA. (2008). *Energy indicators, worldwide trends in energy use and efficiency.* France/Paris. Retrieved 2008 from http://www.iea.org

Marlino, L. (2005). Government's role in energy efficient transportation. *IEEE Power Electronics Society Newsletter, 17*(4), 7–9.

National Energy Policy Development Group. (2001). *Reliable, affordable, and environmentally sound energy for America's future.* Retrieved 2008 from http://www.whitehouse.gov

Chapter 2
Storage and Usage of Energy

OVERVIEW

A necessity to store energy in purpose to use it later in time arises in different occasions. For example, in uninterruptible power supplies systems examined in Chapter 10, stored energy is required when a drop off of the electrical network system happens. In systems used to obtain energy from renewable sources examined in Chapter 9, the necessity to store energy also arises. This is explained by the fact that energy is not always possible to be consumed at the moment of its generation and of the same quantity. For example, the periods of strong sun lightening or strong wind may not coincide with the periods of maximum consumption.

Nowadays, the following methods to store energy are developing and using:

- Storage of energy as electrochemical energy
- Storage of energy as electromagnetic energy
- Storage of energy as electrostatic energy
- Storage of energy as mechanical energy

DOI: 10.4018/978-1-61520-647-6.ch002

Figure 1. Comparison among different methods of energy storage

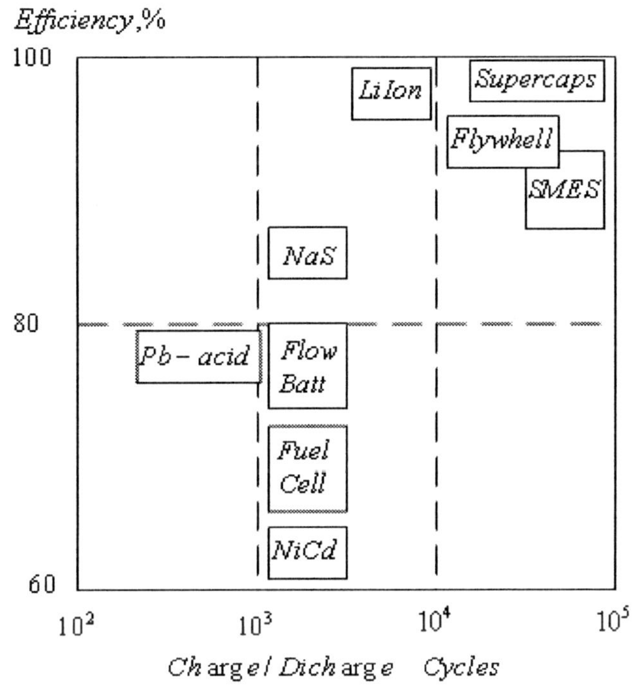

As a consequence of all the above mentioned energy types, electrical energy, which is used by consumers, can be obtained.

Basic indicators of the processes of storage and giving energy are efficiency coefficient and long-lasting time – number of cycles charge/discharge. Figure 1 displays a comparison among the different methods of energy storage.

The figure includes: accumulator batteries –lead-acid, nickel-cadmium (NiCd), sodium NaS, lithium-ion (Lilon); Fuel Cells and Flow Batteries; Supercapacitors (Supercaps); Flywheel batteries; storage in magnetic field under super conductivity (SMES).

It appears in Figure 1 that supercapacitors and Flywheel batteries have the best indicators considering long-lasting time and efficiency coefficient.

Besides the efficiency coefficient and number of cycles, other important indicators are specific power and specific energy. Comparative data about different methods for energy storage also exist for the last mentioned indicators (Ter - Gazarian, 1994).

STORAGE OF ENERGY AS ELECTROCHEMICAL ENERGY

The most widely spread method for energy storage is its storage as electrochemical energy. Accumulators are devices used to realize that storage. The first Pl-acid accumulator was made by Gaston Plante in 1859. Basic principle of operation and also chemical processes of Pl-acid accumulator have remained during the time. The researches in Pl-acid accumulators are orientated towards their seal hermetically and changes in their construction.

Figure 2. Data about specific power and specific energy of different types of accumulators and fuel cells

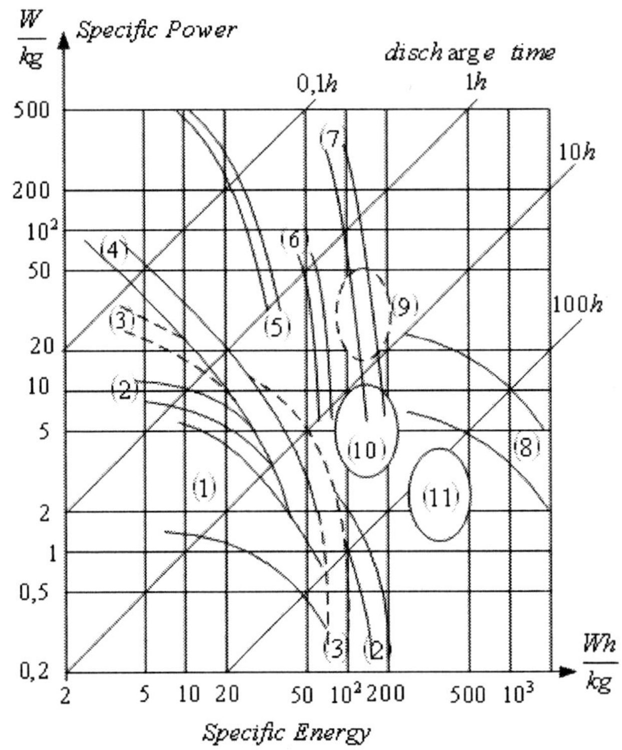

To choose accumulator, its specific energy (Wh/kg), its specific power (W/kg) and their change dependent on the duration of energy giving at discharge are used as general indicators. Also, choosing an accumulator, the following parameters are taken in consideration – efficiency coefficient, expiring period, necessity of accumulator serve, technical safeness, price, etc.

Figure 2 illustrates the change of the specific energy and specific power of different accumulators at different discharge time – form 0.1 to 100 hours. Also, data for fuel cell is presented. An idea about the possibilities of the different elements may be obtained using this data.

Curves 1, 2 and 3 are for primary electrochemical elements (so-called unchargable batteries). Curve 4 is for Pl-acid accumulators, curve 5 –NiCd accumulators, curve 6 - Fe-Ni accumulators, curve 7 – silver-zinc accumulators, curve 9 – NaS accumulators, curve 10 – Al accumulators, curve 11 – Li accumulators and curve 8 – fuel cells. Accumulator efficiency decreases at lower discharge times, i.e. higher discharge currents. Figure 2 shows the advantage of the fuel cells regarding the specific energy, as well as this of NiCd accumulators regarding the specific power.

For the time being Pl-acid accumulators have the widest spreading. That is the reason why their special features will be examined first. Hermetically sealed maintenance free accumulators are used in contemporary systems. PlO_2 is used on the positive plate as an active substance, and spongy lead on the negative plate. The plates are mechanically separated by a separator which secures high permeance of electrolyte and prevents short-circuits into the cell. The electrolyte in these batteries is connected as gel substance (Exide Technologies, 2008). It is a mixture of H_2SO_4, H_2O and SiO_2. The water losses are irretrievable in conventional batteries because the components of chemically decomposed water (hydrogen and oxygen) do not connect again while they are freely liberated in the atmosphere. At maintenance free accumulators, the developing process of recombination on the negative plate almost fully compensates the quantity of decomposed. This happens due to the possibility of transferring hydrogen towards the negative plate through the jelly electrolyte. Water losses are regulated and minimized in such a way that they do not influence the expiring period of the batteries. Each cell of the battery is hermetically sealed and closed with a single-way valve. The valve opens in the case of increased inner pressure and it leaves the additional hydrogen and oxygen out of the cell. After the gases emit the cell, the valve closes again and protects the cell of atmosphere oxygen penetration. Ordinary polypropylene or special polymer is usually used to implement a battery body.

Nominal voltage of an accumulator cell is set $U_N = 2V$. Nominal voltage of the battery is calculated by multiplication of the nominal voltage cell value by the cell number. For example, usually the cell number is 6, therefore, the nominal voltage of the whole battery is 12V.

Ultimate discharge voltage is determined by the value under which the cell voltage should not be decreased at discharge with the corresponding discharge current. The value of the ultimate discharge voltage depends on the magnitude of the discharge current. If the cell voltage becomes lower than the value of the ultimate discharge voltage then an area of deep discharge is reached.

Capacity of an accumulator battery C is measured in ampere-hours Ah. Besides, it is also indicated at how many hour discharge this capacity is given – usually it is in mode of 10- or 20-hour discharge C_{10} or C_{20}.

Nominal discharge current is determined by the capacity data:

$$I_{10} = \frac{C_{10}}{10h}$$
$$I_{20} = \frac{C_{20}}{10h} \qquad (2.1)$$

The value of ultimate discharge voltage of a cell is usually 1.8V in 10-hour discharge and 1.75V in 20-hour discharge.

Deepness of discharge is the detracted battery capacity during discharge. For example, if 80% of the battery capacity is used then the deepness of discharge is 80%. The maximum permissible deepness of discharge depends on the magnitude of the discharge current and on the temperature.

Magnitude of discharge current is given as value divisible to I_{10} or I_{20} - for example $4 \cdot I_{10}$ or $20 \cdot I_{20}$. Value of maximum 5-second discharge current is usually given. It is used at design of the battery fuse.

Self-discharge of a battery is given in percentage of day capacity. It is extremely low in *non-served* batteries – usually below 0.05%.

Inner resistance of a battery is a sum of resistances of all cells and resistances of inner battery connections. Value of inner resistance for a cell at ambient temperature 20-25C in *m ΩAh/cell* is usually put at manual data. The inner resistance of the whole battery is calculated using the equation:

$$R_i = \frac{N.R_{ik}}{C} \qquad (2.2)$$

where in n is cell number, C – battery capacity and R_{ik} - inner resistance of a cell. For example, if the battery capacity is 6.5 Ah, it has 6 cell and the inner resistance of a cell is 57 *mΩAh*, then the inner resistance of the whole battery is 52.6 *mΩ*.

The producers offer not only table relationships but also graphical ones of basic parameters of batteries. These relationships are used during the design process to choose an accumulator battery. Besides, the relationships are also used in determining requirements to operational mode and parameters of a charge device. These characteristics are different in the different producers and series accumulator batteries. The relationship use has to be carefully performed, especially taking in consideration that the relationships are frequently dependent on the ambient temperature.

As an example, the basic characteristics of maintenance free accumulator batteries series dryfit of Sonnenschein may be examined (Exide Technologies, 2008).A feature is a dependence of battery capacity on temperature and on discharge current – Figure 3.

With an increase of the discharge current, shown as divisible one to I_{10}, as well as with a decrease of the temperature, the usable part of the total battery capacity decreases. As it appears in Figure 3, the total capacity is guaranteed at temperatures above 20°C and at discharge current equal to the nominal 10-hour discharge. The temperature decrease into the area of negative values, where a zone of freezing of the electrolyte is, influences especially unfavorably these accumulator batteries.

Some producers offer so-called discharged curves for an initial design. The discharge curves show the dependence of the discharge voltage on the discharge current value and discharge time. These curves are limited from below by ultimate discharge voltage and the trespassing of the limitation line leads to entrance the deep discharge area. This is not advisable to happen because it may lead to reduction of the expiring period and to failure of the battery. Monitoring of the ultimate discharge voltage is provided in the devices using accumulators to protect it from a deep discharge. Reaching the ultimate discharge voltage, the consumer turns off from the accumulator.

Since the design based on the discharge curves has several special features requiring higher classification, recently, detailed table data is presented. The data includes ultimate discharge voltage and the time at discharge with a constant current or constant power (Exide Technologies, 2008). A particular type and operational mode of an accumulator battery may be chosen based on that data.

Figure 3. Dependence of battery capacity on temperature and on discharge current (Source: Industrial Batteries- Network Power Sonnenschein A400 ©Exide Technologies. Used with permission)

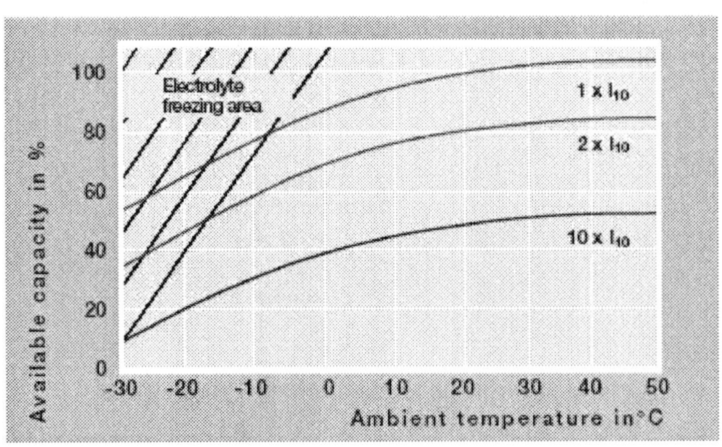

The following ways of charging accumulator batteries are well known: charge with a constant voltage (U-characteristic), charge with a constant current and voltage limitation (IU-characteristic), charge with IUI-characteristic.

Figure 4 displays a charge with U-characteristic. This is the simplest way of charging batteries. Here, the charge current depends on the state of the battery itself. The current decreases with the increase of the charged degree.

Figure 5 illustrates a charge with a constant current. At the beginning of the charge, the battery is charged with a constant current till the set charge voltage is reached. After that moment, the charge voltage is maintained constant while the charge current is limited by the battery itself. With the increase of charged degree the current decreases.

Figure 6 depicts a charge with IUI - characteristic. This charge is recommended at cycling running. The charge begins with a constant current I_1 till a determined value of the accumulator battery voltage is reached $U_B{}'$, for example at t_1. After that moment, finishing-up charge begins with a constant voltage. The characteristic is extended with an additional current branch I_2, when the charge current decreases under a set value. This value is usually about 25% of I_{20}, for example from t_2. This mode is maintained till it is terminated by a charge device, for example, at reaching a maximum charge voltage.

The last mode has a modification called "*accelerated charge*", shown in Figure 7. The charge is made at two stages. The first one is with a constant current which is within the range of 0.05 to 0.2 from the battery capacity till the charge voltage reaches 2.4V per cell, for example, at t_1. The second stage is undercharge with a constant voltage of 2.25V per cell. The charge current decreases and reaches its minimum value at a fully-charged battery.

Figure 4. Charge with U-characteristic

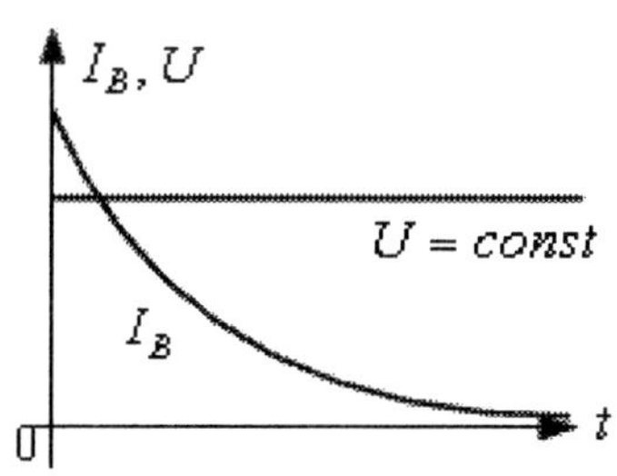

Figure 5. Charge with a constant current and voltage limitation

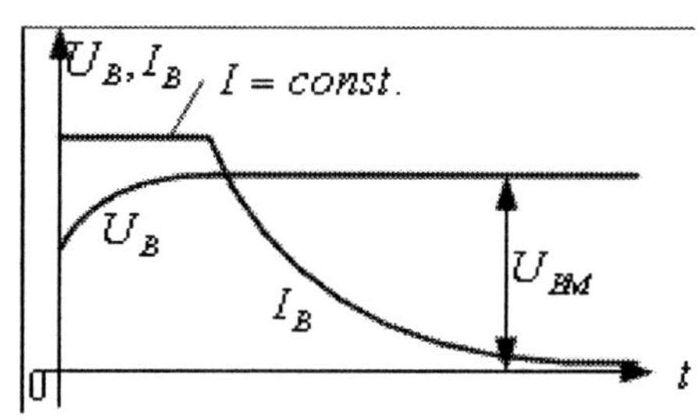

Figure 6. Charge with a constant current and finishing-up charge with a constant voltage

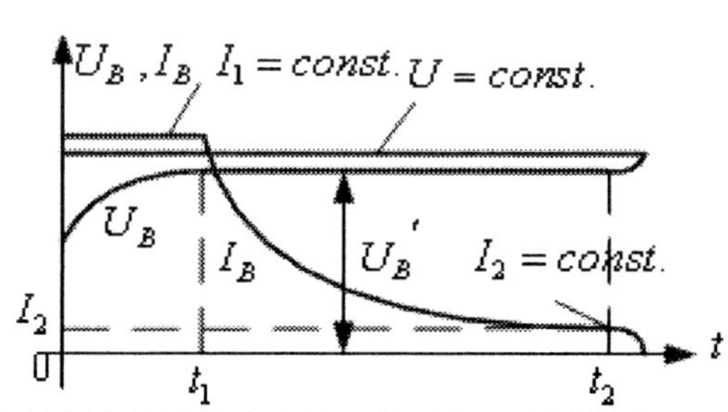

Figure 7. Accelerated charge

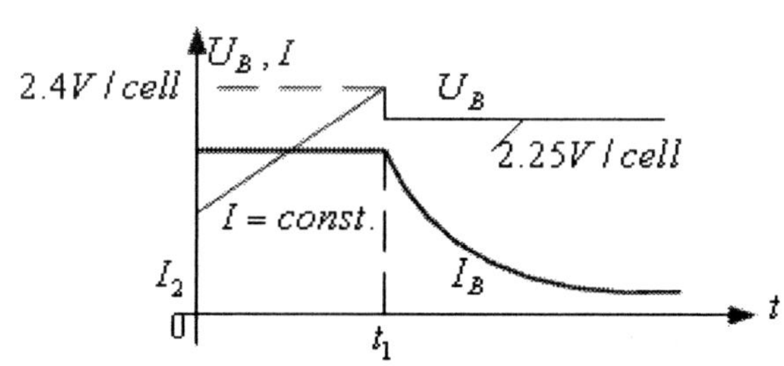

Figure 8. Dependence of the charge voltage per cell on the temperature (Source: Industrial Batteries-Network Power Sonnenschein A400 ©Exide Technologies Used with permission)

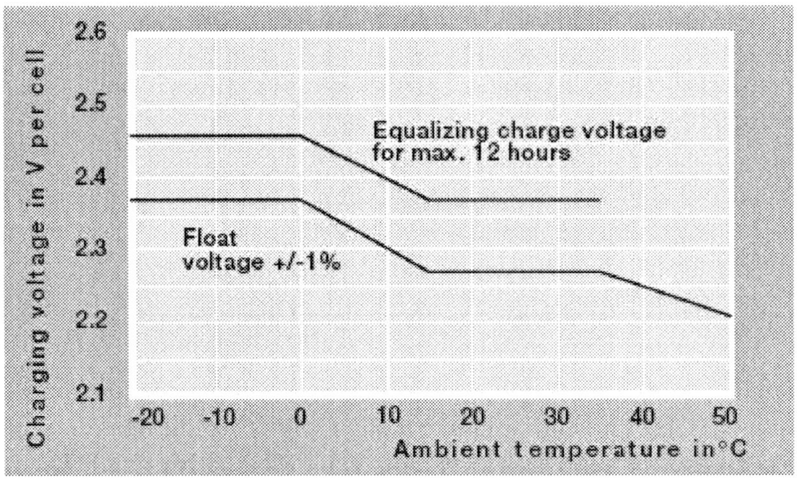

Figure 8 illustrates recommended values of charge voltage per cell dependent on the temperature. The relationships are interesting from charge device point of view. It is visible that the recommended voltage values stay constant in the ranges below 0 and above 15°C.

It is more often assumed that the value of the charge current has to be equal to the nominal discharge current at 10-hour discharge – I_{10}. The charge time depends on the deepness of the previous discharge, on the value of the charge voltage and on the magnitude of the initial charge current. These characteristics are depicted in Figure 9. It appears in them, that the accumulators sustain also an accelerated charge mode with a value of current several times higher than the value of the nominal discharge current.

Ripples of the charge current additionally cause a harmful increase of the battery temperature. AC component of the charge current should not increase 10% of the battery capacity under a cycling running

Figure 9. Dependence of the charge time on the charge current (Source: Industrial Batteries- Network Power Sonnenschein A400 ©Exide Technologies. Used with permission)

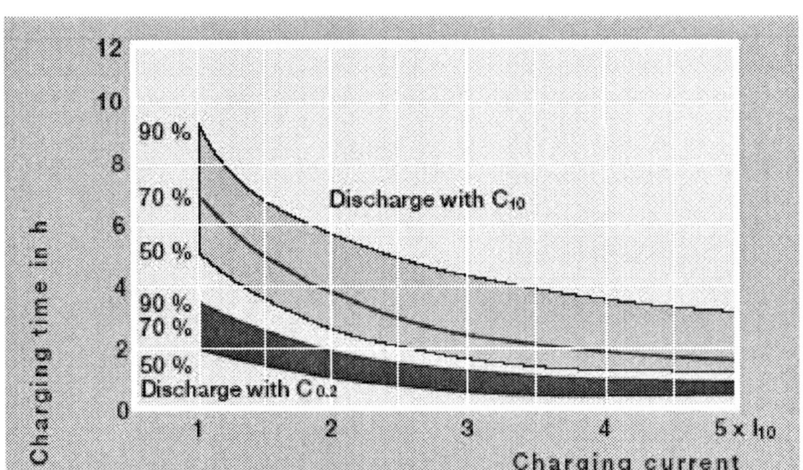

Figure 10. Dependence of the capacity on the storage time and temperature (Source: Industrial Batteries- Network Power Sonnenschein A400 ©Exide Technologies. Used with permission)

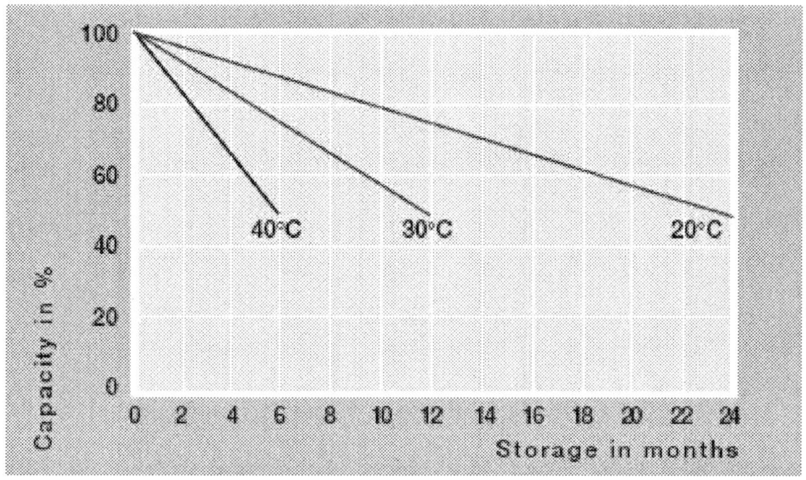

- mode charge-discharge. For example, for battery of 24Ah – the current has to be lower than 2.4A. At undercharge mode – 5% of the capacity, or concerning the example - 1.2A. In all cases, the temperature during the charge should not increase with more than 3°C.

Expiring period of the batteries depends on the temperature. The producers guarantee 5 or 10-year expiring period. It has to be taken into consideration that at storage the battery capacity decreases. The stage of decrease depends on the temperature. As an example, a graphical relationship for batteries of A600 type is shown in Figure 10.

Increase of the temperature leads to decrease of capacity of Pl accumulators.

Other reason of decreasing the capacity is boosted number of charge-discharge cycles of the accumulator. Specially designed batteries with improved indicators in this regard are offered. These batteries are meant to operate in systems with renewable energy sources and they allow increased number of charge-discharge cycles. Figure 11 shows a relationship between their capacity and cycle number as an example of the allowance.

Recently, the implementation of NiCd accumulator increases. A comparison of their capabilities to those of Pl accumulators is frequently required to be made. For this, some of NiCd accumulator special features will be examined.

The nickel-cadmium secondary cell was developed around 1899 by the Swede Waldemar Jungner. New things in Ni-Cd accumulators were the choice of electrode material and the principle of alkaline electrolyte. Since this type of electrolyte only serves as an ion conductor and does not take part itself in the chemical reactions within the cell, it is subject to hardly any change, which has a positive effect on the ageing process of the accumulator.

Since 1950, nickel-cadmium accumulators have also been produced in a gas-tight design, making them maintenance-free and also prolonging their service life. A safety valve is integrated in the gas-tight Ni-Cd cells as an overcharge protection as one already explained in lead accumulators.

NiCd rechargeable batteries display excellent qualities such as a high energy density, ruggedness, resistance to cold and service life as well as a favorable price and long storage life. However, account of the toxicity of the cadmium they contain numbers their days. In some countries corresponding bans are already in effect and the replacement of this type of accumulator has begun, primarily through nickel-metal hydride accumulators.

The basic parameters of Ni-Cd accumulators are:

- **Cell terminal voltage:** 1.2 volts
- **Service life in charge cycles:** 1,000 to 1,500 (2,000)
- **For use in ambient temperatures:** (-40°C) -30°C to +50°C (+70°C)
- **Typical energy density:** 50 - 70 Wh/kg (theoretically approx. 240 Wh/kg)
- **Typical volumetric energy density:** 130 Wh/l
- **Typical power density:** 150 - 200 W/kg
- **Typical spontaneous discharge:** To 20% residual capacity within 3 months

Disadvantages of Ni-Cd batteries are the relatively high spontaneous discharge, relatively high susceptibility to the so-called memory effect and the toxicity of the cadmium.

NiCd cells have extremely diverse fields of use: millions of individual small cells can be found in a wide variety of devices such as torches, portable phones or CD players. Assembled into battery packs (Power-Packs) they are used as energy sources in mobile phones, notebooks or cordless screwdrivers. As plate accumulators with peak currents of 500-1,000 amperes they are used, for example, in the aerospace industry, as traction batteries for industrial trucks or as starter batteries in very cold environments.

Other element to store energy and to generate electrical energy using electrochemical way is a fuel cell (Ramani, 2006). It is demonstrated in a primary mode even earlier than lead accumulator – in 1839 from Grove in Britain. If a strict differentiation has to be made, in spite of accumulators, the fuel cells are not energy storage elements as accumulator batteries, because they do not store theirs own fuel. Fuel cell requires hydrogen and oxygen to be able to operate. Usually, hydrogen is stored in reservoirs and oxygen is supplied by the air. Solutions based on regeneration processes into the cell exist (Walter, 1997).

Using the fuel cell, a direct receiving of electrical energy as a result of chemical processes is made. Figure 12 shows the way of obtaining electrical energy. Generation of electrical energy using Carnot-

Figure 11. Dependence of the capacity on the cycle number (Source: Sonnenschein Solar Safe Storage capacity for renewable energy ©Exide Technologies. Used with permission)

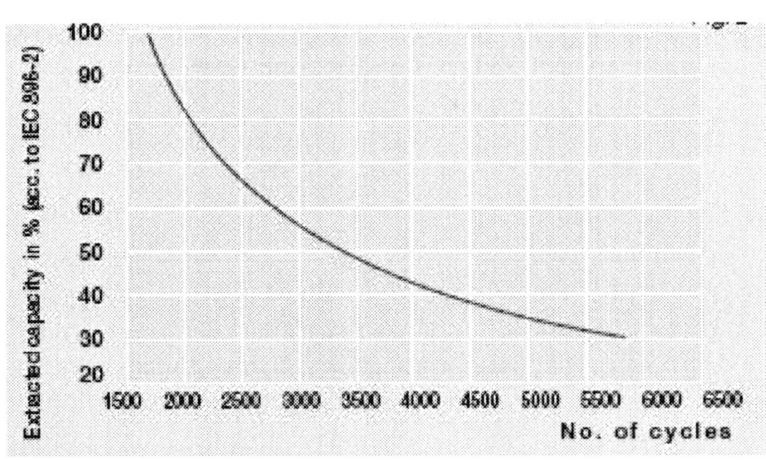

Figure 12. Ways to generate electrical energy using chemical energy

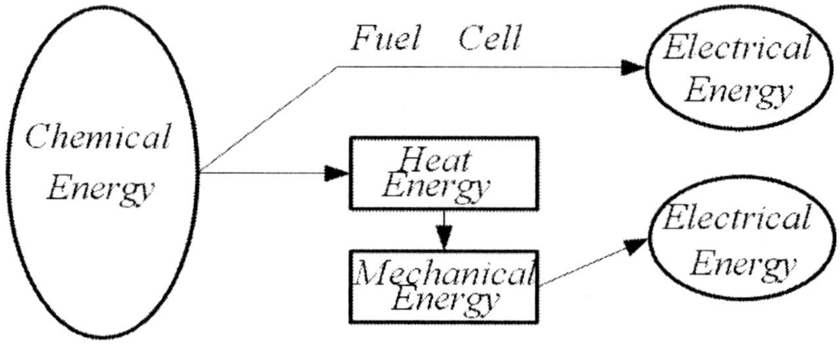

cycle machine (motor with inner combustion) is also shown in the same figure for a comparison. In Carnot-cycle machine, at first, chemical energy is converted into heat energy then into mechanical energy and finally into electrical one.

Operational principle of a hydro-oxygen fuel cell is illustrated in Figure 13. As it appears in the figure, both electrodes have to be able to let gas through them. The process consists of three phases – passing of gas (gas phase), conduction of electrons (solid phase), conduction of ions (liquid phase). Hydrogen is passed under pressure to the anode and under catalyst it is decomposed. Electrons go to the cathode, while ions penetrate the electrolyte. Oxygen is passed under pressure to the cathode and as a result of chemical reaction taking place there, water is generated. The entire process is accompanied with evolve of heat energy. So, if hydrogen and oxygen are inputs of the system, then the output is electrical energy and additional products are water and heat energy. Fuel cells with solid or liquid electrolyte are available. System for the electrolyte circulation and dragging out the water is also required in the liquid electrolyte cells. In the fuel cells with solid electrolyte, the water formed is directly passed from the electrolyte into the cathodic gas compartment and removed there.

Figure 13. Operational principle of a fuel cell

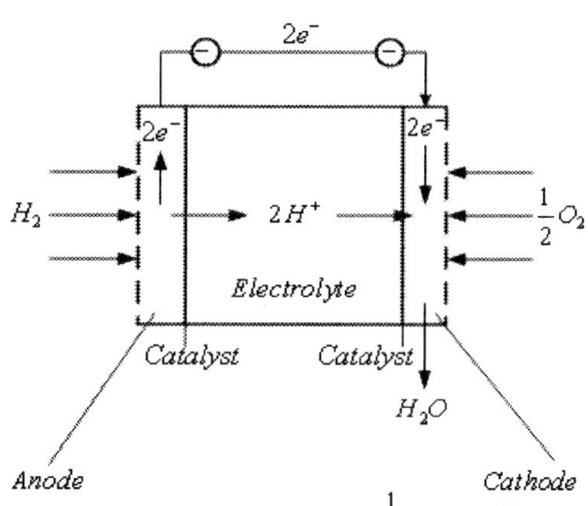

Anode

$$H_2 \rightarrow 2H^+ + 2e^-$$

Cathode

$$\frac{1}{2}O_2 + 2H^+ + 2e^- \rightarrow H_2O$$

Total cell reaction

$$H_2 + \frac{1}{2}O_2 \rightarrow H_2O$$

Efficiency coefficients of Carnot-cycle machine and fuel cells are defined in a different way. A comparison between the two coefficients is made in (Julich Forschungscentrum, 2008). For most types of fuel cells the efficiency coefficient is about 50%, but if the emitted heat is used, it reaches up to 80%.

Various types of fuel cells exist dependent on the two gases, used materials in their design and their construction. Figure 14 classifies these types (Julich Forschungscentrum, 2008).

Alkaline Fuel Cell (AFC)

This is one of the oldest designs for fuel cells; the United States space program has used them since the 1960s. The AFC is very susceptible to contamination, so it requires pure hydrogen and oxygen. It is also very expensive, so this type of fuel cell is unlikely to be commercialized.

Polymer Exchange Membrane Fuel Cell (PEM)

The PEM has a high power density and a relatively low operating temperature (ranging from 60 to 80 degrees Celsius, or 140 to 176 degrees Fahrenheit). The low operating temperature means that it does not take very long for the fuel cell to warm up and begin generating electricity.

Figure 14. Comparison among different types of fuel cells (Source: Basic terms and operating principle of a fuel cell electricity generation system ©Institute of energy research, Julich Forschungscentrum. Used with permission)

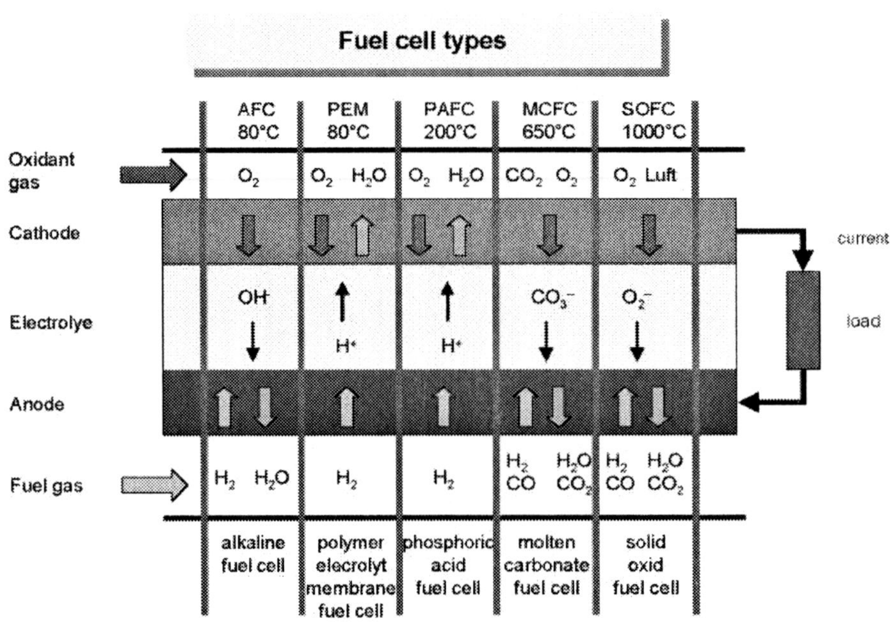

Phosphoric-Acid Fuel Cell (PAFC)

The phosphoric-acid fuel cell has potential for use in small stationary power-generation systems. It operates at a higher temperature than polymer exchange membrane fuel cells, so it has a longer warm-up time. This makes it unsuitable for use in cars.

Molten-Carbonate Fuel Cell (MCFC)

Like the solid oxide fuel cells (SOFC), these fuel cells are also best suited for large stationary power generators. They operate at 600 degrees Celsius, so they can generate steam that can be used to generate more power. They have a lower operating temperature than SOFC, which means they do not need such exotic materials. This makes the design a little less expensive.

Solid Oxide Fuel Cell (SOFC)

These fuel cells are best suited for large-scale stationary power generators that could provide electricity for factories or towns. This type of fuel cell operates at very high temperatures (between 700 and 1,000 degrees Celsius). This high temperature makes reliability a problem, because parts of the fuel cell can break down after cycling on and off repeatedly. However, solid oxide fuel cells are very stable when in continuous use. In fact, the SOFC has demonstrated the longest operating life of any fuel cell under certain operating conditions. The high temperature also has an advantage: the steam produced by the fuel cell can be channeled into turbines to generate more electricity. This process is called co-generation of heat and power (CHP) and it improves the overall efficiency of the system.

Researches on the so-called regenerative cell are presented in (Ramani, 2006). Figure 15 illustrates its idea. In fuel-cell mode, hydrogen and oxygen are input products. Under shown interaction, electrical energy and water are generated. In electrolyzer cell mode, the water is input product and also electrical source is connected to the cell. Output products are hydrogen and oxygen. The described regenerative cell is developed and used in cosmic devices.

So-called flow batteries are realized based on the regenerative fuel cells. A polysulfide-bromide flow battery is shown in Figure 16. Besides the regenerative fuel cells, two reservoirs, in which electrolyte solutions are put, are used. Circulation into one or other directions is maintained by the pumps. In generation mode, electrical consumer is connected to the electrodes and in regeneration mode – electrical energy source is connected.

STORAGE OF ENERGY AS ELECTROMAGNETIC ENERGY

In a Superconducting Magnetic Energy Storage (SMES) system energy is stored in magnetic field in superconductivity conditions (Luongo, 1996). In these processes no chemical or mechanical energy is used. SMES systems use liquid helium to maintain appropriate temperature to the winding of the inductance which is often made from niobium-titanium alloy. More often the processes take place in low temperatures (LTS – low temperature system) – about 4K (about $-270°C$). Installations for superconductivity in high temperatures (HTS – high temperature system) also exist. Since their operational temperature is closer to this of the ambient air, the thermal efficiency coefficient is increased causes an increase of the whole system efficiency. At the widest spread installation, constant current of several hundred amperes flows through the windings of the inductance. In the superconductivity conditions, resistance of the wire is approximately 0 and the losses in the windings are minimum. Additional cryogen installation is required for the operation. Basic installations are for 1-6 MJ energy. This energy can be delivered for a significant short time – for example, 1 to 60 s. Some of the possible applications of these systems are in disturbances of energy quality of the network voltage – spikes and sags, interruptions, i.e. in the UPS systems, studied in Chapter 10. Their use also in maintaining the transmission line stability in systems studied in Chapter 8 is typical, too. Sometimes the pike loadings of power consumed by the loads, when it exceeds several times the power of the state mode, have to be covered (Akhill, 1993). Such type of installations is used in USA. Their feature is their capability to be connected and to operate in parallel.

Figure 15. Processes in a regenerative cell a) fuel-cell mode, b) electrolyzer-cell mode

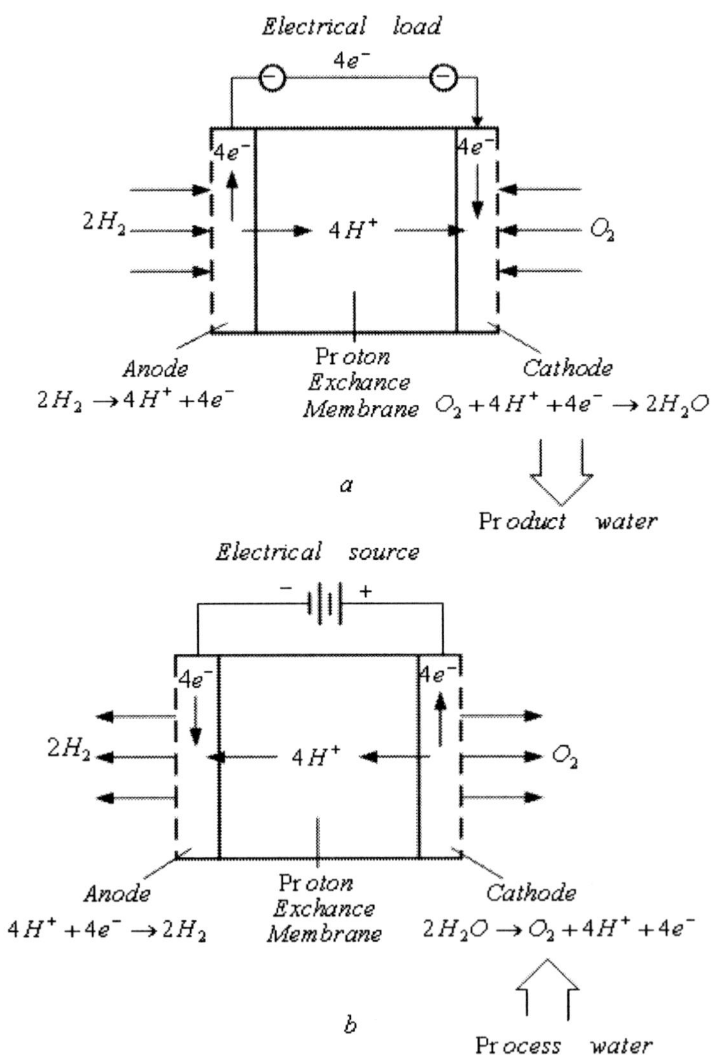

Figure 16. Polysulfide-Bromide flow battery

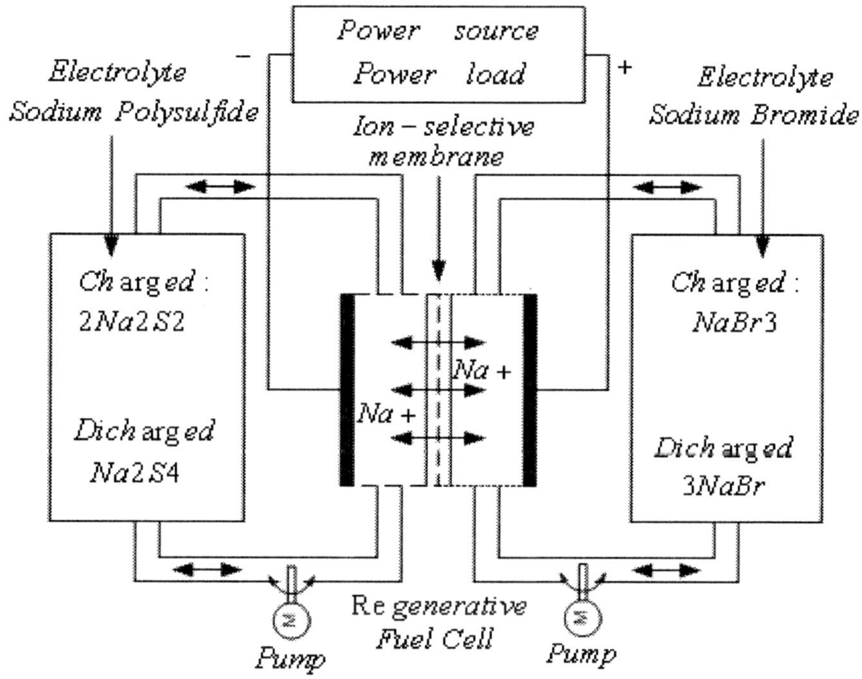

Figure 17 displays a block-schema of Superconducting Magnetic Energy Storage (SMES) system (Henry, 2007). The system operates in the following way. In the conditions of energy storage, the cold switch is normally closed to maintain the circulation of the flowing current through the superconducting coil. A bidirectional converter (see Chapter 4) secures in its output only a little energy value to replace losses of the part of its output circuit, which is not in superconductivity conditions. In disturbances of the quality of AC voltage the cold switch opens when energy should be delivered to the network. In this mode, the bidirectional converter operates as inverter (see Chapter 7) and the control system controls the converter in such a way to secure normal quality of AC voltage to the connected consumers. This systematic transfer of energy from the magnet to the load keeps the load interruption free for optimum performance of critical processes. In the past decade studies on the application of high-temperature superconductivity have been carriet out. Study to determine the optimal resistive value of a superconducting fault-current limiter for enhancing the transient stability of a power system is presented (Sung, 2009).

Energy storage in the inductance is within the range of several minutes and the number of storage/delivering cycles may reach thousands without degradation of the super-magnet. A special feature of this type of installations is a lack of pollution of the environment – there are no harmful chemical materials or gases which are omitted during the operation. Additionally, the installations are mobile and their life is estimated about 20 years. Researches in the field of energy storage as electromagnetic energy in superconductivity conditions are towards reaching higher temperature of its storage and towards study of new materials with decreased losses. This would permit also a realization of systems of lower powers. These systems would be more popularly used.

Figure 17. A block schema of Superconducting Magnetic Energy Storage (SMES) system

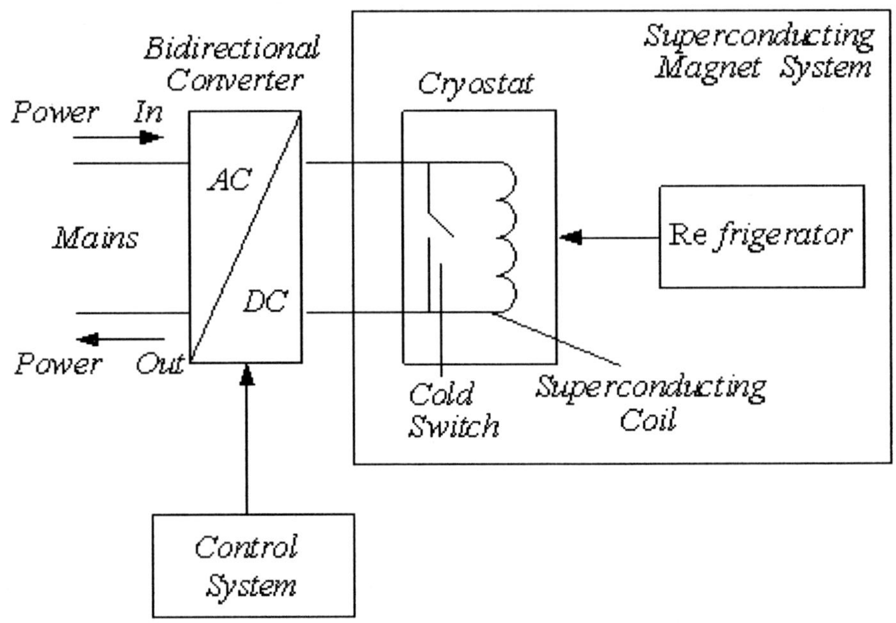

STORAGE OF ENERGY AS ELECTROSTATIC ENERGY

Storage of energy in electrical field is used in electrical capacitor long ago. For example, electrolyte capacitors are designed for this purpose. Because of their low values of the capacity, the electrolyte capacitors have found applications mainly as components in different electrical schematics and very rarely as a short-time supply source for memories.

In this chart, the basic attention is paid on so-called electrochemical double-layer capacitors (EDLC). Further down, EDLC is used as a common name, besides, as we will see, different producers use others names as well.

Accumulation of electrical charges in the metal-electrolyte boundary is known long time ago, but practical use of double-layer capacitors started in 1957, when General Electric registered the use of electrolyte capacitor with porous carbon electrodes. Later on, in 1966, Standard Oil Company, Cleveland, Ohio, made a patent of a device for energy storage in the boundary between two layers. In 1971 Nippon Electron Corporation (NEC) offered for a commercial use the first double-layer capacitors as supercapacitors. In 1980 1980 Matsushita Electric Industrial Co. (Panasonic) offered Gold Capacitor, which had been developed in 1978. In 1992 Maxwell Laboratories started a program for a design of ultracapacitors. Nowadays, there are rather many companies, which offer EDLC under different names.

Figure 18 depicts a schema with whose assistance the operational principle of those capacitors is illustrated (Maher, 2006). EDLC stores energy electrostatically, as polarizing electrolyte solution. The presence of the solution is made the capacitor close to electrochemical devices, but here no chemical reaction lays in the mechanism basis of its energy storage. Its storage mechanism is reversible and it allows the capacitor to charge and discharge millions of times.

Figure 18. Operational principle of EDLC

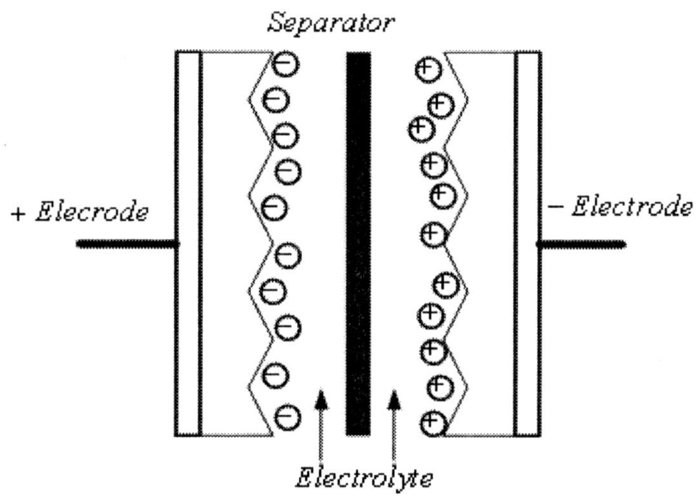

EDLC may be examined as two non-reactive porous plates, suspended in electrolyte, and a voltage applied between the plates. Negative ions are piled up around the positive electrode, and positive ions – around the negative electrode. Thus, two layers of charges separated from each other are generated. As it is known, the capacitor value is proportional to the plate area and in inverse proportion/ratio to the distance between the plates. In ordinary electrolyte capacitors, their values are up to thousands micro-farads. In spite of them, the capacity value in ADLC reaches several hundred farads.

EDLC can give very high values of energy for a short time, which makes them very attractive to be used in means of transportation.

Electrodes in EDLC are from porous carbon material. The porous structure allows the surface area to be about 2000 square meters per gram, far bigger than this of the conventional capacitors. The distance between the charges in EDLC is mainly determined by the size of the ions accumulated around the electrodes into the electrolytes. Since this distance is below 10 angstroms, very higher values of the capacities are obtained.

Figure 19 displays a comparison among EDLC, conventional capacitor, accumulator batteries and fuel cells per their specific power and specific energy.

Popularly used, for the presence, carbon electrodes and organic electrolytes allow the operation in potential difference of 1 to 3V. Monolithic electrodes with a decreased resistance, as well as, ionic liquid electrolytes are known. Their combination allows the operation in 6V. Researches in the field of EDLC are oriented towards three elements of the capacitor structure – electrodes, electrolyte and body. The purpose is to obtain bigger surface of the electrodes and an appropriate for the corresponding electrolyte diameters of the pores. Such electrodes are *composite* ones combining the big surface of the carbon electrodes with electro-conductivity of some polymers.

Figure 19. Comparison per specific power and specific energy

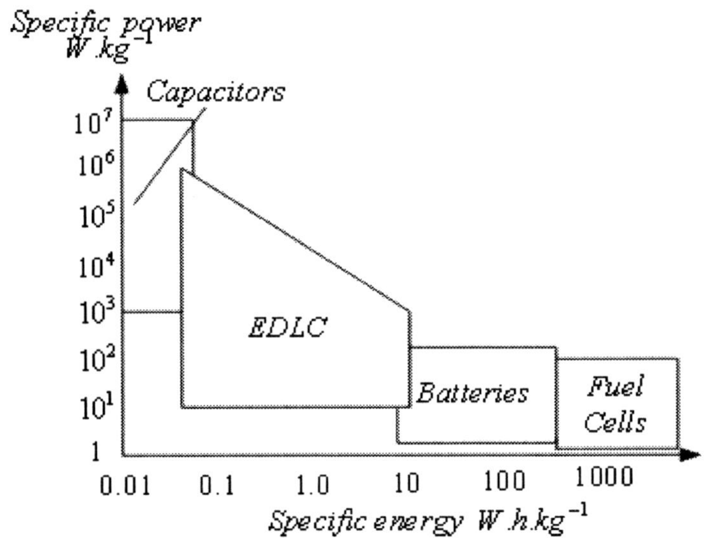

STORAGE OF ENERGY AS MECHANICAL ENERGY

Storage of energy as kinetic energy is realized within rotating flywheels. This idea is comparatively old one. Recently, the interest towards it has been increased due to the development in the field of new materials, development of control motions, and progress in the field of Power Electronics. When a question is about flywheel battery, it is connected with rotating mechanical elements.

One of the elements of a flywheel battery is a flywheel rotor, which for simplification is called very frequently flywheel. Actually, this flywheel is a rotating element with a particular volume and it also store energy mechanically as a kinetic energy.

Kinetic energy stored into a rotating mass is proportional to moment of inertia of the mass of material and the square of the angular speed:

$E_C = \frac{1}{2}.J.\omega^2,$ Where in J is moment of inertia, and ω - angular speed. (2.3)

On the other hand, the following equation is known:

$$j = k.M.R^2, \tag{2.4}$$

where in M is mass, and R- radius. Inertial constant k depends on the construction and the type of the rotor.

Therefore, at a constant speed to store higher value of kinetic energy, the mass has to be increased. This is obtained using high density materials. On the other hand, it is good to increase the speed of the rotor as the energy depends on it second power. I.e. certain optimization connected with energy

Figure 20. Schema, illustrating the usage of kinetic energy

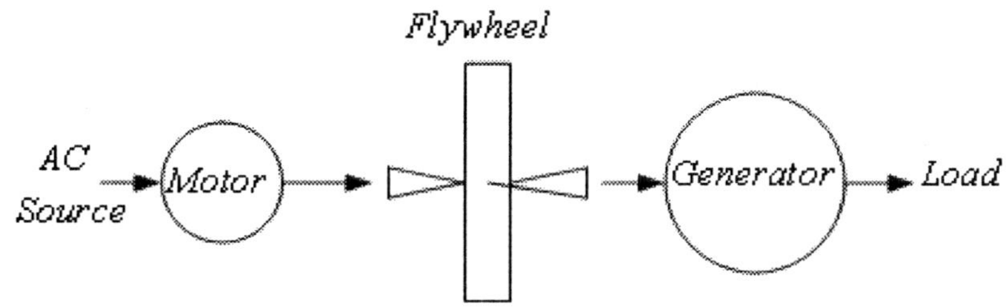

– mass – speed is required. The development in the field of composite materials gives the opportunity to obtain higher speeds. Usage of contemporary high-strength composite materials allows reaching ultra high speeds of the flywheels.

The main idea of storage and usage of kinetic energy is illustrated in Figure 20.

Mechanical connection between motor and flywheel is made in energy storage mode, while a connection between generator and flywheel misses. The motor swings the flywheel to the maximum possible speed under the influence of AC energy source and then they are mechanically separated. Thus, the flywheel continues to rotate as its energy decreases due to the friction with the ambient air and bearings. If the environment is a vacuum, then the friction with the bearings has to be minimized in case to minimize the losses of mechanical energy. If a necessity to deliver the stored energy occurs, a mechanical connection between the flywheel and the generator is realized and thus the load is supplied and the mechanical energy is transferred into electrical one.

Figure 21. Principle structure of a low-speed flywheel system

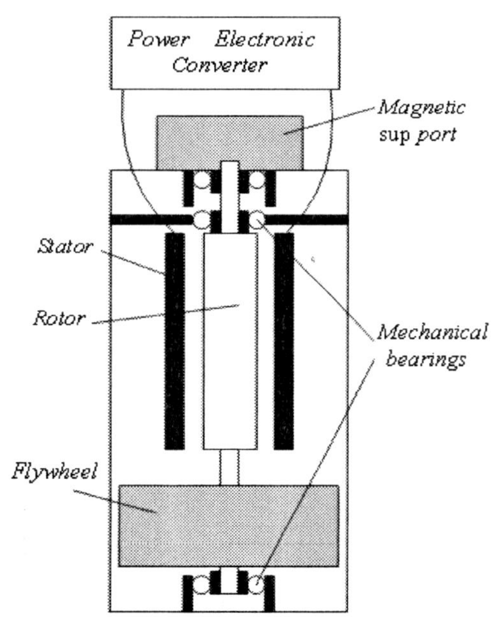

Figure 22. Principle design of a high-speed flywheel system

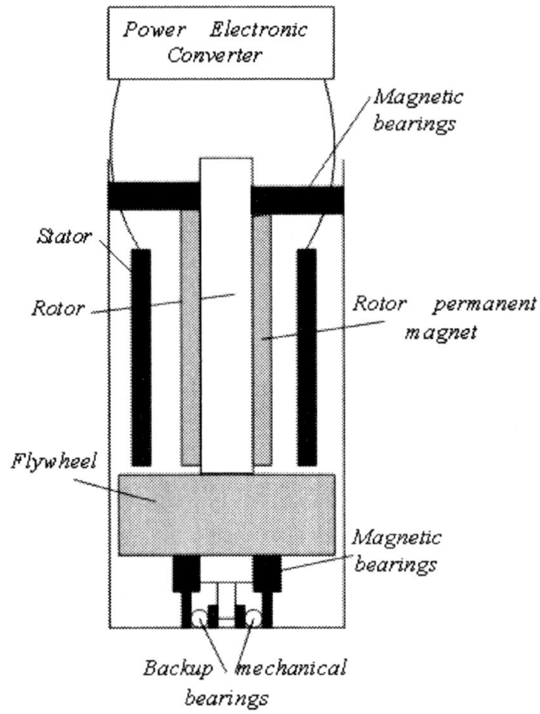

Since AC energy is at the input and the output of the system, the functions for energy transformation can be jointed. This is made using power electronic converter in the contemporary flywheel battery. Flywheels can be classified as low and high speed ones dependent on the speed of the rotor rotation (Emadi, 2005).

A principle structure of a low-speed flywheel system is shown in Figure 21.

Since the stored kinetic energy is proportional to the square of the speed and the speed is limited to not very high levels, a considerably larger inertia is necessary for low-speed flywheels, which results in a large weight. Consequently, magnetic support for the conventional bearings is advisable to increase the bearings' life. Low-speed flywheels are usually designed for short discharge times and also referred to as power wheels.

Low-speed flywheels rotate at up to 600 rpm. The output frequency of the generator is between 100 and 200 Hz.

Compared to the high-speed flywheels, weight and space requirements of the low-speed storage systems are doubled; but, the simpler construction results in lower cost, approximately five times lower. Low-speed, systems are used only in stationary applications where cost considerations are of primary concern.

Figure 22 shows a schema illustrating a structure of a high-speed flywheel system.

High-speed flywheels rotate at speeds above 10000 rpm up to 100000 rpm. Such high speeds are possible because of the relatively small mass moment of inertia of the rotor. It is wound from high-tech compound materials with a specific strength, five times higher than the steel. It is expected that, in the near future, with the advancement of material science, the cost of high-tech composite materials will go down considerably. In addition, conventional bearings at high speed are of no use; therefore, magnetic bearings have to be employed. Furthermore, at such high speeds, aerodynamic losses are considerable. Then, since the rotor has to run in a vacuum, no effective cooling can be provided.

The high-speed flywheels have a low weight, small size, and high energy density, which make them suitable for vehicular applications. However, the technology is relatively new and not widely proven in the field.

Advantages of the flywheel systems compared to the other methods of energy storage are as follows (Boyes, 2000):

- Very fast recharge and low temperature dependence
- Easer control and monitoring of stored energy because to realize these it is enough to measure the speed
- The systems requires comparatively lower expenses for maintenance
- They are suitable for higher powers dependent on power of the power electronic converter

Systems designed for powers for dozens of kW to dozens of MW are known.

Figure 23. Usage of electrical energy

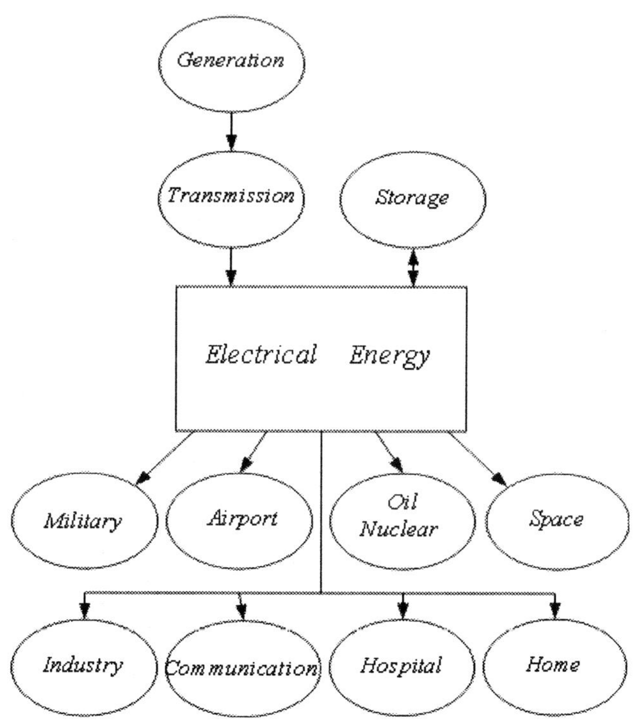

USING THE ENERGY AS ELECTRICAL ENERGY

As it has been mentioned in Chapter 1, a part of energy is generated using fossil fuels. Recently, the part of energy used as electrical energy has increased and the trend is to extend this increase. Figure 23 shows the place of electrical energy in the processes of its generation, storage and usage.

Electrical energy is required to be generated. This happens either using fossil fuels or using renewable energy sources. Next stage is energy transmission to the consumers. Since it is not possible to consume entire quantity of energy in its generation moment, energy storage is also required. This is also imposed in the cases when temporary interrupts of distributed energy from network are probable.

Figure 23 also displays the fields of electrical energy usage. In some fields, energy is used directly, while in others it is converted in other type – mechanical and heat energies, etc.

Power Electronics takes an important role in all processes connected with generation, transmission and usage of electrical energy (Ahmed,1999),(Ericson,2001). Chapter 3 is devoted to these topics.

REFERENCES

Ahmed, A. (1999). *Power electronics and technology.* UK: Prentice Hall.

Akhill, A., Butler, P., & Bickel, T. (1993). *Battery energy storage and superconducting magnetic energy storage for utility applications: A qualitative analysis.* (Tech. Rep. SAND93-2477). USA: Sandia National laboratories.

Boyes, J., & Klark, N. (2000, July). *Flywheel energy storage and super conducting magnetic energy storage systems.* Paper presented at IEEE PES 2000 summer meeting, Seattle, WA.

Emadi, A., Nasiri, A., & Bekiarov, S. B. (2005). *Uninterruptible power supplies and active filters.* Florida: CRC Press.

Erickson, R., & Maksimovic, D. (2001). *Fundamentals of power electronics.* Norwell, MA: Kluwer Academic.

Forschungscentrum, J. (n.d.). *Basic terms and operating principle of a fuel cell electricity generation system.* Retrieved December 2008 from www.fz-juelich.de

Henry, L., & Strunz, K. (2007). Superconducting magnetic energy storage (SMES) for energy cash control in modular distributed hydrogen – electric energy systems. *IEEE Transactions on Applied Superconductivity, 17*(2), 2361–2364. doi:10.1109/TASC.2007.898490

Luongo, C. A. (1996). Superconducting storage systems: An overview. *IEEE Transactions on Magnetics, 32*(1), 2214–2223. doi:10.1109/20.508607

Maher, B. (2006). Ultracapacitors provide cost and energy savings for public transportation applications. *Battery Power Products and Technology Magazine, 10*(6).

Ramani, V. (2006). Fuel cells. *The Electrochemical Society's Interface, 15*(1), 41–44.

Sung, B., Park, D., Park, J., & Ko, T. (2009). Study on a Series Resistive SFCL to Improve System Transient Stability: Modeling, Simulation, and Experimental Verification. *IEEE Transactions on Industrial Electronics*, *56*(7), 2412–2419. doi:10.1109/TIE.2009.2018432

Technologies, E. (n.d.). *Industrial batteries – Sonnenschein A600. Premium quality for uninterrupted communication.* Retrieved April 2008 from http://www.sonnenschein.org

Technologies, E. (n.d.). *Industrial batteries- Network power Sonnenschein A400. Outstanding cost efficient. Specifications.* Retrieved April 2008 from http://www.sonnenschein.org

Technologies, E. (n.d.). *Handbook for Gel-VRLA - batteries.* Retrieved April 2008 from http://www.sonnenschein.org

Ter-Gazarian, A. (1994). *Energy storage for power systems.* UK/London: Peter Peregrinus Ltd.

Walter, K. (1997).The unitized regenerative fuel cell. *Science & Technology Review*, May, 12-14.

Chapter 3
Power Electronics and Its Role in Effective Conversion of Electrical Energy

OVERVIEW

As it has been clarified by information provided in the previous chapters, there is a necessity to convert electrical energy from one input kind into other output kind of energy.

Power electronics is a technology to create power electronic converters associated with conversion of electrical energy using electronic means. Cumulative electronic means covers the following groups:

1. **Power schema made on the basis of power electronic devices (switches):** Diodes, thyristors (silicon controlled rectifier - SCR), insulated gate bipolar transistors (IGBT), metal–oxide–semiconductor field-effect transistors (MOSFET), etc.
2. **Control and regulation system made on the basis of analogue and digital electronic elements:** Analogue and digital integral circuits, microprocessors, microcontrollers, digital signal processors, etc. Additionally, software is included here, if a processor is applied in the control system.
3. **Sensors used to monitor the basic variables:** Current, voltage, power, temperature, etc.

DOI: 10.4018/978-1-61520-647-6.ch003

Figure 1. General schema of electrical energy conversion

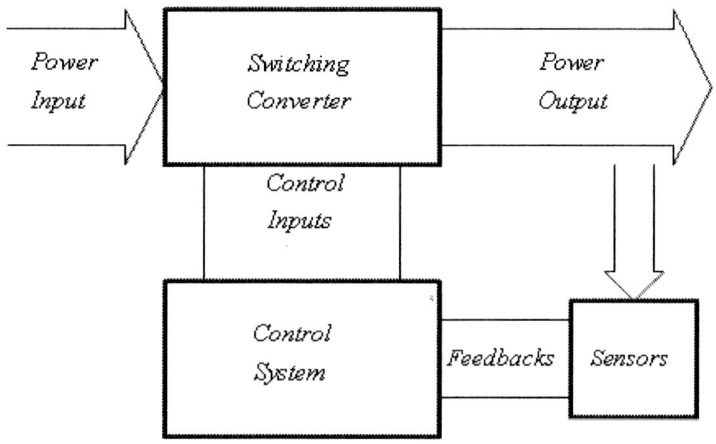

Figure 1 shows the basic schema to explain the definition. The conversion is made under the action of control signals, which are obtained from the control system. The most frequently, sensors are applied to monitor parameters referred to the output energy. Also, it is possible to monitor the parameters referred to the input energy, as well as variables in the converter itself. The sensor output signals are employed as feedbacks to the control system.

In Figure 2 the schema is extended with showing several of the possible sources of electrical energy as input ones, as well as several of the possible loads demanding electrical energy as output energy of the conversion.

Dependent on the input and output type of the electrical power, power electronic converters can be divided in several basic groups as follows:

1. Converters of AC into DC power – they are called rectifiers
2. Converters of AC into AC power

Figure 2. Sources and consumers of electrical power

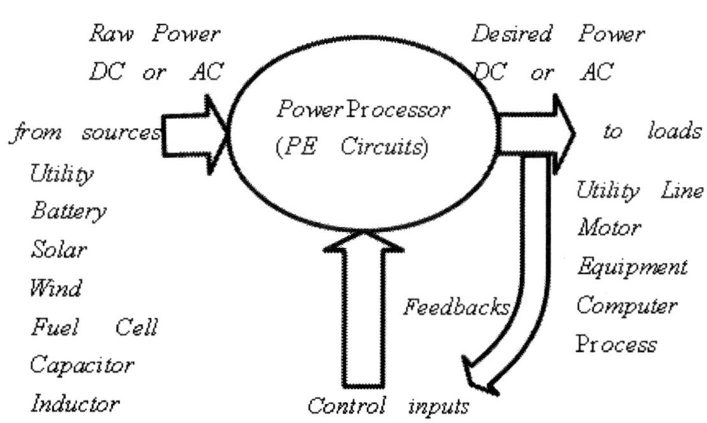

3. Converters of DC into DC power
4. Converters of DC into AC power – they are called inverters

In this chart, as well as in the whole book, international accepted symbols for the type of energy AC (alternating current) and DC (direct current) are used.

The Power Electronics is applicable within the whole field of electrical power systems. The power range of its applications is from a few VA/Watts to several MVA/MW.

The Power Electronics demands knowledge many and also different fields, thereby it is interdisciplinary (Peter, 1994). Figure 3 displays these fields, which are:

1. Structure, operational principle, characteristics and parameters of the power semiconductor devices
2. Electronic circuit theory – analogue, digital and microprocessor – required to synthesize the control system
3. System and regulation theories are required because the power electronic converters operate in closed loop for automatic control
4. Signal theory, because different signals in type – analogue, digital, and in shape – sinusoidal waveforms, saw waveform, find application in the Power Electronics
5. Microprocessors, digital signal processors, programming logic – increasingly are approved as a kernel in the control systems. It might be as well to know programming languages and developing systems for them
6. Electrical machines and mechanisms – a lot of power electronic converters are designed for control of different motors – synchronous, asynchronous, DC and AC
7. Electromagnetism and electromagnetic compatibility – power electronic converters emit EMC disturbances caused by the switching processes in them onto the supply network and onto the ambience space

Figure 3. Interdisciplinary nature of the Power Electronics

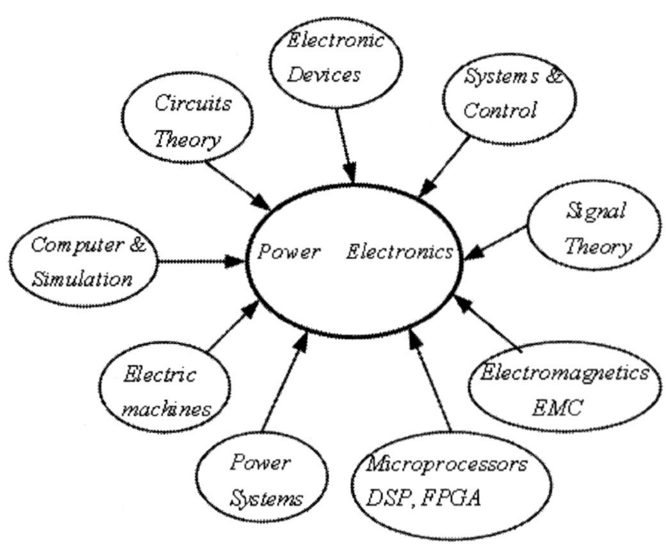

8. Software for simulation of the power electronic circuits, schemas and control systems - they are required in the processes of analysis and design, as well as during the put in action process and adjustment of the power electronic converters
9. Generation and distribution systems of electrical power

The experience in power electronics study is an exciting and challenging one. The expanse of power electronics applications is growing rapidly (Lander,2003). New devices keep coming into the market, sustaining development work in the Power Electronics (Bose,1992).

Power electronic converters are applied in many and different fields; the most important of them are examined in Section 3 in this book. They are:

- Conversion under generation, distribution and transmission of electrical energy
- Uninterruptible power source systems
- Systems for power supply and distributed generation of electrical power using renewable energy sources
- Systems with electrical engineering technology application
- Home appliances
- In automobile, railway and air transportation

Nowadays, about 33% of total energy is consumed as electrical energy of worldwide importance. This share is predicted to increase to 67% towards 2040 year. In this way, the role of the Power Electronics increases significantly not only in the processes of generation, distribution and transmission, but also in generation of electrical power using renewable energy sources - sun, wind, etc. (Blaabjerg, 2005; Bose, 2007).

Figure 4. Saving power through the Power Electronics. Source: Saving Energy Through Innovation and Technology ©2008Infineon Technologies North America Corp., Page 10, Figure 2.1

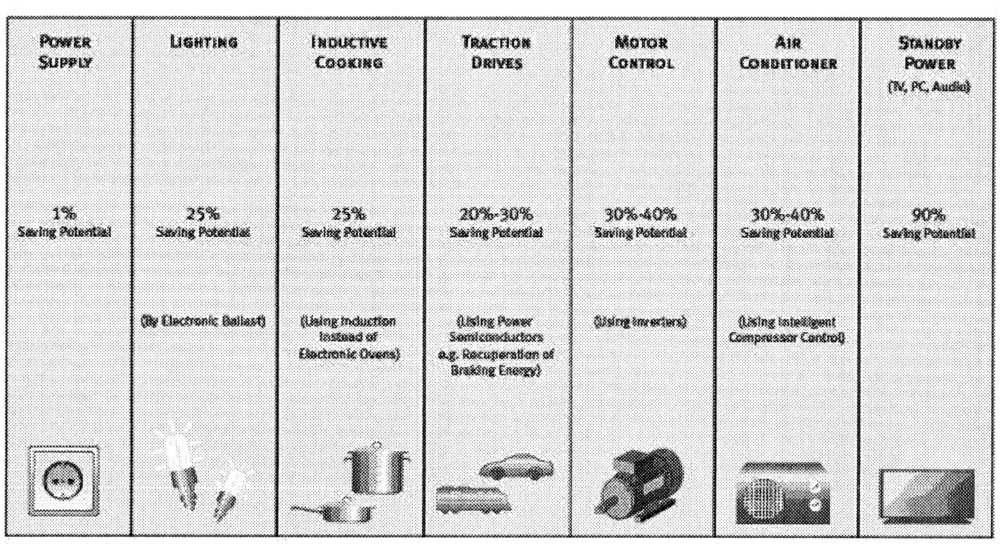

Source: Eupec GmbH: BVG- Berlin: Siemens / ECPE.

According to several researches, remarkable achievements in the field of the Power Electronics can lead to energy savings up to 50%. Figure 4 shows the possibilities for the savings in several more fields (Infineon, 2008).

According to already made researches, electrical motors consume about 65% of total electrical energy; another 20% are used up for lightning. Control of the electrical motors using the power electronics means is a significant step towards the increase of energy efficiency. For example, only in the fields of air conditioning and pressure pumps, the energy saving can reache up to 25% (IEEE Board of Directors, 2008). In the lightning field, fluorescent lamps are imposed. Their control is high frequency control and they are approximately 4 times more effective than incandescence wire lamps. New prospects are formed in the use of LEDs. Generally about 15% can be the energy savings in USA if the contemporary achievements of the Power Electronics are used (IEEE Board of Directors, 2008).

The quicken interest about the printed transactions in the field of the Power Electronics also confirms the increased influence of Power Electronics (Santana Communications International, 2007).

Because of the complexity of the power electronic converters, their design demands to joint forces of specialist in different fields. An example is the establishment of the Center for Power Electronics Systems (CPES) in USA. One of its perspective operation is the integration of the power electronic systems (Lee, 2008).

PRINCIPLES OF CONVERSION OF ELECTRICAL ENERGY

The universal classical principles, on which the operation of power electronic converters of electrical power is based, are examined in this chart (Bird,1993),(Mohan, 1995). The converters are shown as blocks because the stress is made upon the waveform of the input and the output signals. The purpose is to clarify the ideas, which serve as bases to implement the different converters.

Operational Principle of Converters of AC into DC Powers: Rectifying Principle

A rectifying principle is explained in Figure 5. The converters using this principle are called rectifiers.

Figure 5. Rectifying principle

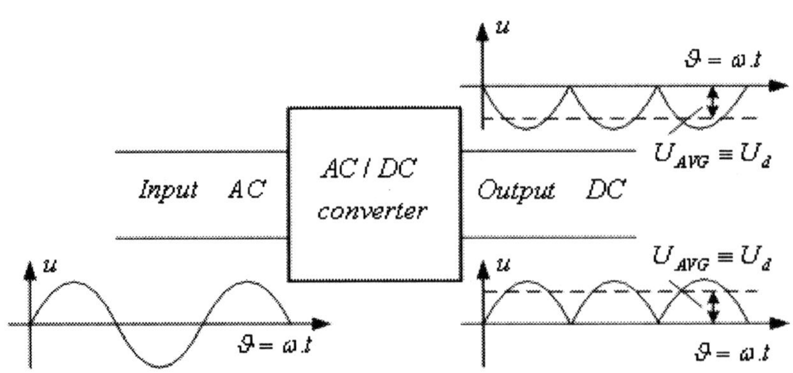

It is seen, that the converter transfers bipolar input voltage into uni-polar output voltage – positive or negative voltage. The output voltage contains constant DC component U_d equal to the average value U_{AVG}, which is necessary to the load requiring DC power. This component is separated from the output voltage by including smoothing filters between the output of the converter and the load.

Figure 5 displays sinusoidal waveform of the input voltage, but the principle is also applicable with others bipolar waveforms – square and saw waveforms, etc.

From the graphics shown in Figure 5 one can see that the average value of the output voltage U_d in the ideal rectifier depends only on the value of the input voltage and it can not be changed in the converter. These rectifiers are uncontrolled rectifiers. Basic element for their implementation is semiconductor diode.

Frequently in practice, the DC loads require the possibility to change the value of the rectifier output voltage. This can be easily achieved within the converter itself– the idea is shown in Figure 6.

The idea is to deliver the voltage across the load only in particular intervals with the possibility to change their duration. In general, the change of the output waveform can be different from that shown in Figure 6.

The rectifiers, in which possibility to change the value of the output voltage by an influence in the converter is available, are called controlled rectifiers. Thyristor SCR is the basic element for implementation of the classical principle shown in Figure 6. Recently, power transistors have been widely used, giving the possibility to improve the indicators regarding the source of the input AC energy.

Figure 6. Control of the output voltage in AC/DC converters

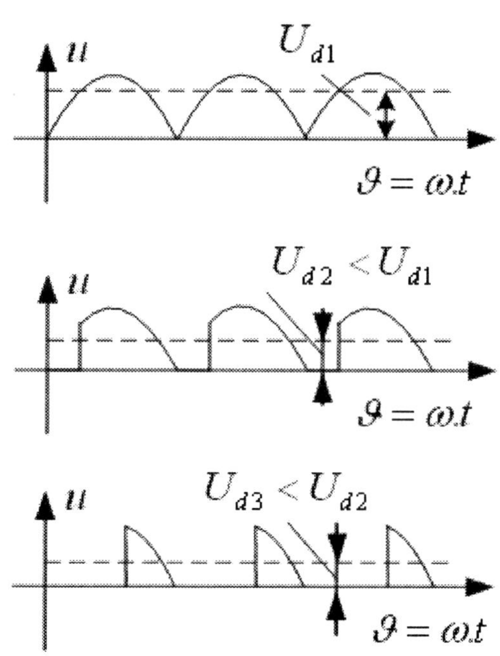

Figure 7. Operational principle of AC/AC converters

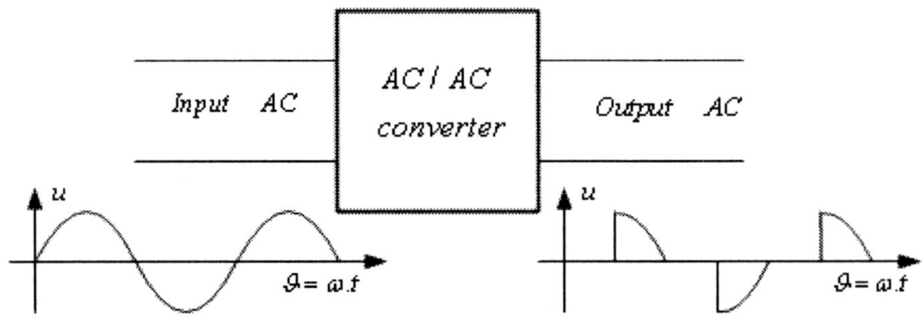

Operational Principle of Converters of AC into AC Powers: AC/AC Converters

The principle of AC/AC conversion is explained in Figure 7. It is applied in AC loads requiring the change of the value of their source voltage.

The output voltage remains bipolar (AC), only its value is changed. The idea is to deliver positive and negative voltages across the load only in particular intervals with the possibility to change their duration. This is made by an influence in the converter. The idea is closed to this of the controlled rectifiers.

In general, other bipolar waveform of the output voltage is possible. They depended on the implementation of the converter and the type of load.

Thoroughly, these converters are used when the input voltage is of sinusoidal waveform. Their important feature is that frequency of the first harmonic of the output voltage is equal to the frequency of the input voltage. Semiconductor elements used here are the same as those used in controlled rectifiers.

Operational Principle of Converters of DC into DC Powers: DC/DC Converters

A schema illustrating the DC/DC conversion principle is shown in Figure 8. This principle is applied in DC loads requiring the change of the value of their source voltage.

The idea is to change the DC input voltage, which can be positive or negative, using the converter into unipolar pulses, which can be either only positive or only negative.

The average value of the pulse voltage is marked in Figure 8 with U_d. The average component is separated from the output voltage by including smoothing filters between the output of the converter and the load.

There are two possible ways to change the value of U_d.

The first is by changing the pulse duration t_p when the period is constant T (frequency f is constant too). This method is called pulse-width modulation.

The second is by maintaining the pulse duration t_p constant and by changing the period T (frequency f). This method is called frequency-pulse modulation.

To implement these converters, power transistors – MOSFET or IGBT are used.

Figure 8. Operational principle of DC/DC converters

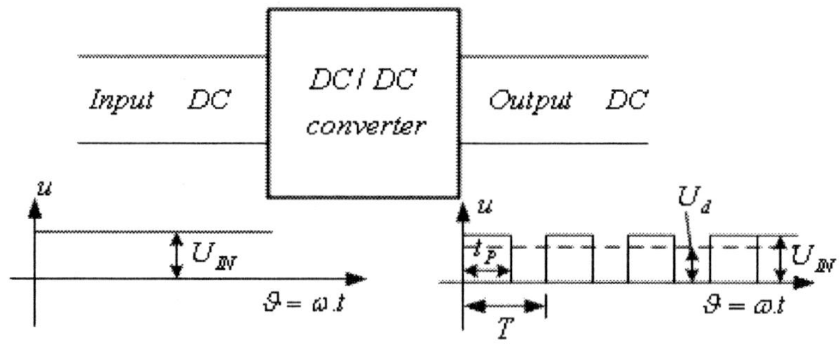

Operational Principle of Converters of DC Into AC Powers: Inverting Principle

The principle is explained in Figure 9. The converters using this principle are called inverters.

The idea is to transfer the DC input voltage which can be positive or negative by the use of the converter into bipolar pulse series. This bipolar pulse voltage can be used to supply AC loads which are not curtailed to the supply voltage waveform. The pulse voltage contains one first harmonic marked as U_1 in Figure 9. If the AC load requires a sinusoidal waveform of the voltage, this fundamental harmonic is separated by a smoothing filter connected between the output of the converter and the load.

A change in the value of the voltage and its first harmonic U_1 is made by a change of the pulse duration t_p under constant period T (frequency f) – pulse-width modulation.

The output voltage in Figure 9 contains one positive and one negative pulses consequently. These single pulses are possible to be a series of pulses.

More frequently, MOSFET or IGBT power transistors are used to implement these converters. There are also inverters implemented using conventional type of thyristors as well as using gate turn-off thyristors (GTO).

Figure 9. Inverting principle

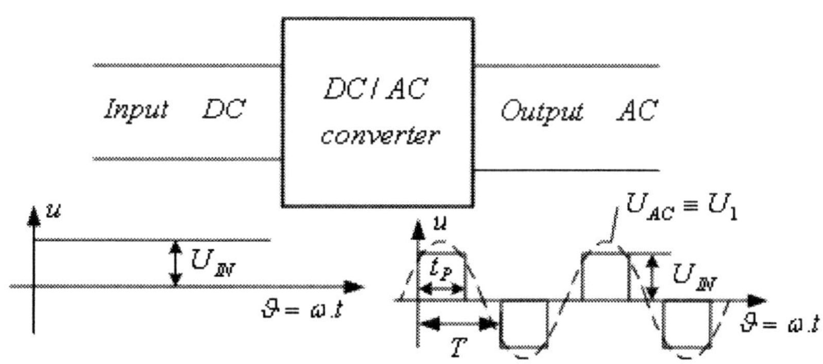

Operational Principle of Matrix Converters

For certain practical needs, the following task has to be solved: input AC voltage with number of phases p_I and frequency f_I is available; and required output AC voltage is with number of phases p_O which may be equal or different from p_I and with frequency f_O in general different from f_I. An additional condition is the possibility to change the effective value of the output voltage. A standard approach to solve the task is to use a double energy conversion – first, into DC energy using rectifier – Figure 5, Figure 6, and then, into AC energy using inverter – Figure 9.

Another possible solution is to use matrix converter, which operational principle is explained in Figure 10. This converter combines rectifiers and inverters functions in one converter.

The figure indicates a case of equality of the phases of the output and input voltages. The waveform of the output voltage of the matrix converter differs from ideal sinewave. In some applications, the frequency of the first harmonic of the output voltage is higher than the frequency of the input voltage, in others – lower. Although in matrix converters the number of the phases of the output voltage may be different from the number of phases of the input voltage, the most frequently used are converters with equal input and output phases.

The power schematics of matrix converters may be implemented using fully-controlled power devices with bi-directional conductivity – bi-directional semiconductor switches.

Only the general principles and ideas, on which are based the operation of the converters of electrical power, has been examined in this chart. It is worthy to be mentioned that these principles can be referred as classical ones. There are also their different varieties but yet the basic idea of the implementation finds just a modification.

After the description of basic conversion principle has been made, the following Section 2 contains the schemas for implementation of the converters and description of their operation, as well as detailed diagrams; moreover, basic relationships describing the processes of energy conversion are put.

As it is seen form the above made review, at the processes of energy conversion, the power electronic converters used are characterized by variables – currents and voltages, which are periodical ones but are non sinusoidal function of time. Appendix contains the harmonic content of those variables. Also, data about which converters from the ones examined in Section 2 are valid is included.

Figure 10. Operational principle of matrix converters

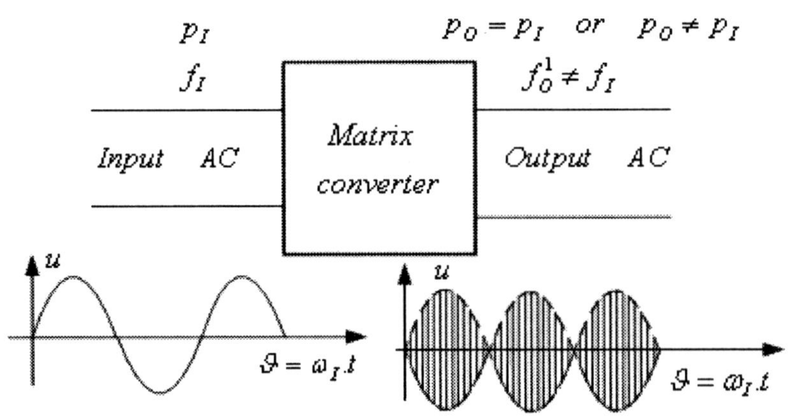

COMPUTER-AIDED DESIGN OF POWER ELECTRONIC CONVERTERS IN POWER ELECTRONICS

An important stage during the synthesis and design of power electronic converters is the usage of computer simulation to study some of the processes developed in them. The importance of the computer simulation is determined by the complexity of a power electronic converter, which consists of power schematic and control system usually operating in a closed automatic regulation system. Through a computers simulation, it is possible to study and forecast some transient processes connected with the initial start of the converter, its faulty modes, as well as to choose the most appropriate way to implement the control system.

Different methods in computer simulation of the power electronic converters have historically developed:

1. Working out a system of differential equations, which describe the operation of the power electronic converter, and solving the system using an appropriate time step through numeral methods
2. Combining the first approach with a next use of special software in solving the system
3. Using of special programs to simulate electronic schematics
4. Using of special programs to simulate power electronic schematics and their control system

The first approach is applied on the assumption that the semi-conducting devices are ideal switches with two possible states – on and off. The differential equation system, which describes the operation of the power schema, is worked out. Then, this system is solved using appropriate methods of numerical integration. To solve the system, a program that is executed by a computer is written at adequate programming language. Graphical change of some variables connected with the operation of power electronic schematic is generated as a result of program execution. Nevertheless, this method has been applied successfully for a long time, there are number of difficulties met using this method. The difficulties are connected with the following: the equation system of faulty mode usually differs from this of normal operation mode; the simultaneous study of the power schematic and its control system is impossible to be performed; the influence of real power elements over the operation of the power electronic converter is difficult to be recognized; the number of graphically obtained results is limited.

The second approach has been developing in the course of time. Instead of writing a program at a programming language, ready-made software having a possibility to solve the differential equation system and to visualize the results obtained is used (Carpita, 1996). Such software is *MATLAB*[1] where the results obtained from the equation system solving are visualized (Barucki, 2007; Toker, 2007). The control system and the operation of a closed outline of automatic regulation may be study separately from the power schematic using *SIMULINK*[1] module (Orizondo, 2006). This method saves quite many efforts and time to obtain the results solving the differential equations.

The disadvantages, connected with the joint study of the power schema and its control system, as well as the influence of the reality of the power semi-conduction devices, remain in the first two approaches using direct solving of the differential equation system.

The third approach uses special programs to simulate electronic schemas. Its level and possibilities are rather higher than those of the first two methods. Libraries of the special program used in the third case frequently contain models of power semi-conductor devices (Mohan, 1992). These models are used at simulation. A typical example of such a program is *ORCAD*[2]. Using *ORCAD*[2], the power schematic is drawn in the *CAPTURE*[2] module. Moreover, in the same module different types of analyses and conditions are set. The simulation itself is made through the *PSPICE*[2] module and then the results obtained may be graphical monitored. Difficulties to obtain stable solution, to choose an appropriate time step, to obtain accuracy of the result generated, etc., are met because the programs are not designed especially for the Power Electronics use (Wu, 2007).

Later on, special programs for computer simulation of power electronic schematics and their control systems appear at the Market. This is the fourth possible approach for computer simulation. This approach is in continuous development. The author knows two software – *PLECS*[3] and *PSIM*[4].

The first of them -*PLECS*[3] (Plexim GmbH, 2002-2008) is toolbox for a high-speed simulation of power electronic schemas and electrical machine using *MATLAB*[1]/*SIMULINK*[1]. A joint simulation of the power schematic and its control system is possible. The user may draw an electrical schema with components. This schema is represented as a subschema in the *SIMULINK*[1] level. The component library comprises different current and voltage sources, switches, passive components, voltmeters and amperemeters, as well as complex components such as electrical machine. Signals for controlled sources and switches are the inputs for the subsystem, while the results measured from voltmeters and amperemeters – outputs. Simulation of thermal modes in power electronic converters is also possible (Hammer, 2008).

The second software – *PSIM*[4](Powersim Technologies Inc., 2006) consists of several modules:

- Motor Drive Module includes modules of electrical machine and mechanical loads used to study motor drive systems.
- Digital Control Module includes functional blocks with their discreet gear functions, digital filters, blocks used to make one signal as a discreet signal and used to analyze discreet regulation systems.
- SimCoupler Module is an interface between *PSIM*[4] and *MATLAB*[1]/*SIMULINK*[1] for a co-simulation.
- Thermal Module permits an analysis of active power losses across the power devices.
- MagCoupler Module is an interface between *PSIM*[4] and software for magnetic field analysis *JMAG*[5] for a co-simulation.
- MagCoupler –RT Module is a connection between *PSIM*[4] and *JMAG-RT*[5] data files.

The schematic is drawn using PSIM Schematics. Some parameters of the semi-conductor devices, such as forward voltage drop, are possible to be set during the schematic draw. Thus, the simulation process comes nearer to the real physic process. Furthermore, ready-made schematics of power electronic converters, such as rectifiers, inverters, etc., may be used. Also, it is possible to connect current and voltage sensors to provide feedbacks. After the schematic is finished, a simulation through PSIM Simulator is made and then the results are graphically presented in *SIMVIEW*[4]. *PSIM*[4] permits a thoroughly simulation of a power electronic schematic, its control system and sensors for the feedbacks, i.e. simulation of the entire power electronic converter as it is presented in Figure 1.

A new approach to study the power electronic converters through computer simulation has arisen with the coming of the digital signal processors in the converter control systems. The approach combines the hardware and software means. A part of the power electronic converter is simulated using an appropriate software while another part of it is physical implemented (Monti, 2003; Jacobs, 2004).

The new product flowSIM[6] made by Tyco Electronics (Obermaier, 2007) is a new step in the simulation field using power electronic modules in the fields of electro-motion, electro welding and electrical energy generation. At determined steps of the simulation process, this product uses actual data of the measurements of real converters instead of simulation modules. Besides, simulation of high-frequency transient processes may be combined with low-frequency temperature simulation.

Let us see an example of a study of a power electronic converter using computer simulation. The example illustrates part of the capabilities of *PSIM*[4]. Figure 11 shows the schematic for the computer simulation (Antchev, 2009). The operation of a single-phase bi-directional AC/DC converter, whose schema is shown in Chapter 4, Figure 27, is studied here. Hysteresis current control (see Section 2) is used. The DC voltage in the output is stabilized by proportional-integral regulator. Using multiplier, the reference sinusoidal waveform for the source current is gained. The transitory values of the source current are monitored by a feedback. The difference between the values of the reference and real sinusoids determines which pair of diagonally connected transistors has to be switched on.

The example illustrates the possibilities of *PSIM*[4] to simulate the power schema together with the process and the sensors for feedbacks.

The computer simulation results are presented in Figure 12 and Figure 13 in rectifier and inverter mode, respectively, with effective value 230 V of network voltage and with value of the DC voltage 350 V.

From the results obtained, it is seen that the source current is in phase with (Figure 12) or displaced with 180° (Figure 13) to the source voltage and also the current is of sinusoidal waveform. Therefore, in both cases, the power factor is much closed to 1. Also, the tracing of the reference current with a certain hysteresis is seen.

The simulation results are confirmed also by experiments in both operation modes (Antchev, 2009).

Using an appropriate mathematical description of power electronic converters in *MATLAB*[1], it is possible to obtain oscilograms characterizing their operation. The following example analyses a three-phase to single phase matrix converter (Kunov, 2009) – see Figure 10 and Chapter 5.

The circuit for mathematical analysis is presented in Figure 14a. Bidirectional switches are used, realized as shown in Figure 14b and controlled by a corresponding algorithm. The converter is sup-

Figure 11. Schematic used for the computer simulation through PSIM software

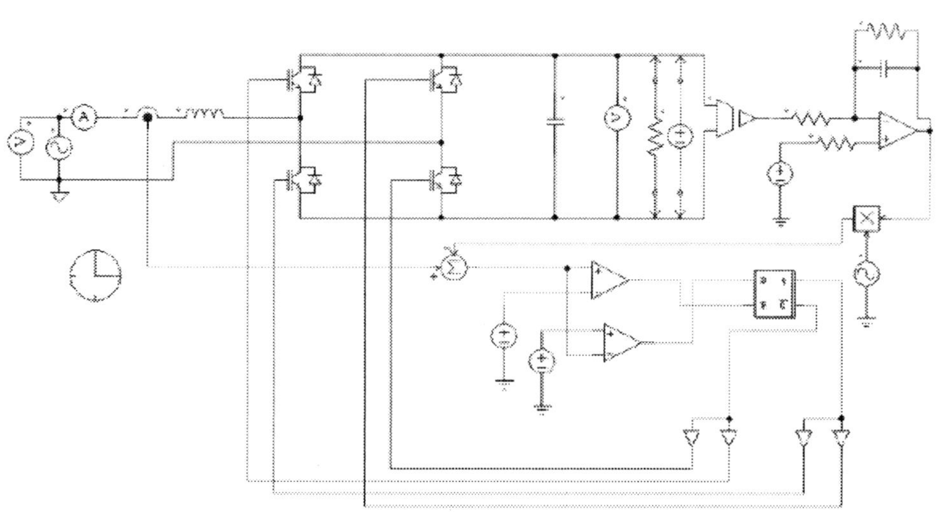

Figure 12. Computer simulation results in rectifier mode of operation – the network voltage and current

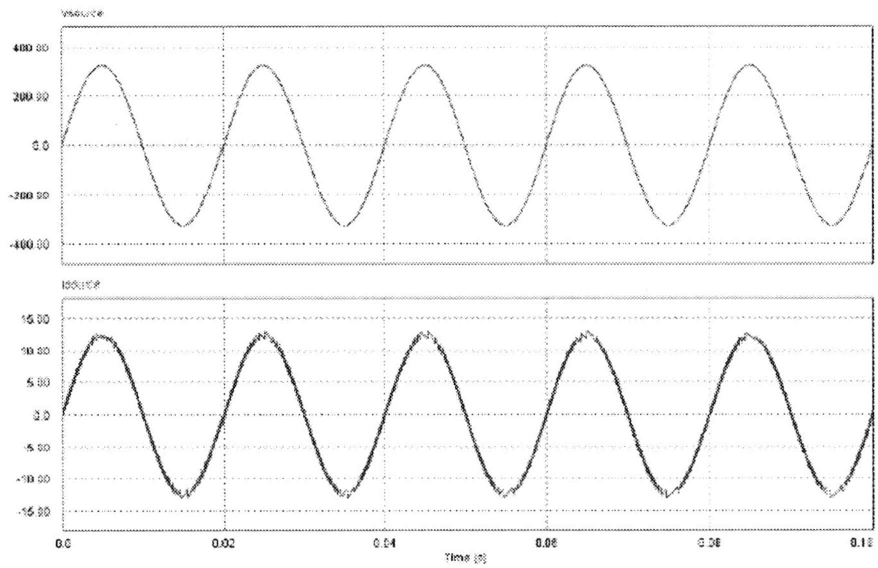

plied directly by the mains network. The three-phase line input voltages are described by the following equations:

$V_R = V_m \mathbf{sin}\theta$

Figure 13. Computer simulation results in inverter mode of operation – the network voltage and current

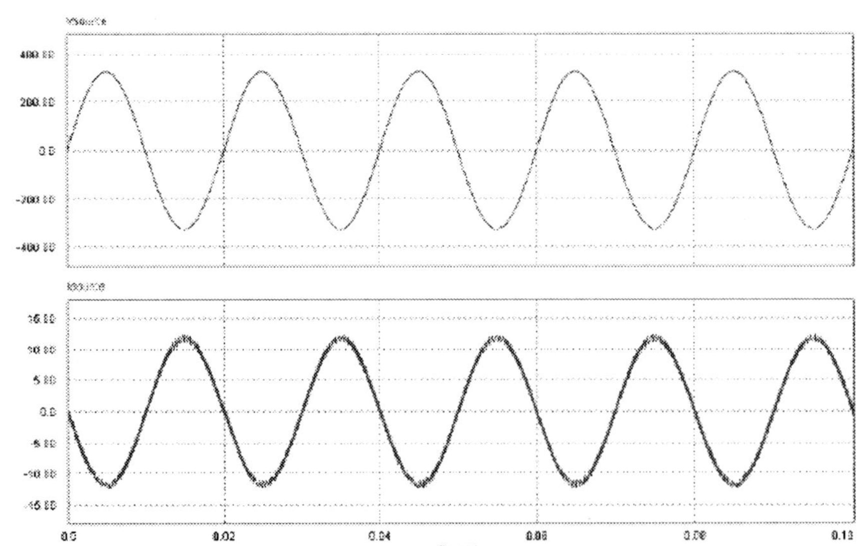

$$V_S = Vm\sin\left(\theta - \frac{2\pi}{3}\right)$$
$$V_T = Vm\sin\left(\theta + \frac{2\pi}{3}\right) \quad (3.1)$$

a. Equivalent circuit
b. Bidirectional power switch

The condition of potentials V_P, V_N regarding the null of the three-phase voltage system may be presented in matrix form as:

$$\begin{bmatrix} V_P \\ V_N \end{bmatrix} = \begin{bmatrix} S1 & S3 & S5 \\ S4 & S6 & S2 \end{bmatrix} \cdot \begin{bmatrix} V_R \\ V_S \\ V_T \end{bmatrix} \quad (3.2)$$

Figure 14. Circuit for investigation

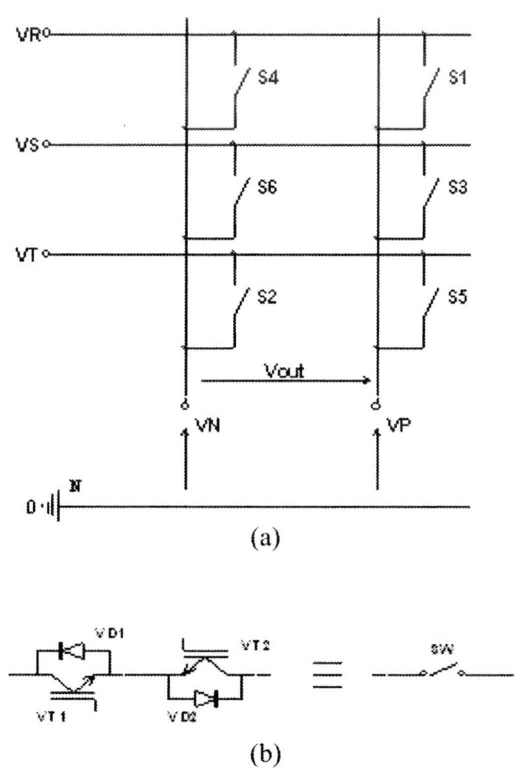

S1-S6 are pulse series with an amplitude of 1. These series are depicted in an appropriate way according to the switching on and off the corresponding switches in Figure 14. Level '1" corresponds to the turned on switch, while "0" – turned off one.

The single-phase output voltage in Figure 14 is

$$V_{out} = V_P - V_N \tag{3.3}$$

and from (3.1) and (3.2) it is written in the form:

$$V_{out} = (S1 - S4) * V_R + (S_3 - S_6) * V_S + (S_5 - S_2) * V_T \tag{3.4}$$

Figure 15. Potentials VS1-VS6 produced by the bidirectional switches.

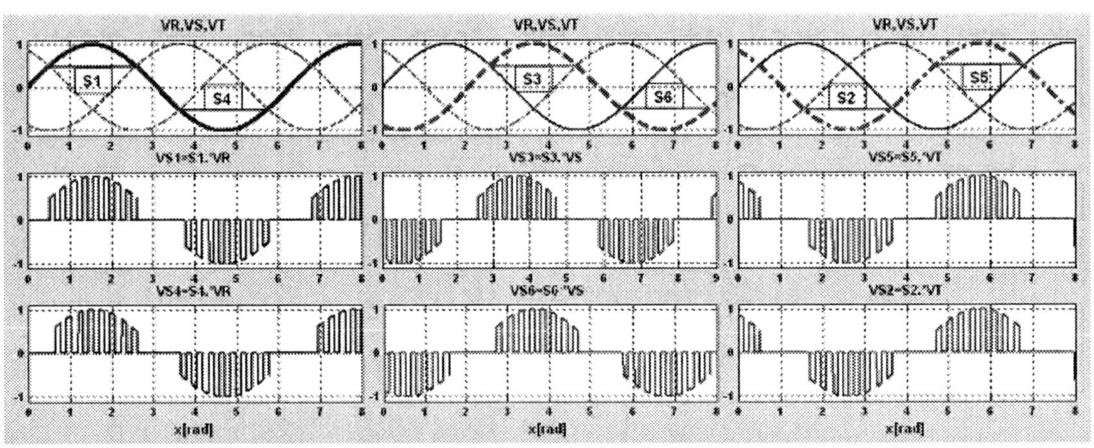

Figure 16. Potentials V_P, V_N and voltage V_{OUT}

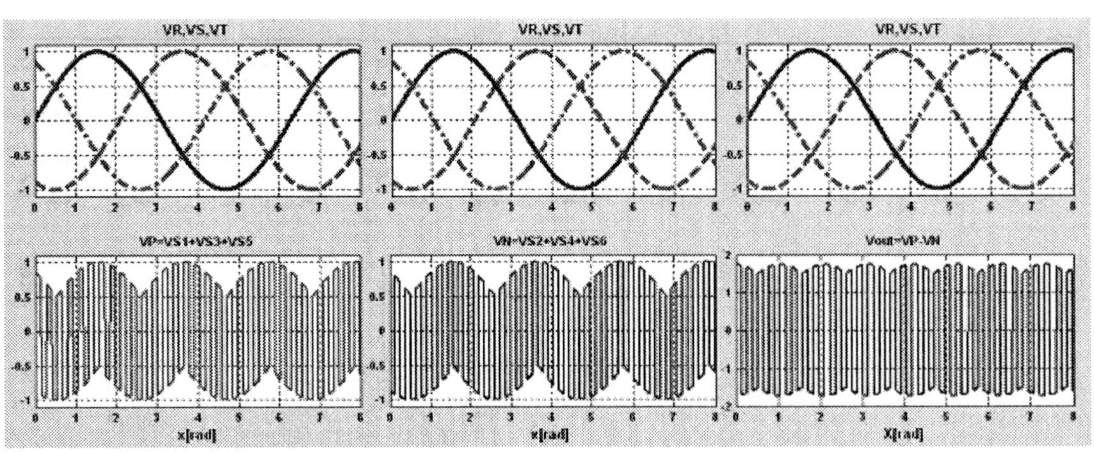

The *MATLAB*[1] program is used for the investigation of the matrix converter by solving the equations found. The potentials *VS1-VS6* produced by the bidirectional switches are shown in Figure15. The potentials V_P, V_N and voltage V_{out} are presented in Figure 16.

REFERENCES

Antchev, M., Petkova, M., Gurgulitsov, V., & Antchev, H. (2009). *Study of a single-phase bidirectional AC/DC converter with a high power factor.* Paper presented at IEEE conf. EUROCON 2009, May 2009, Russia, Saint- Petersburg.

Barucki, T. (2007, November). Analysis tool covering wide application range. *Bodo's power systems* (pp. 40-41).

Bird, B., King, K., & Pedder, A. (1993). *An introduction to power electronics.* Chichester, NY: John Wiley & Sons.

Blaabjerg, F., Consoli, A., Ferreira, J. A., & Van Wyk, J. D. (2005). The future of electronic power processing and conversion. *IEEE Transactions on Power Electronics, 20*(3), 715–720. doi:10.1109/TPEL.2005.846516

Bose, B. (1992). *Modern power electronics, evolution, technology, and application.* Institute of Electrical and Electronics Engineers. Piscataway, NJ: IEEE Press.

Bose, B. K. (2007). Power electronics – Why the field is so exciting. *IEEE Power Electronics Newsletter, 19*(4), 11–19.

Carpita, M., Ricerche A., & Monti, A. (1996). Voltage source converters and drives simulation at system level for control design applications. *European power Electronics and Drives Journal, 5*(3/4), 49-55.

Hammer, W. (2008). Combined thermal and electrical simulation of power electronic systems with PLECS. *IEEE Power Electronics Society Newsletter, 20*(2), 22–23.

IEEE Board of Directors. (2008). *Energy Efficiency.* Retrieved 2008 from http://www.ieeeusa.org

Infineon (2008). *Saving energy through innovation and technology.* Retrieved 2008 from http://www.techonline.com

Jacobs, J., Detjen, D., Karipidis, C. U., & De Doncker, R. W. (2004). Rapid prototyping tools for power electronic systems: demonstration with shunt active power filter. *IEEE Transactions on Power Electronics, 19*(2), 500–597. doi:10.1109/TPEL.2003.823240

Kunov, G., Antchev, M., & Gadjeva, E. (2009). *Computer modeling of three-phase to single- phase matrix converter.* Paper presented at 15th Symposium Power Electronics Ee2009, October2009, Novi Sad, Serbia.

Lander, C. (2003). Power Electronics (3rd Ed.). London/New York: McGraw-Hill.

Lee, F. C., Boroyevich, D., Van Wyk, J. D., Ngo, K., Lu, G. Q., & Xu, M. (2008). Power electronics system integration – a CPES perspective. *IEEE Power Electronics Society Newslwtter, 20*(2), 14–20.

Mohan, N. (1992). *Power Electronics: computer simulation, analysis and education using PSPICE.* Minneapolis, MN: Minesota Power Electronics Research and Education.

Mohan, N., Undeland, T. M., & Robbins, W. P. (1995). *Power electronics converters, applications and design.* New York: John Wiley & Sons.

Monti, A., Santi, E., Dougal, R. A., & Riva, M. (2003). Rapid prototyping of digital controls for power electronics. *IEEE Transactions on Power Electronics, 18*(3), 915–923. doi:10.1109/TPEL.2003.810864

Obermaier, W. (2007, July). A new class in power electronics simulator. *Bodo's Power Systems,* pp. 16-20.

Orizondo, R., & Alves, R. (2006).*UPFC simulation and control using the ATP/EMTP and Matlab/ Simulink programs.* Paper presented at the IEEE PES Transmision and Distribution Conference and Exposition Latin America. Venezuela.

Peter, J. M. (1994). Education in power electronics, the engineers point of view. *European Power Electronics and Drives Journal, 4*(1), 47–49.

Plexim GmbH (2002-2008). *PLECS Users Manual.* Switzerland, Zurich.

Powersim Technologies Inc. (2006). *PSIM Users Guide Version 7.0 Release 4.* Retrieved 2008 from http://www.powersimtech.com

Santana Communications International. (2007). European power electronics reader survey. Retrieved 2008 from http://www.powersystemsdesign.com

Toker, O., & Carroll, E. (2007, November). PLECS – the user-friendly simulation program for power electronics. *Bodo's Power Systems,* pp.48-51.

Wu, X., Wong, S. C., Tse, C. K., & Lu, J. (2007). Bifurcation behavior of SPICE simulations of switching converters: a systematic analysis of erroneous results. *IEEE Transactions on Power Electronics, 22*(5), 1743–1752. doi:10.1109/TPEL.2007.904207

ENDNOTES

1. MATLAB and SIMULINK are registered trademarks of the MathWorks Inc.
2. ORCAD, CAPTURE, PSPICE and PROBE are copyright by the Cadence Design Systems, Inc.
3. PLECS is registered trademark of the Plexim GmbH.
4. PSIM and SIMVIEW are registered trademarks of the Powersim Technologies Inc.
5. JMAG and JMAG-RT are copyright by the Japan Research Institute Ltd.
6. flowSIM is registered trademark of the Tyco Electronics.

Section 2
Electronic Energy Converters

Chapter 4
AC/DC Conversion

BASIC INDICATORS IN RESPECT TO THE SUPPLY NETWORK

Figure 1 displays a power electronic converter connected to the mains. In general, a power electronic converter is an electrical power converter – controlled or uncontrolled rectifier, AC regulator, compensator of reactive power, converter of phase number, active power filter. The converter supplies a load with power Pout, and in the same time it loads the mains with active power P and total power S.

Power factor is defined as a ratio of active power P to total apparent power S:

$$K_p = \frac{P}{S} \tag{4.1}$$

If the voltage and current of the supply network are with non-sinusoidal waveform, they contain DC component and they can be presented in Fourier series, then the active power is given as:

$$P = U_0 \cdot I_0 + \sum_{k=1}^{n} U_k \cdot I_k \cdot \cos\phi_k \tag{4.2}$$

DOI: 10.4018/978-1-61520-647-6.ch004

Figure 1. Powers in conversion of AC electrical energy

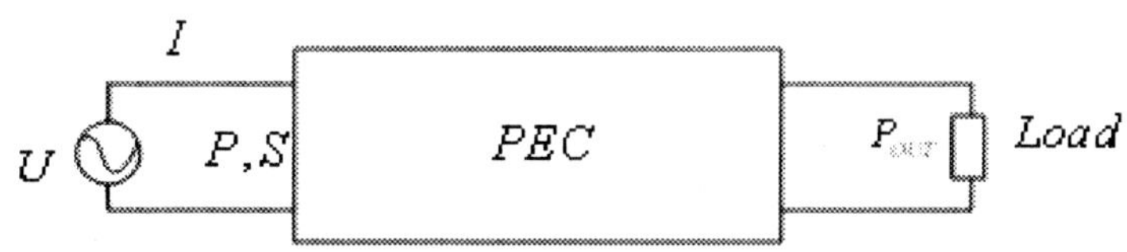

where in U_k and I_k are the effective values of the kth harmonic of the voltage and current, respectively, and ϕ_k is the displacement angle.

Total power is a product of the effective values of the source voltage U and the source current I:

$$S = U \cdot I \tag{4.3}$$

where in

$$U = \sqrt{\sum_{k=0}^{n} U_k^2} \text{ and } U = \sqrt{\sum_{k=0}^{n} U_k^2} \tag{4.4}$$

After substituting (4.4) in (4.1), it is found:

$$K_p = \frac{U_0 \cdot I_0 + \sum_{k=1}^{n} U_k \cdot I_k \cdot \cos\phi_k}{\sqrt{\sum_{k=0}^{n} U_k^2} \cdot \sqrt{\sum_{k=0}^{n} I_k^2}} \tag{4.5}$$

The mains voltage is usually accepted to be of a pure sinusoidal waveform and it does not contain a DC component, and the source current is usually accepted to be of non-sinusoidal waveform. So, it is derived:

$$K_p = \frac{U_1 \cdot I_1 \cdot \cos\phi_1}{U_1 \cdot \sqrt{\sum_{k=1}^{n} I_k^2}} = \frac{I_1 \cdot \cos\phi_1}{\sqrt{\sum_{k=1}^{n} I_k^2}} = \frac{I_1}{I} \cdot \cos\phi_1 = \nu \cdot \cos\phi_1 \tag{4.6}$$

From (4.6) it is seen, that the power factor is a product of two variables, the highest value of each of them can be equal to 1. The two variables are:

- ν : The distortion factor
- $\cos\phi_1$: The displacement factor, where in ϕ_1 is the angle of the displacement between the sinusoidal source voltage and the first current harmonic.

Besides the distortion factor, the current non-sinusoidal waveform is also characterized by harmonic distortion factor or a total harmonic distortion defined as:

$$K_H = \frac{\sqrt{\sum_{k=2}^{n} I_k^2}}{I_1} \tag{4.7}$$

It is easy to find the relationship between the two coefficients:

$$K_H = \frac{\sqrt{I^2 - I_1^2}}{I_1} = \frac{1}{\nu} \cdot \sqrt{1 - \nu^2} \tag{4.8}$$

or

$$\nu = \frac{1}{\sqrt{1 + K_H^2}} \tag{4.9}$$

As a result from (4.6) and (4.9), the power factor can be present in two ways using different variables, as follows:

$$K_P = \nu \cdot \cos\phi_1 = \frac{1}{\sqrt{1 + K_H^2}} \cdot \cos\phi_1 \tag{4.10}$$

The ratio of the output power P_{out} to the active power P is the efficiency coefficient:

$$\eta = \frac{P_{out}}{P} \tag{4.11}$$

Complete idea about the converter efficiency is given by efficiency factor:

$$K_E = \frac{P_{out}}{S} = \eta \cdot \nu \cdot \cos\phi_1 = \eta \cdot \frac{1}{\sqrt{1 + K_H^2}} \cdot \cos\phi_1 \tag{4.12}$$

The ways to increase the efficiency factor up to 1 are seen from (4.12). The three methods are:

- Increasing of the efficiency coefficient of the power electronics converters $\eta = 1$
- The consumed source current has to be with low harmonic contains $K_H = 0$ or $\nu = 1$
- The first harmonic of the source current has to be in phase with the source voltage - $\phi_1 = 0$

Let assume that the source current has pure sinusoidal waveform but it is displaced at angle ϕ to the source voltage. A vector diagram for powers is presented in Figure 2. Besides the total power S and active power P, Figure 2 presents also the reactive power Q.

In this case $\nu = 1$, and the following equations are valid:

$$S^2 = P^2 + Q^2 \qquad K_P = \cos\phi \tag{4.13}$$

Figure 2. Powers in the case of linear loads

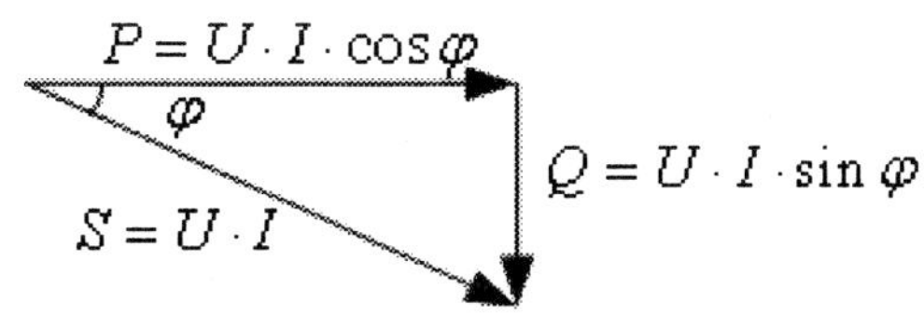

A vector diagram for powers, in the case when the source current has different waveform from the pure sinusoidal waveform, is shown in Figure 3. The first harmonic of the current is displaced at angle ϕ_1 to the voltage. The diagram presents also the apparent fundamental power S1 and distortion power D.

In this case $\nu = \cos\theta$ and the following equations are valid:

$$S^2 = P^2 + Q^2 + D^2 \qquad K_P = \nu \cdot \cos\phi_1 = \cos\theta \cdot \cos\phi_1. \tag{4.14}$$

It has to be mention that both the consumer and the distribution electrical company have benefits from the use of converters whose power factor is close to 1.

Figure 4 illustrates the benefit for the consumer. A single-phase network 230V is available. It is possible to connect total load with 15A. At minimum source voltage 207V, the total possible power is 3105VA. If the supply of a computer has to give power 280W but it works with power factor 0.6 (typical value without power factor correction) it will load the network with power equal to 470VA. This means that t it is possible to connect only 6 computers to this network.

If the consumer has computers which supplies are with a power factor correction and its value is 1 (typical values are above 0.96), each of the computers will load the network with power equal to 295VA. Then, the consumer may connect 10 computers to the same network.

Figure 5 illustrates the benefit for the power distribution company. In both cases the network voltage is the same. In the first case, the converters incorporated in the supplied building have an insufficient power factor, which means that they needs higher total power S, i.e. higher effective current value. Consequently, the power supply wires should have higher section.

If the connected converters at the supplied building are with power factor closed to 1, they will require less total power S, less effective current value. As a result, the section of the wires in this case can be decreased.

Figure 3. Powers in the case of nonlinear loads

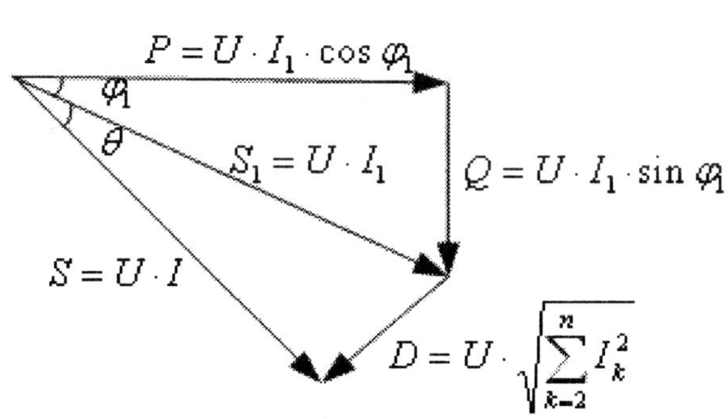

Figure 4. Benefit for the consumer from operation with a high power factor

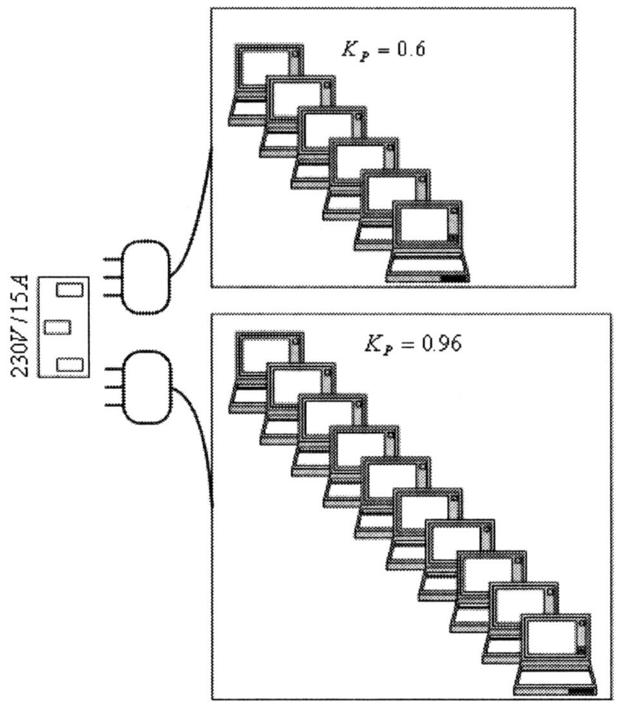

Figure 5. Benefit for the power distribution company from operation with a high power factor

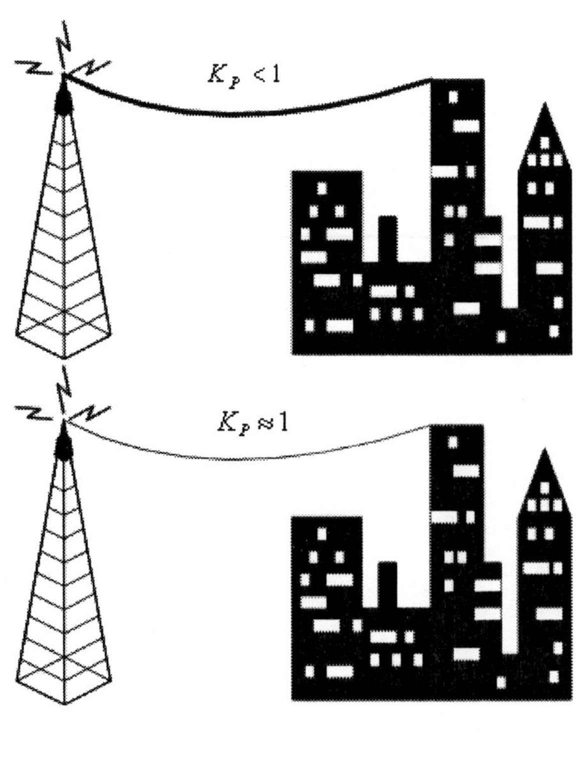

SINGLE-PHASE AND THREE-PHASE UNCONTROLLED RECTIFIERS

As it has been written in Section 1, the rectifiers are uncontrolled and controlled ones dependent on the absence or presence of the possibility to change the output voltage value using influence in rectifiers, respectively.

Besides this characteristic, the rectifiers also differ in:

- **Number of phases:** More often single-phase and three-phase rectifiers
- **Type of the current of the supply source for each phase:** Half-wave rectifiers (current flows only during one of the half-periods of the source frequency – either negative or positive half-period) and full-wave(current flows during both half-periods of the source frequency)
- **Type of load:** More often – active, active-capacitive and active-inductive types

It has to be mentioned that the active load is met very rarely in practice.

Semiconductor devices with one-direction conductivity – diodes, are used to implement the uncontrolled rectifiers. Semiconductor devices with a control electrode – thyristors or transistors, are used to implement the controlled rectifiers.

Single-Phase Uncontrolled Rectifiers

The operation of a single-phase bridge rectifier with active load, whose principal electrical schematic and its typical waveforms are shown in Figure 6, has the closest operation to the described rectifier operational principal in Section 1. The single-phase bridge rectifier is the most propagated one in practice. On its basis the fundamental concepts connected with the rectifiers will be introduced.

During a conventional positive half-period of the AC voltage, when the upper end of the secondary winding in the schematic has higher potential than the lower one, diodes VD1 and VD3 conduct, and during the negative half-period – VD2 and VD4 conduct. Thus, the current through the secondary winding, and through the primary winding, flows during both half-periods – a fully-wave rectifier. As the current through the load always flows from point A towards point B, point A has higher potential than point B. The waveform of the output voltage u_d in Figure 6 is made on the assumption that the point B has potential equal to zero volts. If the potential of the point A is assumed to be zero, than the output voltage will be negative. During the study, the diodes usually are assumed to be ideal switches – with infinitely low resistances during conductivity and with infinitive high resistances in the turned off state. Consequently, the shape of the output voltage traces the voltage of the secondary winding but it is unipolar – rectified.

The waveform of an ideal DC voltage is a straight line above or below X-axis dependant on its polarity. As it appears in Figure 6, the waveform of u_d differs from the ideal one. It is a periodical even function, which in accordance with Fourier series has a DC component and even harmonics. They are determined by the corresponding coefficients a_0 and a_K from the Fourier series.

The DC component is the ideal DC voltage and the put over it harmonics are voltage changes – they are called ripples of the output voltage.

The DC component is equal to the average value of the output voltage:

Figure 6. Electrical schematic and typical waveforms of a single-phase bridge rectifier with active load

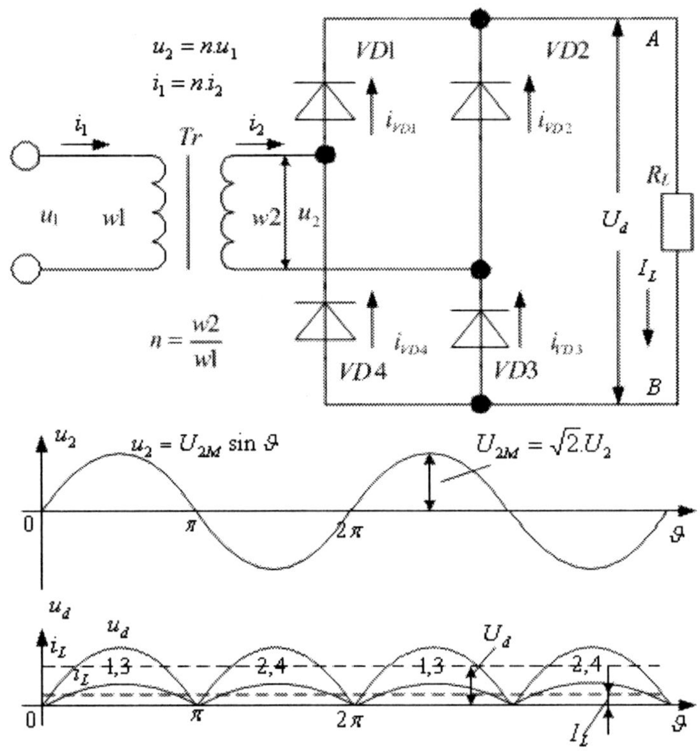

$$U_d = \frac{a_0}{2} = \frac{1}{\pi}\int_0^{\pi} \sqrt{2}.U_2.\sin\vartheta.d\vartheta = \frac{2.\sqrt{2}}{\pi}.U_2 = \frac{2\sqrt{2}}{\pi}.n.U_1 \tag{4.15}$$

It is visible in Figure 6, that in one period of the supply voltage u_2, two periods of the output voltage are hold. Therefore, the frequency of the first harmonic of the output voltage is twice higher than the frequency of the source voltage.

The ratio of these frequencies in all rectifiers – uncontrolled and controlled, is marked as m and it is called a rate frequency of ripples.

For the most widely propagated rectifiers there is a following relationship:

$$m = p.l\,, \tag{4.16}$$

where in p is the number of phases and $l = 1$ in half-wave and $l = 2$ in full-wave rectifiers. In single - phase full - wave rectifiers $m = 2$.

The maximum value of the first harmonic of the output voltage is derived from the Fourier series:

$$U_{dM}^{1} = \frac{2}{\pi}\int_0^{\pi} \sqrt{2}.U_2.\sin\vartheta.\cos.2.\vartheta.d\vartheta = \frac{2}{3}.\frac{2\sqrt{2}}{\pi}.U_2 \tag{4.17}$$

In all rectifiers – uncontrolled and controlled, the ratio of the maximum value of the first harmonic of the output voltage to the DC component is ripple factor:

$$S_1 = \frac{U^1_{dM}}{U_d} \tag{4.18}$$

This coefficient shows the quality of the output voltage; the smaller the coefficient is, the closer to the ideal DC voltage the output voltage is.

After substituting (4.15) and (4.17) in (4.18), for the single-phase full-wave rectifier is derived that $S_1 = \frac{2}{3} \approx 66.7\%$.

As we will see further down, the number m is different for the different rectifiers. The higher it is, the higher the frequency of the first harmonic of the output voltage is and the easier its filtering is. Moreover, the ripple factor is smaller because the following relationship for the uncontrolled rectifiers is provable:

$$S_1 = \frac{2}{m^2 - 1} \tag{4.19}$$

The waveform of the load current i_L in Figure 6 and this of the voltage u_d are of the same form because the load is active. The relationship between the average values of the current and voltage in the output of the rectifier is:

$$I_L = \frac{U_d}{R_L} \tag{4.20}$$

As it appears in Figure 6, in each half-period, the waveform of the current i_L has sinusoidal form coinciding with the one of the supply voltage u_2 but unipolar. In each moment, the current value is equal to the value of the secondary current i_2. From the rectifier operation is obvious that the current i_2 flows into different direction in the different half-periods – during the one it comes out from the upper end of the secondary winding through VD1, and during the other current goes in the winding through VD4. Therefore, the current waveform is as this of the secondary voltage – sinusoidal and with the same phase. Consequently, also the current through the primary winding (from the AC source) i_1 is sinusoidal one in phase with the voltage u_1. In accordance with the stated in the first chart in this chapter, the single-phase full-wave rectifier with active load has theoretically a power factor equal to 1, on assumption of ideality of the elements.

As it has been clarified, the aim is the ripple factor of the output voltage to be decreased. For this purpose, smoothing filters are used. More often they contain passive elements – capacitors and inductances, and they are connected to the output of the rectifiers.

The simplest smoothing filters, but also the most widely used, contain only one capacitor or only one inductance. The capacitor is connected in parallel to the active resistance R_L in the output of the rectifier, while the inductance – in series with the resistance. The filter connection changes the type of the processes in the rectifier, thereby, in these cases it is accepted to be talked about active-capacitive and active-inductive loads of the rectifiers.

Figure 7. Electrical schematic and waveforms of a single-phase half-wave rectifier with active-capacitive load

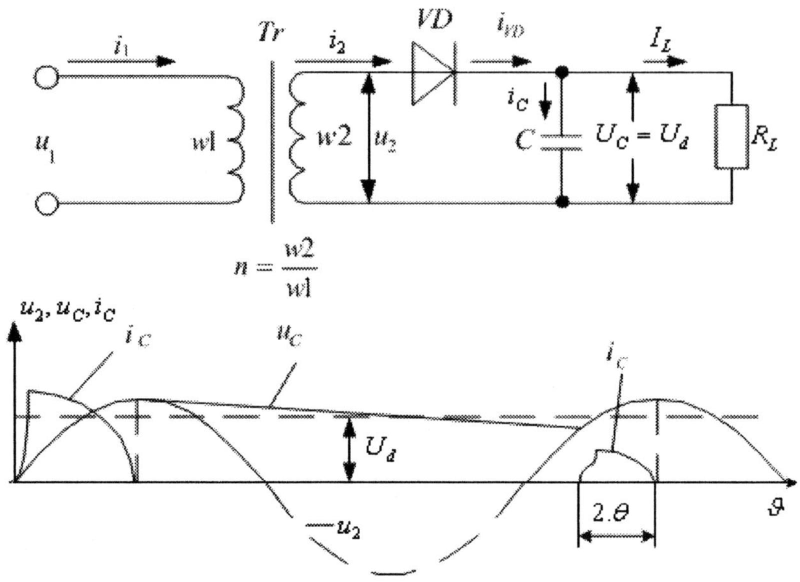

For better understanding, first the operation of a single-phase half-wave rectifier with active-capacitive load will be examined. It is shown in Figure 7. Its implementation is limited in practice.

At first, when the capacitor voltage is 0V, the diode begins to conduct as soon as the voltage applied to its anode becomes positive. The capacitor begins to charge. When the capacitor voltage is 0V, higher charging current flows through the diode. With the increase of the capacitor voltage, the current decreases. When the capacitor voltage becomes equal to the maximum value of the secondary voltage, the current through the diode stops flowing; the charge of the capacitor is terminated. During the rest of the half-period of the source voltage, the secondary voltage decreases. At the beginning of the next half-period, to the anode of the diode is applied negative voltage. The diode is blocked.

When the diode is blocked, the capacitor discharges through the load R_L and its voltage decreases. If the capacitance of the capacitor C and the load R_L are small, the capacitor discharges fast and the load current stops flowing before the above explained process starts again. If the capacitance of the capacitor C and the load R_L are high enough, the capacitor discharges slowly. Therefore, at the beginning of the next positive half-period, its voltage is higher than 0V. As a result, the anode of the diode is more negative than its cathode and the diode remains blocked.

Therefore, the current through the secondary winding of the transformer i_2, respectively the source current i_1, flows only during one of the half-periods of the source voltage – half-wave rectifier. This feature is typical for all half-wave rectifiers and it causes low values of the power factor. Besides this feature, due to the waveform of the current, particularly to its DC component, the operation of the supply transformer worsens.

The above features basically define the limited practical implementation of the half-wave rectifiers. Their implementation is in low powers. Because of that, comparatively little attention is paid to them in this book.

The diode of the rectifier shown in Figure 7 begins to conduct after the secondary voltage has become higher than the capacitor voltage. Therefore, the current through the diode flows less than one fourth of the period. The time duration of the current flowing is equal to angle 2θ. In practice, the discharge time constant of the capacitor $\tau_{dis} = C \cdot R_L$ is chosen to be much higher than the period of the source voltage T. Also, the charge time constant of the capacitor $\tau_{char} = C \cdot R_D$ is chosen to be much smaller than 0.25T, where in R_D is the inner resistance of the conducting diode. That is why the current through the diode flows during the following time $2 \cdot \theta = \left(0.1 \div 0.2\right) \cdot T$.

The above explained process is repeated continuously. It is obvious that the capacitor causes two effects in the output voltage – the increase of the DC output voltage U_d and the decrease of the ripples of the output voltage in comparison with the active load case. The higher the capacitor is and the smaller the load current is, the higher the two effects in the output voltage are. The frequency of the ripples of the output voltage is equal to the source frequency - $m = 1$ - a half-wave rectifier.

If the load misses ($R_L = \infty$), the charged capacitor will not discharge and its voltage will be:

$$U_{CM} = \sqrt{2}.U_2 = U_{2M}. \tag{4.21}$$

The reverse voltage of the diode is:

$$U_{DR} = U_{2M} + U_{cM} = 2 \cdot U_{2M}. \tag{4.22}$$

One of the disadvantages of the single-phase half-wave rectifier circuit is the higher reverse diode voltage.

Figure 8 depicts the operation of a single-phase bridge rectifier, known from Figure 6, but with active-capacitive load. Knowing the operation of the rectifier shown in Figure 7, the following features should be mentioned. The current $i_{d'}$ charging the capacitor C, flows only during $2.\theta$ angle in both half-periods around the maximum of the secondary voltage u_2. The ratio $\frac{2.\theta}{T}$ dependent on the power is from 0.08 to 0.2.

During the same intervals $2.\theta$, current i_2 with different direction in both half-periods, flows through the secondary winding. Waveforms of the current and voltage of the primary winding have the same shape as those of the secondary winding. The current waveform shown is characterized by a wide frequency spectrum and a significantly increased third harmonic. Possible values of the ratio of the higher harmonics to the fundamental harmonic are shown in Table 1(Mohan, 1995).

The worse harmonic content is the basic reason for lower power factor of the rectifier in respect to the supply network. Typical values dependent on the power are from 0.55 to 0.65. The attention paid to the low power factor is connected with the wide propagation of the rectifier. It is included in the supply

Table. 1. Values of the ratio of the higher harmonics to the fundamental harmonic

n	3	5	7	9	11	13	15	17
$\frac{I_n}{I_1}.\%$	73.2	36.6	8.1	5.7	4.1	2.9	0.8	0.4

Figure 8. Electrical schematic and typical waveforms of a single-phase bridge rectifier with active-capacitive load

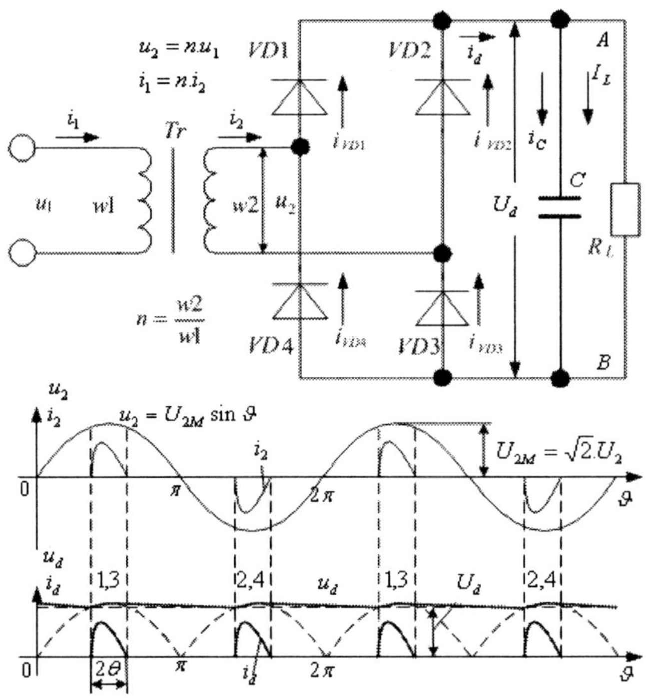

blocks of the audio and video equipment, communicational devices, personal computers and computers systems, electronic ballast for lightning, etc.

In particular seldom met cases, the single-phase full-wave rectifier can be implemented by the use of two diodes and by a middle point of the secondary winding of the supply transformer. The electrical schematic and waveforms of the rectifier with active-capacitive load are shown in Figure 9.

The waveforms of the voltages of the two halves of the secondary winding u_{2A} and u_{2B} are drawn in accordance with the middle point potential, which is assumed to be zero. During the positive half-period of u_{2A}, the current i_{2A} flows coming out from the beginning of the upper half of the secondary winding; during the positive half-period of u_{2B} the current i_{2B} flows coming out from the end of the lower half of the secondary winding.

Therefore, the current i_1 has the same waveform as in the single-phase bridge rectifier and all about its power factor is valid here, too. To obtain the output voltage with a value U_d equal to the value of the single-phase bridge rectifier, the voltage of each half of the secondary winding is necessary to be equal to the secondary voltage of the bridge rectifier. When a diode shown in Figure 9 conducts, the voltage applied across the other diode (that does not conduct) is equal to the voltage of the whole secondary winding. Therefore, in comparison to the voltage of the bridge rectifier, here the reverse diode voltage is twice higher. This, as well as the middle point of the transformer, makes this rectifier less propagated.

Another filter used to decrease the ripples of the load voltage is an inductance connected in series with the load. As it has already been mentioned, in this case it is said that the rectifier operates with active-inductive type of load. In particular cases, the load itself can have its own inductance.

Figure 9. Electrical schematic and waveforms of a single-phase full-wave rectifier with a middle point of the transformer

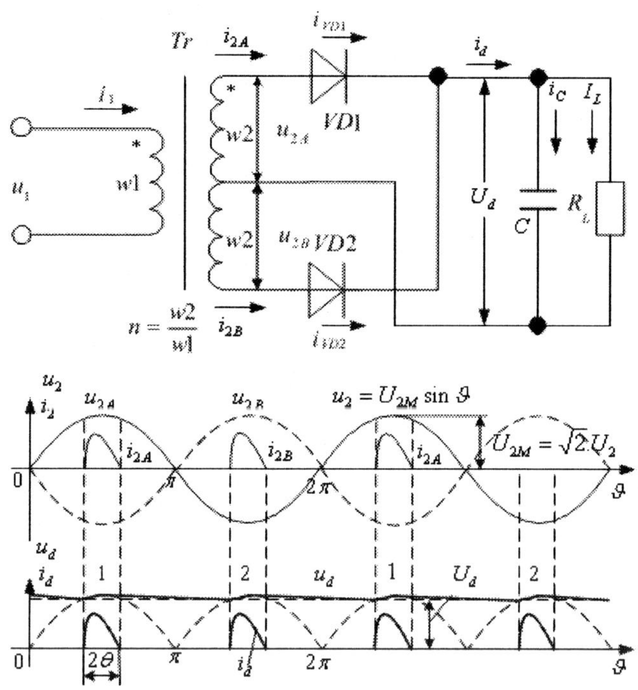

Figure 10 displays power schema of a single-phase bridge uncontrolled rectifier and a double-phase half-wave uncontrolled rectifier, both are with active-inductive load. Figure 11 depicts the current and voltage typical waveforms for the in the studied schemas.

The load has equal parameters for both schemas. The equality of the secondary transformer voltage of the bridge rectifier to each voltage of the secondary half-winding of the transformer of the double -phase half-wave rectifier results in equal output voltages for the two rectifiers.

The shape of the voltage u_d waveform is identical to this of the rectifier with active load – Figure 6. Therefore, the average value of the load voltage is:

$$U_d = \frac{2 \cdot \sqrt{2} \cdot U_2}{\pi} = \frac{2.\sqrt{2}}{\pi}.n.U_1 \qquad (4.23)$$

Because of the inductance L_L, the waveform of the current i_L is not the same as the output voltage u_d. As it appears in Figure 11 from the i_L and u_d waveforms, the current has a DC component I_L, also, because of the smoothing inductance, the ripples of the output current are lower than those of the output voltage. Flowing through the load resistance R_L, this current will produce a voltage drop across it. The voltage drop will have lower ripples compared to these of the output voltage. This is the smoothing effect that the inductive filter causes.

If the value of the load inductance is infinitive large, the load current is almost constant and its value is I_L - the dotted line in Figure 11.

The last diagram in Figure 11 shows the waveform of the primary current. The waveform differs from the sinusoidal one, which leads to decrease of the power factor.

Three-Phase Uncontrolled Rectifiers

Three-phase half-wave rectifiers have found very limited implementation in practice, thereby they are not examined in this chart.

Figure 12 displays electrical schematic and waveforms of a three-phase bridge rectifier with active load. This type of load is comparatively seldom met but its study is made to facilitate the understanding of operation of the further down stated three-phase rectifiers – uncontrolled and controlled.

The circuit in Figure 12 contains two groups of diodes – anode (VD4, VD6, VD2) and cathode (VD1, VD3, VD5). At each moment one of the diodes of the anode group, whose cathode is the most negative, with one of the diodes of the cathode group, whose anode is the most positive, conduct. A change in the conducting pair of the diodes happens at a point of natural commutation, when the phase voltages become equal to each other– points a, b, c, etc. Furthermore, assuming that the diodes are ideal, at each moment one of the right ends of a secondary windings of the transformer is connected to point A, and a right end of one of the rest windings is connected to point B. Thus, the phase-to-phase voltage for a pair of phases is applied across the load. Therefore, the waveform of the output voltage u_d has the shape depicted in Figure 12 with the assumption made that the potential of point B is zero. The subsymbols

Figure 10. Electrical schematics of single-phase full-wave rectifiers with active-inductive load

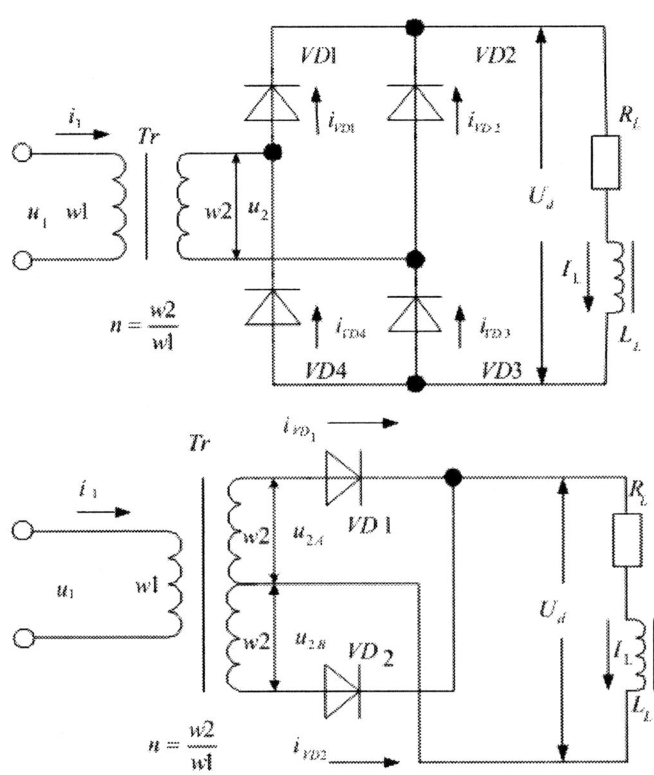

Figure 11. Waveforms for single-phase rectifiers with active-inductive load

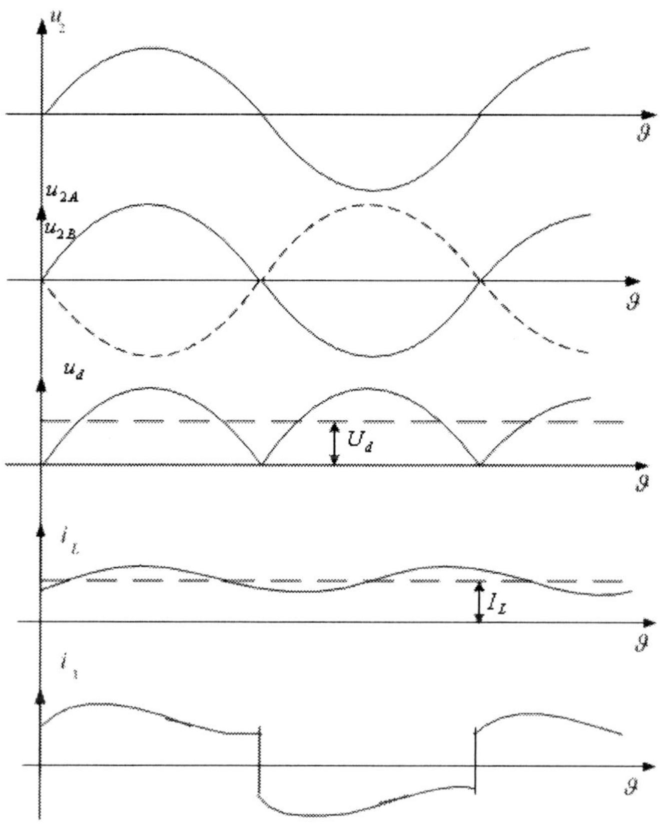

of the switches conducting during all intervals are also marked. For example, if VD1 and VD6 conduct, the phase-to-phase voltage u_{RS} is applied across the load.

The DC component of the output voltage is equal to the average value. For convenience the beginning of the coordinate system is moved in the maximum of the phase-to-phase voltage. Thus, the following equation is derived:

$$U_d = \frac{6}{\pi}\int_{-\frac{\pi}{6}}^{\frac{\pi}{6}} \sqrt{6}.U_2 \cos\vartheta.d\vartheta = \frac{6}{\pi}.\sqrt{6}.U_2.\sin\frac{\pi}{6} \approx 2.34.U_2 = 2.34.n.U_1, \qquad (4.24)$$

where in U_1 and U_2 are the effective values of the phase voltages of the primary and secondary windings, and n is the turns ratio.

The shape of the waveform of the load current is identical to the shape of the output voltage, and its average value is:

$$I_L = \frac{U_d}{R_L} \qquad (4.25)$$

Figure 12. Electrical schematic and waveforms of a three-phase bridge rectifier with active load

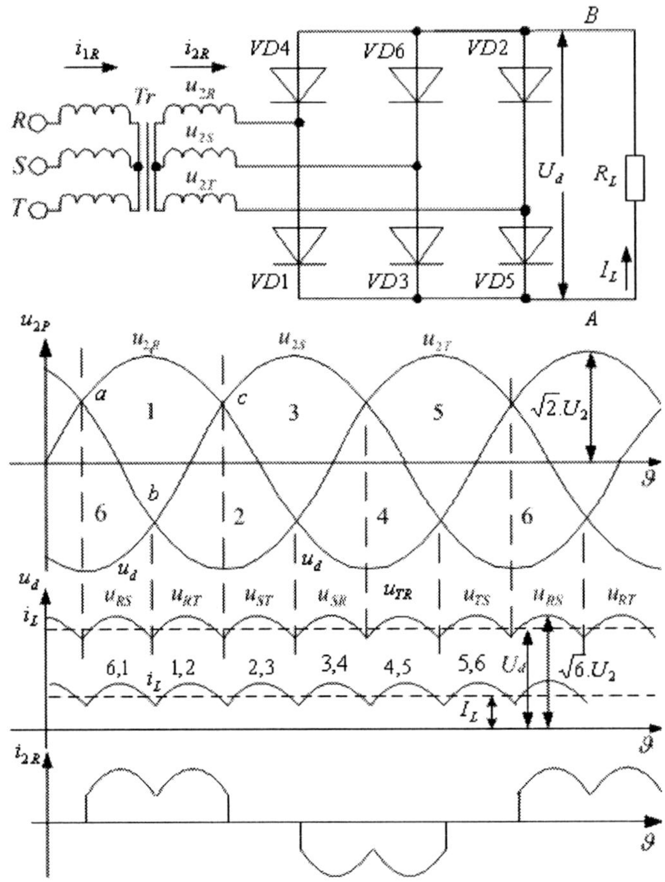

For this rectifier in accordance with (4.16) $m = 6$ because $p = 3$, $l = 2$. This also can be seen in Figure 12 where the number of the ripples of the output voltage within a period of the source frequency is 6. Therefore, the frequency of the first harmonic of the output voltage is 6 times higher than this of the source voltage.

Consequently, from (4.19) the ripple factor is $S_1 = \frac{2}{35} \approx 5.7\%$. It is considerably smaller than that of the single-phase full-wave rectifiers. The smaller ripple factor is a significant advantage.

The last waveform shown in Figure 12 is the source current for phase R – i_{2R}. It differs from the sinusoidal one and in contrast to the single-phase rectifier shown in Figure 6, nevertheless, the active type of load the power factor of the three-phase rectifier is less than 1.

As the operation of the three-phase bridge rectifier with active type of load is already known, it is easy to understand also the operation of the same rectifier with active-capacitive type of load. The electrical schematic and waveforms for this case are depicted in Figure 13. It has to be mentioned that this rectifier is widely spread in practice. Taking also in consideration the operation of the single-phase rectifiers with active-capacitive load, it is obvious that pairs of diodes conduct only around the maximums of the phase-to-phase voltages. The current i_d consists of unipolar pulses and flows during those intervals. During the remaining time the capacitor discharges through the load. From a comparison of the waveforms of the output voltages shown in Figure 12 and Figure 13, it is seen that the value of the first harmonic, and correspondingly the ripple factor, are smaller in the case of active-inductive load than in the case of active load. The influence of the capacitor as a smoothing filter explains the difference of the ripple factors. The average value of the output voltage is higher than the value defined with (4.24). The difference between the two values depends on the values of C and R_L.

The last waveform in Figure 13 shows the shape of the consumed current for one of the phases - R. Here the problem with the smaller than 1 power factor also exists.

Figure 14 displays electrical schematic and waveforms of a three-phase bridge rectifier with active-inductive load. As it is already known from the similar case in the single-phase rectifiers, the inductance is a smoothing filter. In contrast to Figure 11, here the inductance is assumed to be indefinitely high, to be able to see its influence over the ripple factor. With this assumption made, the load current i_L has only a DC component I_L. As a consequence, the voltage across the resistor R_L is an ideal DC voltage without ripples.

From the waveform of the output voltage u_d it appears that this voltage is the same as that in the case of active load – Figure 12. Therefore, its average value is defined with (4.24).

Figure 13. Electrical schematic and typical waveforms of a three-phase bridge rectifier with active-capacitive load

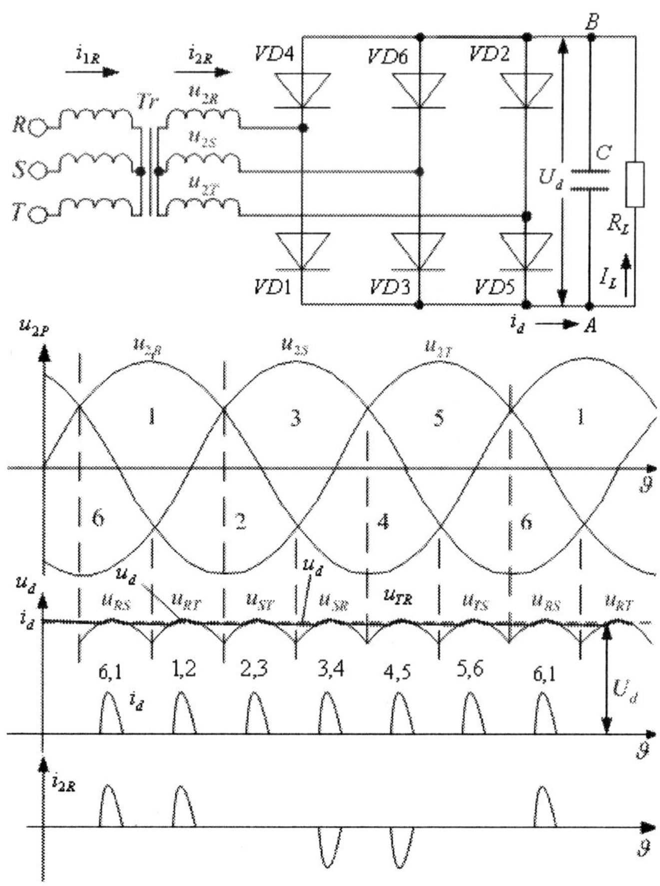

Figure 14. Electrical schematic and waveforms of a three-phase bridge rectifier with active-inductive load

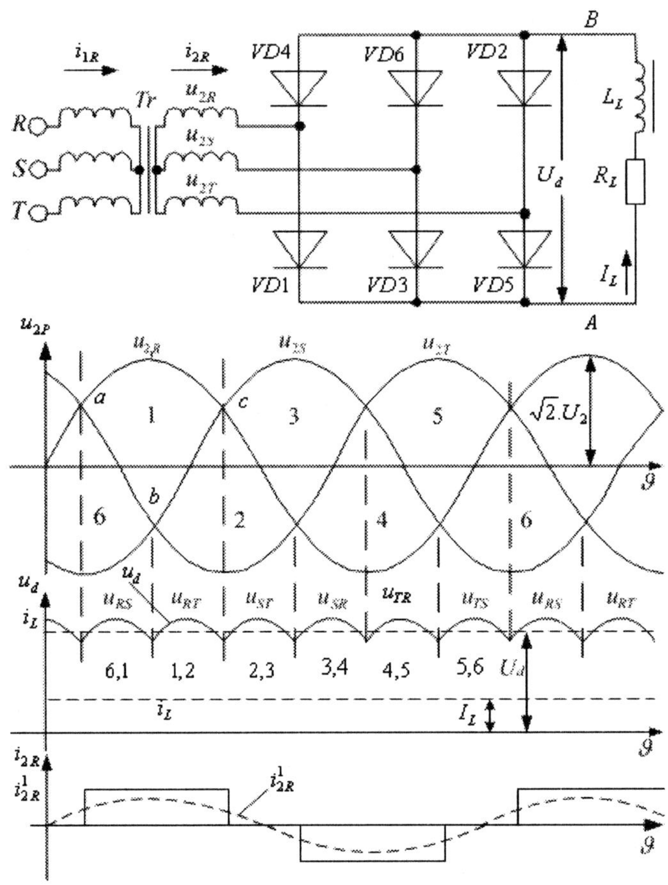

The source current is non-sinusoidal (see the lowermost waveform in Figure 14) and it consists of series of square pulses. Its first harmonic i^1_{2R} is in phase with the source voltage. Therefore $\cos\phi = 1$. However, the current contains higher harmonics and in accordance with the stated in the beginning of this chapter and (4.6), the power factor of the rectifier is less than 1.

SINGLE-PHASE AND THREE-PHASE CONTROLLED RECTIFIERS

Single-Phase Controlled Rectifiers

The controlled rectifiers are characterized by the use of controlled power devices – most often thyristors. As it is described further down, the use of transistors with a purpose to improve energy indicators is possible. To switch on thyristors, it is necessary not only to apply forward voltage to them, but also to deliver a firing pulse to their third electrode – control electrode or gate. Once a thyristor of a standard type is switched on, it is impossible the thyristor to be switched off by the use of the control electrode.

Switching off happens when the current through the devices becomes less than so called "holding current", which is one of the parameters of the thyristor. Gate turn-off thyristors are thyristors turning on and off using the gate electrode. They also find place in the controlled rectifiers (Skvarenina, 2002). The controlled rectifiers are characterized by the possibility to change the effective value of their output voltage by the change in respect to the rectifier input supply AC voltages of the moment at which the firing pulses are passed to the thyristors.

The firing pulses are supplied from the rectifier control system. Their displacement in respect to the moment of natural commutation – the moment when the diodes begin to conduct in the uncontrolled rectifiers, is called firing angle and is marked as α .

Additionally, to the already known types of rectifiers, in the controlled rectifiers another classification exists – symmetrical and asymmetrical rectifiers. The symmetrical rectifiers contain only thyristors, and the asymmetrical rectifiers – both thyristors and diodes. More often the controlled rectifiers operate with active-inductive type of load, by that reason their study is made with this type of load. The inductance is assumed to be infinitive high.

Figure 15. Electrical schematic and waveforms of a single-phase full-wave controlled rectifier with a middle point of the transformer with active-inductive load

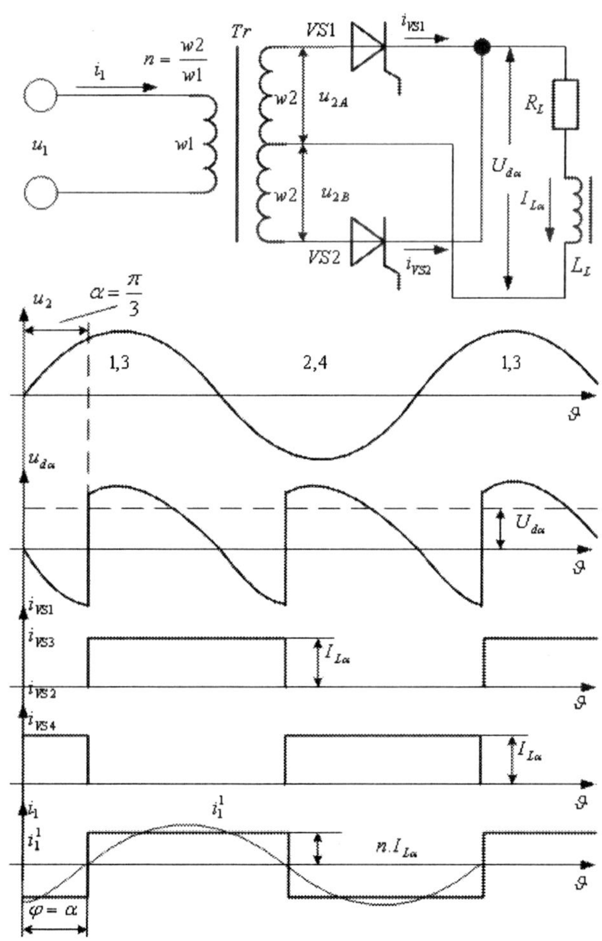

Figure 15 and Figure 16 display circuits of double-phase half-wave rectifiers with a middle point of the source transformers without and with a commutating diode, respectively.

The firing pulses are passed to the thyristors at delay angle α - firing delay angle. The delay is in respect to the moment at which the source voltage crosses 0. The currents through the power devices are also shown. For example, the thyristors of the rectifiers shown in Figure 15 continue to conduct during part of the next half-period because of the stored energy in the load inductance. As a result, the output voltage contains sections below the X-axis. Let see Figure 16, at the moment, in which the source voltage crosses 0, the commutating diode begins to conduct. Thus, the stored energy in the load inductance is delivered through it to the output. During the time of energy transfer, the output voltage is 0V. The lower waveforms in both figures show the current consumed by the rectifiers from the mains. The current has non-sinusoidal waveform and its first harmonic is displaced at angle φ to the source voltage. The displacement causes the effect of loading the mains with reactive power, and also causes the emission of higher current harmonics on the grid. These harmonics are connected with distortion power. To eliminate them, active power filters are connected in several cases in UPS. The higher the angle φ is,

Figure 16. Electrical schematic and waveforms of a single-phase full-wave controlled rectifier with a middle point of the transformer, a commutating diode and active-inductive load

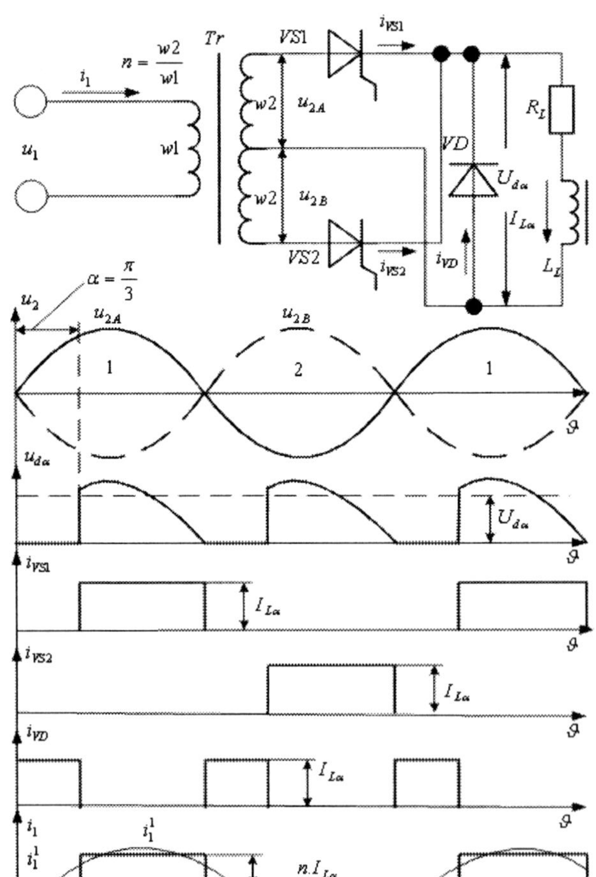

the higher the reactive power is. So, the circuit shown in Figure 16 has higher power factor and better indicators in respect to the network than this in Figure 15.

Control characteristic is the relationship between the average value of the output voltage $U_{d\alpha}$ and the firing delay angle. For the rectifier schema of Figure 15 is valid the following equation:

$$U_{d\alpha} = \frac{2.\sqrt{2}}{\pi}.U_2.\cos\alpha = U_{d0}.\cos\alpha \tag{4.26}$$

where U_2 is the effective value of the secondary winding voltage, U_{d0} – the value of the output voltage of the uncontrolled rectifier (if α=0). Therefore, the maximum firing delay angle at which the output voltage becomes equal to 0 is $\frac{\pi}{2}$. The schema shown in Figure 16 has maximum firing delay angle equals to π, and the control characteristic as follows:

$$U_{d\alpha} = U_{d0}\frac{1+\cos\alpha}{2} \tag{4.27}$$

A single-phase symmetrical rectifier (fully-controlled rectifier) (see Figure 17) has four thyristors. The output voltage has the same waveform as this of the output voltage of Figure 15. As it has been already mentioned in the chart of uncontrolled rectifiers, the voltage of each half of the secondary winding of the double-phase half-wave rectifier is necessary to be equal to the secondary voltage of the bridge rectifier to produce output voltage with a value U_d in both cases. For a single-phase symmetrical rectifier, the maximum firing delay angle is $\frac{\pi}{2}$, and the control characteristic is (4.26).

Sometimes, an asymmetrical rectifier is used because of its better power factor in respect to the mains, as well as because of easier implementation of the control of its two thyristors, which have common cathodes.

Figure 18 displays a schema and waveforms explaining the operation of a single-phase bridge asymmetrical rectifier (half-controlled rectifier), containing diodes in its anode group, and thyristors in its cathode group.

It is obvious from the waveforms that during the intervals in which the output voltage is 0V, the stored energy in the load inductance dissipates via the devices connected in one leg, for example, VS1, VD2 or VS2, VD1. The source current diagram is the same as this of the rectifier from Figure 16. The displacement angle between the source voltage and the first current harmonic is equal to the half of the firing delay angle. The equation for the control characteristic of the single-phase asymmetrical controlled rectifier is (4.27).

Three-Phase Controlled Rectifiers

Figure 19 displays the most widespread schema of a three-phase rectifier – three-phase bridge symmetrical controlled rectifier whose output voltage may contain pauses (values equal to 0) depending on the load characteristic, the firing delay angle, presence or absence of the commutating diode. The rectifier consists of two groups of thyristors – anode group (VS2, VS4, VS6) and cathode group (VS1, VS3, VS5). Each of the thyristors connected to one of the phases of one of the groups can conduct together

Figure 17. Electrical schematic and waveforms of a single-phase symmetrical controlled rectifier with active-inductive load

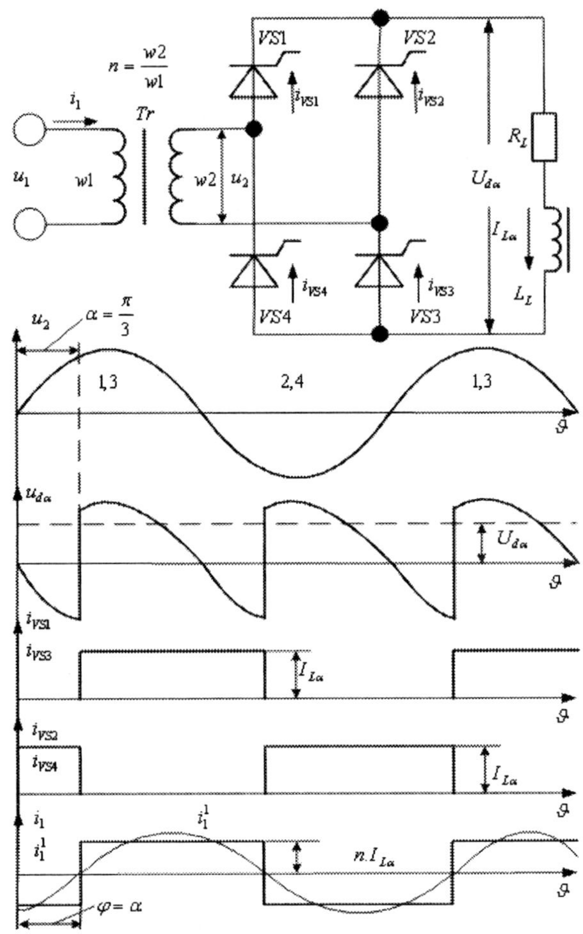

with a thyristors of the other group connected to the other two phases. The value of the firing delay angle, whose further increase leads to appearance of pauses, is so called crucial angle. For this rectifier the crucial angle is $\frac{\pi}{3}$. For angles smaller than the crucial angle, the waveform of the output voltage does not depend on the load type and on the presence of the commutating diode (see Figure 20b and Figure 20c). When the load is pure active load, after the crucial angle pauses appear in the waveform of the output voltage, also, the current through the load is discontinuous Figure 20d. When the load includes an inductance, the waveform of the output voltage contains also negative values because of the stored energy in this inductance which maintains the current trough the conducting thyristors higher than the holding current after the change of the sign of the corresponding phase-to-phase voltage – Figure 20e. If a commutating diode and active-inductive type of load are available, the stored energy is dissipated in the load through the commutating diode, and the conducting thyristors switch off after the change of the sign of the corresponding phase-to-phase voltage – Figure 20f. The output voltage contains pauses, and also, the current through the load is discontinuous.

Figure 18. Electrical schematic and waveforms of a single-phase asymmetrical controlled rectifier with active-inductive load

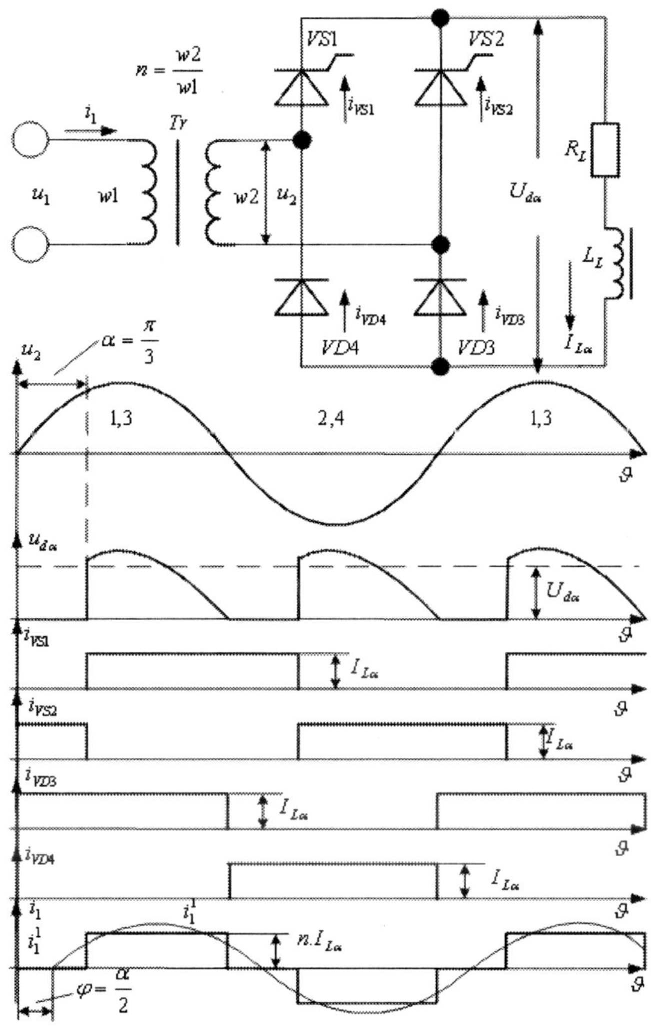

Figure 20g displays the waveform of the voltage across the first thyristor. The shape of the voltage depends on the firing delay angle. The maximum value of the reverse voltage is equal to the maximum value of the phase-to-phase voltage.

The control characteristics are as follows:

- For pure active load and for active-inductive load with commutating diode and $\frac{\pi}{3} < \alpha \leq \frac{2.\pi}{3}$

$$U_{d\alpha} = \frac{3.\sqrt{6}}{\pi} U_2 \left[1 - \sin\left(\alpha - \frac{\pi}{6}\right)\right] = 2.34.U_2.\left[1 - \sin\left(\alpha - \frac{\pi}{6}\right)\right] = U_{d0}\left[1 - \sin\left(\alpha - \frac{\pi}{6}\right)\right] \tag{4.28}$$

the maximum firing delay angle is equal to 2π/3.when $0 \leq \alpha \leq \frac{\pi}{3}$

Figure 19. Electrical schematic of a three-phase symmetrical controlled rectifier with active-inductive load

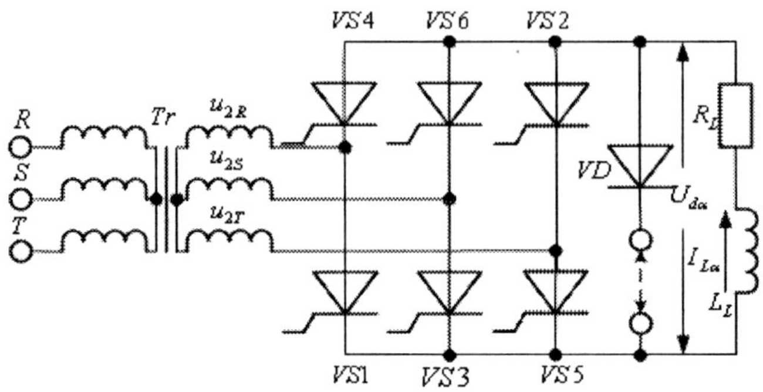

Figure 20. Corresponding waveforms for the schema shown in Figure 19

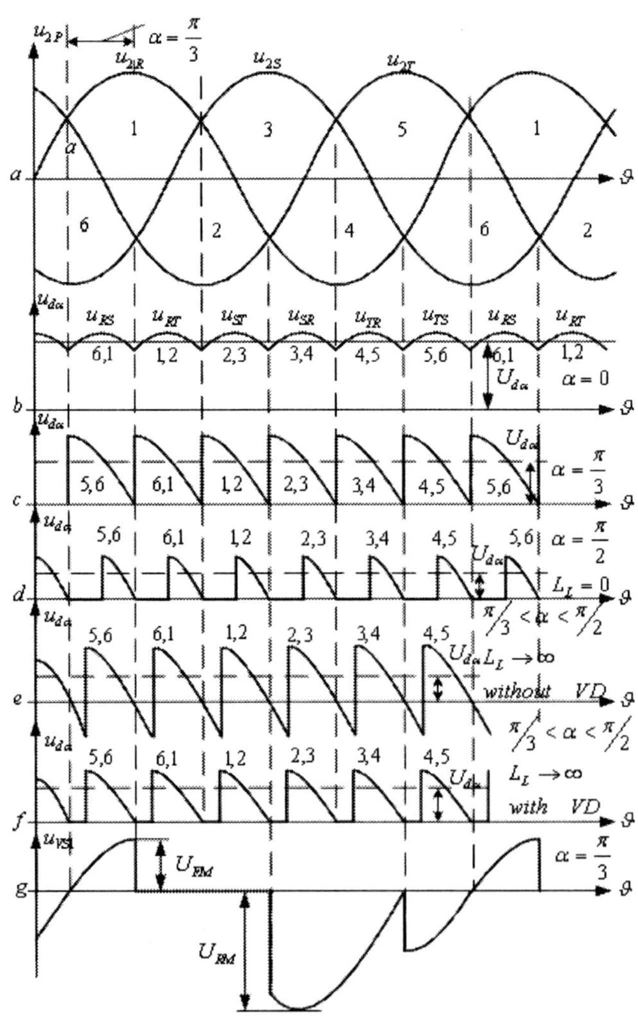

$$U_{d\alpha} = U_{d0} \cos \alpha \tag{4.29}$$

the maximum firing delay angle is equal to π/2, U_2 is the effective value of the phase voltage of the secondary winding of the supply transformer.

- For active-inductive load without commutating diode:

The equation (4.29) is valid for the whole regulation range and maximum firing delay angle is equal to $\frac{\pi}{2}$.

Figure 21 displays the schema of an asymmetrical three-phase bridge rectifier, which contains diodes in the one group, and thyristors in the other. It appears in the waveforms- Figure 22, that the output voltage contains pauses despite of the load type – active or active-inductive types. During the pauses the thyristor, which has been conducting till this moment, and the diode connected in the same leg are used as a commutating diode. The diagrams shown in Figure 22 clarify which devices conduct during the different intervals. The control characteristic is:

$$U_{d\alpha} = U_{d0} \cdot \frac{1 + \cos \alpha}{2} \tag{4.30}$$

where the maximum firing delay angle is equal to π. The advantage of the rectifier is the improved power factor in respect to the supply network and the easies implementation of the control system.

Figure 23 displays the control characteristics of the examined three-phase controlled rectifiers.

Output characteristics of the rectifiers – Figure 24, show the relationship of the output voltage and the output current. The ideal output characteristics are straight lines parallel to the X-axis. The real characteristics declines in respect to the X-axis, and the change of the output voltage is given by the equation:

$$\Delta U_{d\alpha} = \Delta U_{d\alpha PD} + \Delta U_{d\alpha R} + \Delta U_{d\alpha Ls} \tag{4.32}$$

Figure 21. Electrical schematic of an asymmetrical three-phase bridge rectifier with active-inductive load

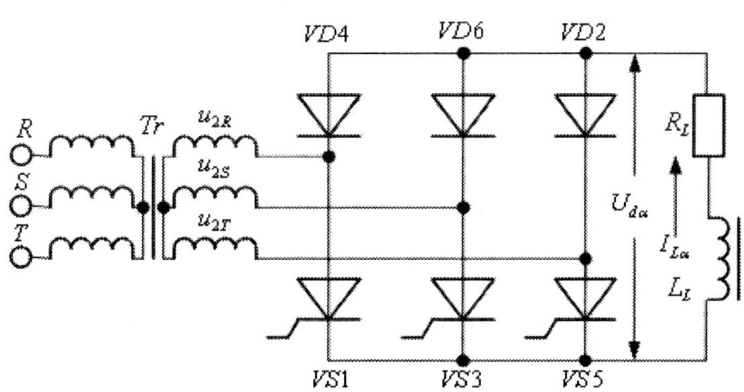

Figure 22.Corresponding waveforms for the rectifier schema shown in Figure 21

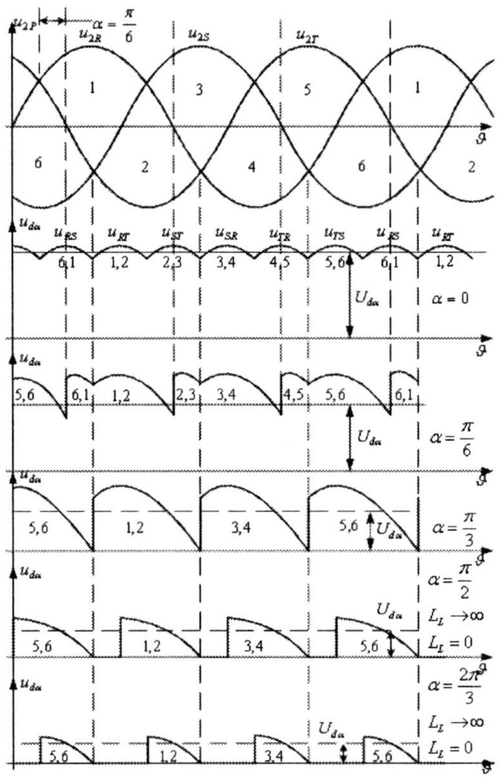

where in $\Delta U_{d\alpha PD}$ is the voltage across the power devices of the rectifier, $\Delta U_{d\alpha R}$ is the voltage across the active resistances (the transformer active resistance, the inductance active resistance, the active resistance of the wires), $\Delta U_{d\alpha L_s}$ is so-called inductive drop voltage. The processes of the commutation of the thyristors result of the inductive dissipation of the supply transformer Ls, which value is defined from the equation:

$$\Delta U_{d\alpha Ls} = m.f.Ls.I_{d\alpha} \qquad (4.33)$$

Figure 23. Control characteristics of three-phase rectifiers

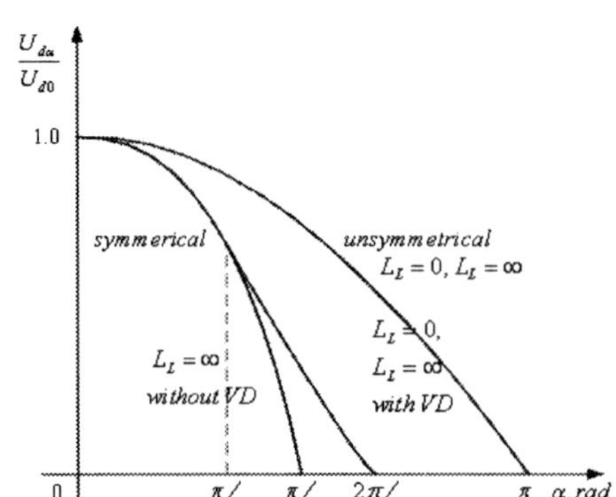

Figure 24. Output characteristics of controlled rectifiers

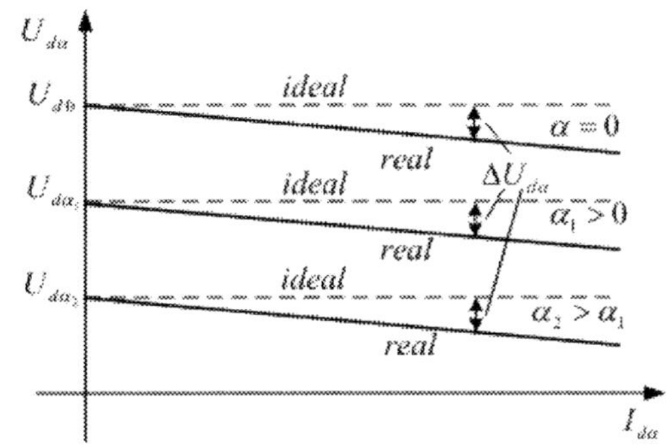

where in f is the frequency of the supply network voltage, m is the rate frequency of the ripples – (4.16).

The rectifiers require a control system to their operation. The basic function of this system is to generate firing pulses for the thyristors with the possibility to change the pulse displacement in respect to the input AC voltage of the rectifier. These pulses have to be passed to the thyristors at the moments in which they could conduct, i.e. a distribution of the pulses is required. Additionally, the control system has also auxiliary functions – protection against break-down regimes, indication of particular signals, regulation of particular parameters, diagnostic of the power schema, etc. In this chart, the main attention will be paid on implementation of the basic functions – generation, formation and distribution of the pulses.

The control systems are implemented using different elements – discreet elements, integral circuit (digital or/and analogue), microprocessors.

The control systems of the rectifiers and AC/AC converters have similar implementation principles. They are synchronous or asynchronous systems.

In the synchronous control systems, the phase angle, at which the following firing pulse is passed, is determined using a particular phase of the voltage supplying the rectifier:

$$\omega.t_i = \phi + \frac{2.\pi}{m}.(i-1) + \theta_i(U_C) \qquad (4.34)$$

where in $\omega.t_i$ is the angle of passing the i-th-firing pulse; ϕ - the primary angle which the delay to the AC voltage is rendered with; θ_i - adjustable delay angle, it depends on the control parameter - here on U_C; m- rate frequency of ripples.

In the asynchronous control systems, the phase angle, at which the following firing pulse is passed, is determined using the moment of the previous pulse:

$$\omega.t_i = \omega.t_{i-1} + \frac{2.\pi}{m} + \theta_i(U_C) \qquad (4.35)$$

The pulse is not directly connected to the parameters $\omega.t$ and ϕ of the AC voltage. This type of systems can operate only in a closed system of automatic control. They are usually applied where big disturbances and distortions of the AC voltage exist and they are rarely implemented.

According to the way of pulse formation, the synchronous systems are single-channel Figure 25.a or multichannel Figure 25.b systems.

The functions of the blocks are as follows:

ISC: An input synchronous circuit – it monitors the change of the AC voltage supplying the power schema and it also synchronizes the operation of the control system to the voltage;

PRC: A phase-regulation circuit – it realizes a change in the phase of the firing pulses to a fixed angle to the AC voltage.

PF: A pulse former – it forms the firing pulses to be of a sufficient duration to be able to fire the thyristors;

PD: A pulse distributor –it distributes the series of the firing pulses to the corresponding thyristors;

Dr: A driver – it secures the required power to fire the thyristors.

As it appears in Figure 25 in the single-channel control systems the phase-regulation circuit is common, while in multi-channel systems each channel has its own circuit. An advantage of the single-channel circuit is the better symmetric of the firing pulses.

The phase-regulation circuits are a specific block in the control systems, therefore, the basic principles of their implementation is studied in the chart. There are two methods to implement the phase-regulation circuits – horizontal Figure 26a and vertical – Figure 26b methods.

The symbols in Figue 26 are:

PRC – a phase - regulation circuit; SLVG–a saw linear voltage generator; C – a comparison device – a comparator.

Figure 25. Structural schemas of the control systems

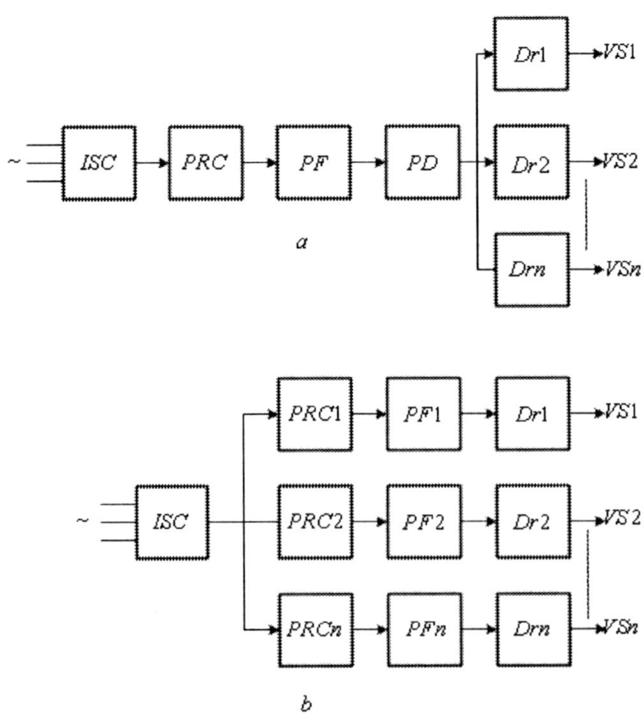

Figure 26. Methods to implement the phase-regulation circuits

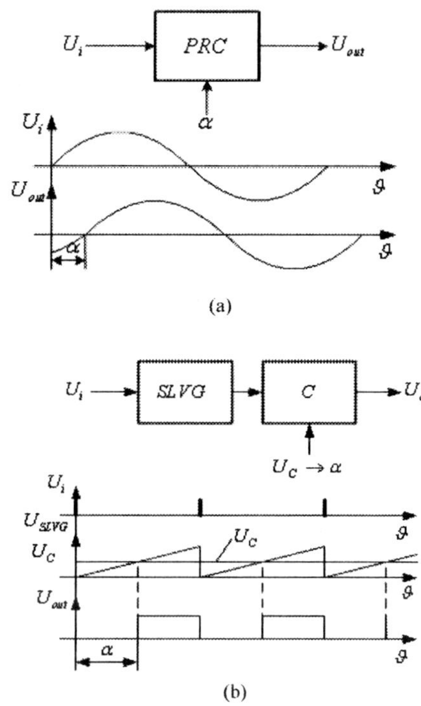

In the first method, the firing pulses are formed at the moment when a sinusoidal voltage U_{out} crosses the zero. The phase of the sinusoidal voltage can be shifted in accordance with U_i. The voltage U_i has a particular amplitude and phase ratio to those of the AC voltage supplying the power schema. In the vertical method, the firing pulses are formed at the moments of equality of a linear changing voltage U_{SLV} and a control voltage U_C. These moments correspond to the front edge of the voltage U_{out}. The comparison is made in the comparison circuit – the comparator. The beginning and the end of the AC voltage U_{SLV} are synchronizes to the AC voltage supplying the power schema.

BIDIRECTIONAL AC/DC CONVERSION

Using power switches, which are possible to switch on and off through the control electrode (more often IGBT), reveals new probabilities to control the flux of electrical power towards and from the source network (Rodriquez, 2005). The basic idea is to make the control of the power system in such a way that the power factor to stay close to 1 regardless of a regulation of the power value consumed from or delivered on the AC mains. Therefore, the current has to be in phase with the source voltage in the rectifer mode, and it has to be displaced with an angle 180° in inverter mode.

Figure 27 displays a single-phase converter with bidirectional transmission of energy.

The input inductance L includes also the mains inductance. Its purpose is to decrease the current high frequency ripples produced from the switching of the transistors. The capacitor C_d is used as an element to store DC energy. The energy storage is also possible to be made in an inductance, but the converters using inductive storage element have more limited implementation. The schema

Figure 27. Electrical schematic of a single-phase bidirectional transistor converter

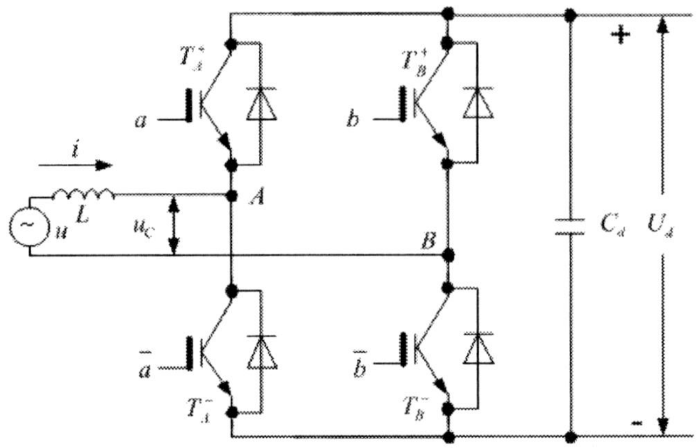

shown in Figure 27 allows the source current to track the preliminarily set curve by using an appropriate control of the transistors. Thus, in the positive half-period of the network voltage, if the current has to increase, transistors T_A^-, T_B^+ turn on, and if the current has to decrease the transistors T_A^+, T_B^- turn on. During the negative half-period, the first pair of transistors turns on for the current to decrease and the second pair turns on for the current to increase. Therefore, if two adjacent current values are compared by their absolute value, in both half-periods when T_A^-, T_B^+ turn on the current increases, and when T_A^+, T_B^- turn on the current decreases. Thus the preliminarily set sinusoidal wave for the source current can be tracked while only the sinusoidal waveform value and its displacement in accordance with the source voltage are changed. At every moment, the following relationships are valid:

$$u(t) = u_L(t) + u_C(t); \quad u_L(t) = L.\frac{di(t)}{dt}, \qquad (4.36)$$

where in u_C is the voltage of the converter between points A and B (see Figure 27), u_L and i_L are the inductance voltage and current, respectively, and u is the network voltage.

Let assume that the network voltage is a pure sinusoidal wave. For the first harmonics the converter variables the following equation is written:

$$U = U_{L1} + U_{C1}; \quad U_{L1} = \omega.L.I_1 \qquad (4.37)$$

A vector diagram corresponding to (4.37) is shown in Figure 28 (Mohan, 1995). The vector of the source voltage U is taken as a basic vector. Inductive reaction of the converter in respect to the mains is examined. The source current lags the source voltage at angle ϕ. The displacement between the source voltage and the first harmonic of the converter voltage U_{C1} is marked as δ. According to the shown vector diagram, the equation for active and reactive powers is written, as follows:

$$P = U.I_1.\cos\phi \qquad (4.38)$$

Figure 28. Vector diagram explaining the operation of the converter shown in Figure 27.

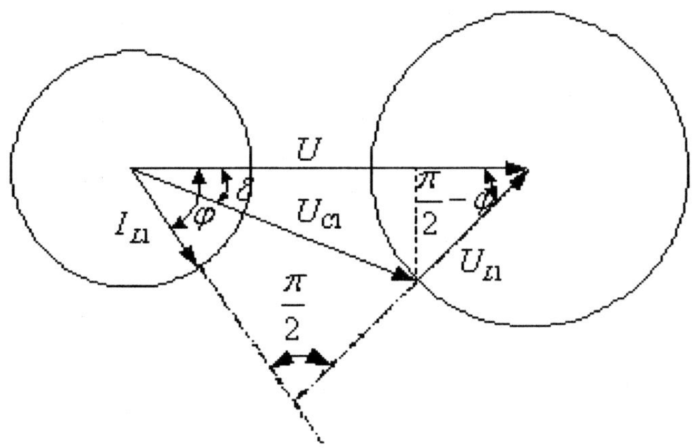

Also, from Figure 28 is clear that:

$$U_{L1}.\cos\phi = \omega.L.I_1.\cos\phi = U_{C1}.\sin\delta \tag{4.39}$$

this leads to:

$$P = \frac{U^2}{\omega.L}\left(\frac{U_{C1}}{U}.\sin\delta\right) \tag{4.40}$$

On the other hand:

$$Q = U.I_1.\sin\phi \tag{4.41}$$

since

$$U_{C1}.\cos\delta = U - U_{L1}.\sin\phi = U - \omega.L.I_1.\sin\phi \tag{4.42}$$

for the reactive power is found:

$$Q = \frac{U^2}{\omega.L}\left(1 - \frac{U_{C1}}{U}\cos\delta\right) \tag{4.43}$$

Besides from (4.35) is valid:

$$I_1 = \frac{U - U_{C1}}{\omega.L} \tag{4.44}$$

The value of the inductance L is constant. From (4.40) and (4.43) is obvious that in particular source voltage value U, the active and reactive power can be changed by the change of the input converter voltage U_{C1} and its displacement to the source voltage δ . The change of this angle leads to a movement of the vector of the source current which leaves 0 or lags the source voltage. It appears in Figure 28, that if the current I_1 is required to be maintained constant during the regulation of the active and reactive powers, as well as during the change of the flux of the powers from or into the mains, the current vector point will describe a circumference the same as the circumference which will be described by the point of the voltage vector U_{C1}.

Figure 29 shows the vector diagrams in rectifier and inverter modes of operation with power factor equal to 1. Figure 30 displays the waveforms of the voltages and currents in the schema shown in Figure 27 when the converter operates in rectifier mode with power factor equal to 1. There is a change made by switching the power devices of the source current around the preliminarily set sinusoidal wave. The input voltage between points A and B is equal to the value of the DC voltage U_d with different sign in particular moments. In accordance with the firing pulses passed to a transistor, either it conducts or its antiparallel diode conducts.

Additional advantage of the converter is the possibility to maintain the DC voltage U_d constant. Figure 31 displays the structural schema which has to be followed during the implementation of the control system of the bidirectional converter. The symbols in Figure 31 are:

PIC – a proportional integral regulator; M – a multiplier; SMCC – a sliding mode current control; Dr – drivers.

The reference current waveform is generated by multiplication of a sinusoidal waveform, in phase (in rectifier mode operation $\varphi = 0$) and in displaced at angle π (in inverter mode operation $\varphi = \pi$) to the source voltage, by the output voltage of the PI regulator. Signal of the comparison between reference voltage U_{dr} value and transitory voltage U_d of the DC voltage values is applied to the regulator input. Furthermore, switching signals for the power devices are generated following the above mentioned algorithm at the continuous comparison of the reference sinusoidal waveform i_r to the transitory source current i value (for example, using a comparator with hysteresis). Then, the drivers come. Additional so-called "dead time" between the switching of the transistors connected in one leg is obligatory to be realized in the drivers

Principle schema of a three-phase bidirectional transistor converter is shown in Figure 32. The instantaneous values of the phase voltages in respect to the star center are marked as u_R, u_S, u_T. The converter voltages in points R, S, T in respect to the same star center are marked as u_{CR}, u_{CS}, u_{CT}. The relationship

Figure 29. Vector diagrams for the converter shown in Figure 17 in both operating modes when power factor is equal to 1

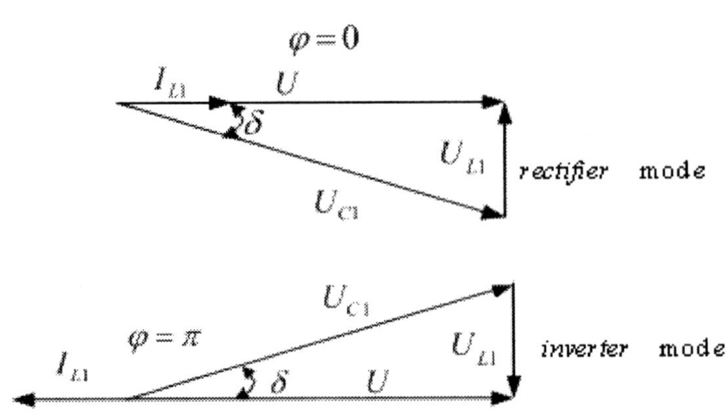

Figure 30. Waveforms explaining the operation of the converter shown in Figure 17 in rectifier mode with power factor equal to 1

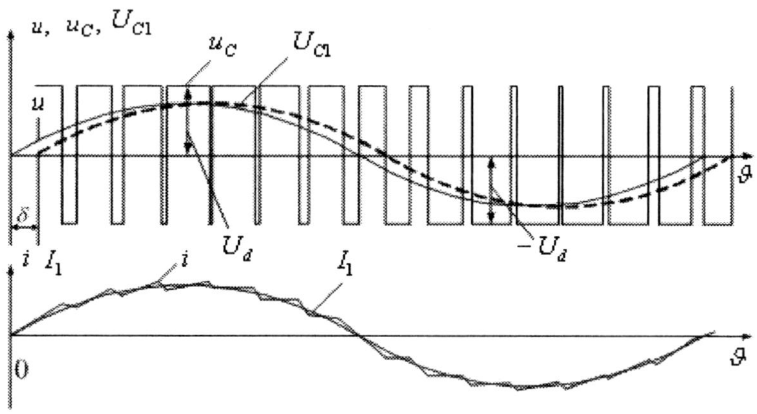

among the corresponding pair of the phase voltage, converter voltage and current through the inductance is given by (4.36). The voltages of the three-phase mains form symmetrical three-phase system:

$$u_R = U_m \sin\vartheta$$
$$u_S = U_m.\sin\left(\vartheta - \frac{2\pi}{3}\right)$$
$$u_T = U_m.\sin\left(\vartheta + \frac{2.\pi}{3}\right) \qquad (4.45)$$

The same system is formed by the phase currents if they are displaced at angle ϕ to the voltages. This angle is 0 in rectifier operational mode and it is π in inverter mode.

$$i_R = I_m \sin\left(\vartheta - \phi\right)$$
$$i_S = I_m.\sin\left(\vartheta - \frac{2\pi}{3} - \phi\right)$$
$$i_T = I_m.\sin\left(\vartheta + \frac{2.\pi}{3} - \phi\right) \qquad (4.46)$$

Figure 31. Structural schema for control system of a single-phase bidirectional transistor converter

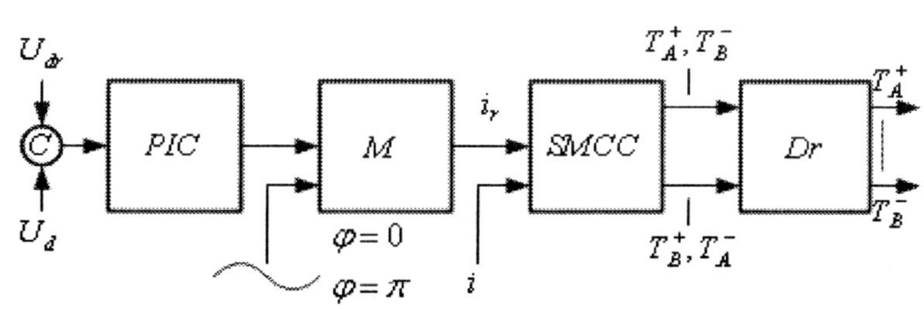

The control signals vector [*r*, *s*, *t*] is used. Each signal is changed using a sinusoidal law with amplitude M and displacement δ (see Figure 30 for single-phase converter).

$$\begin{aligned} r &= M.\sin\left(\vartheta - \delta\right) \\ s &= M.\sin\left(\vartheta - \frac{2\pi}{3} - \delta\right) \\ t &= M.\sin\left(\vartheta + \frac{2.\pi}{3} - \delta\right) \end{aligned} \tag{4.47}$$

The voltages u_{CR}, u_{CS}, u_{CT} are written as follows:

$$\begin{aligned} u_{CR} &= \frac{1}{2}.U_d.r = \frac{1}{2}.U_d.M.\sin\left(\vartheta - \delta\right) \\ u_{CS} &= \frac{1}{2}.U_d.s = \frac{1}{2}.U_d.M.\sin\left(\vartheta - \frac{2\pi}{3} - \delta\right) \\ u_{CT} &= \frac{1}{2}.U_d.t = \frac{1}{2}.U_d.M.\sin\left(\vartheta + \frac{2.\pi}{3} - \delta\right) \end{aligned} \tag{4.48}$$

The corresponding variables from the equations (4.45), (4.46) and (4.48) are connected with each other by the equation (4.37). The connection is the same as in the single-phase converter.

The output current I_d can be written in matrix form:

$$I_d = \begin{bmatrix} r & s & t \end{bmatrix} \cdot \begin{bmatrix} i_r \\ i_S \\ i_T \end{bmatrix}. \tag{4.49}$$

After substituting (4.46) and (4.47) in (4.49) and several transformations, it is derived:

Figure 32. Principle schema of a three-phase bidirectional transistor converter

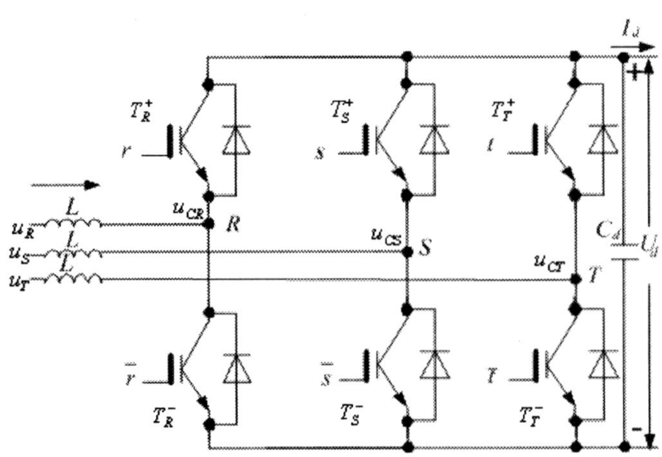

$$I_d = M.I_m.\cos(\delta - \phi) \tag{4.50}$$

The relationship between the output voltage and output current is linear:

$$U_d = R.I_d \tag{4.51}$$

where in R is the resistance in the DC side. A conclusion is made that the value of the DC voltage depends on the maximum value of the current, the amplitude of the control signal and the angle δ (considering that $\phi = 0 \quad or \quad \phi = \pi$). Therefore, the control system of the three-phase bidirectional transistor converter is similar to the already examined system in Figure 31. It is necessary to generate three reference sinusoidal waveforms for each phase current in phase or anti phase with the corresponding phase voltages. Tracking of each sinusoidal waveform leads to switching off the power devices connected to the same phase. The regulator of the output voltage is common for the three phases. Using multiplier, regulator output signal changes simultaneously the amplitudes of the three currents. It should be mentioned that both the single-phase and the three-phase bidirectional transistor converters are boost converters operating in condition of transmitting energy from the mains to the DC side as a result of a periodical storage of energy in the inductances L. Energy is transmitting to the DC side during the switching off the transistors.

The examined bidirectional transistor converters are in the basis of the power active filters – the contemporary means to improve the efficiency of already installed equipment.

METHODS TO IMPROVE POWER EFFICIENCY IN AC/DC CONVERSION

Active Power Factor Correction Techniques in Uncontrolled Rectifiers

As it has been mentioned during the study of the uncontrolled and controlled rectifiers, they consumed from the AC source voltage a current of a shape that differs from the sinusoidal shape and they are also characterized with low values of the power factor. The consumed current contains higher harmonics, which cause harmful influence to the rest of the consumers connected to the same supplying source. There are standards specifying the contents of the harmonics when consumed current is below 16A. These standards are of the International Electrotechnical Commission (IEC1000-3-2), as well as of the European Organization of Standardization in Electrotechnique (EN61000-3-2). In accordance with the standards, the equipment is divided in the following groups:

- **Class A:** Three-phase symmetrical equipment and all that is not included of the rest groups
- **Class B:** Mobile equipment (mobile devices)
- **Class C:** Lightening equipment
- **Class D:** All devices consuming a current with a specific square pulse waveform and are not included in the upper three classes and also are not meant for electrical motors
- For each of those classes the content of the higher harmonics (odd and even) of the consumed current is specified in tables as a maximum absolute value for class A, as a percentage of the fundamental harmonic for class B and as absolute value for a watt consumed power and a maximum absolute value for class C.

More often the converters using uncontrolled rectifiers are of one of these groups. The current consumed by them does not meet the requirements of the above stated standards.

There are two methods to correct their power factor – passive and active methods. The passive method is based on the use of LC-filters and it is effective only if the voltage and the frequency of the supplying source and also the load of the rectifier are changed in a very narrow range. These filters have higher volume and weight, but they cause less EMI disturbances than the active methods to correct power factor.

As a basic method for correction, the active power factor correction is imposed (Andrieu,1996; Garsia, 2003). Its principle is described in Figure 33.

This method to correct the power factor is characterized by including a DC/DC converter between the single-phase bridged rectifier and the load. The converter is controlled in an appropriate way securing that the source current is of sinusoidal waveform and in phase with the source voltage (SGS – Thomson, 1995). Among all schemas of DC/DC converters – buck, boost, buck-boost, etc, the boost converter has the widest implementation (see Chapter 3 of the current Part). The basic elements of the boost converters are inductance L, transistor VT, diode VD and the output capacitor Cd. In contrast to the classical boost converter in which to its input is applied DC voltage, here, the input voltage is a pulse voltage which is a result of a full-wave rectifying in active load. The input voltage is marked as |U~|.

Basically, when the transistor VT is turned on, energy is stored in the inductance L. This energy recharges the capacitor Cd through the diode VD when the transistor is turned off. This process of recharging can be controlled with a purpose to obtain the power factor close to 1. Figure 34 explains the controlled method in the range of a half-period of the source voltage. Figure 34 shows the input voltage of the boost converter $|k.U\sim|$, the reference sinusoidal wave for the inductance current i_R, which is made in the power factor controller, the inductance current i_L, and the source current $i\sim$ during this time period. A diagram of the control pulses for the transistors is shown in the lower position.

With the switching on of the transistor VT, the current through the inductance increases. When its value reaches the value of the reference curve i_R that fact is determined into the controller, and the transistor is turned off. The inductance current decreases; the diode VD conducts; the capacitor Cd recharges. When the inductance current reaches 0, which is determined in the controller with the auxiliary winding W2 (Figure 33), the transistor is turned on again. And the process continues. Figure 34b displays several time intervals of the schema operation. Thus, the average value of the source current $i\sim$ before the EMI filter

Figure 33. Principle of the active correction of the power factor

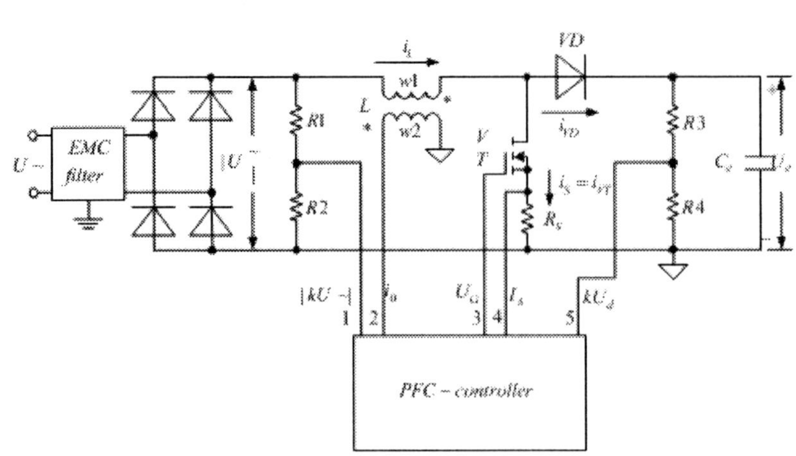

(which removes the higher harmonics), becomes of a sinusoidal waveform and in phase with the source voltage. The described operational principle is characterized by the inductive current fall to zero.

It should be mentioned that there are plenty of and also different integral circuits of PFC – controllers, but to operate in the above set method several basic signals are required.

Figure 33 displays the several basic signals required for the operation of the correction schema, which are used in the power factor controller. They are as follows:

- **Signal 1:** The input voltage of the boost converter $|k.U\ \sim|$, required to generate the reference sinusoidal wave
- **Signal 2:** Signal for determining the decrease of the inductive current to $0 - i_0$
- **Signal 3:** Control pulses for transistor VT $-U_G$
- **Signal 4:** Signal for monitoring the transistor current, as well as the inductance current when the transistor is turned on, for determining its reaching of the reference sinusoidal wave – I_S. This signal is obtained as a voltage drop across a resistor with little value (tenths of 1Ω)

Figure 34. Diagrams of the active power factor correction

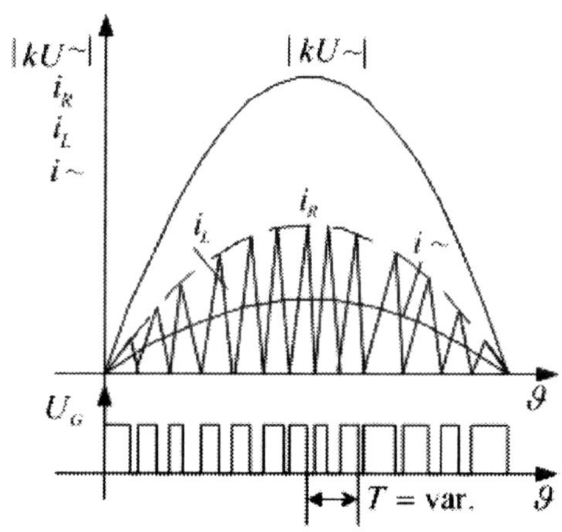

(a)

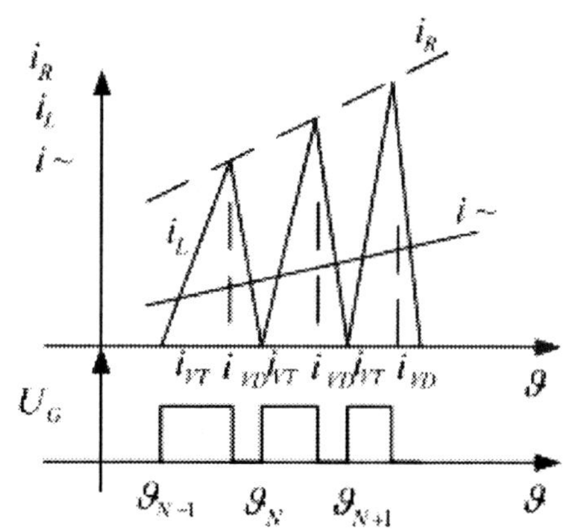

(b)

- **Signal 5:** Feedback of the instant value of the output voltage, which is required for its stabilization – $k.U_d$

The block schema of the inner structure of a PFC-controller, containing the basic blocks used by all companies, applying this method, is shown in Figure 35.

The symbols in the schema are:

RVS – a reference voltage source; EA – an error amplifier; AM – an analogue multiplier; C – a comparator; Dr – a driver.

Two signals are applied to the input of the error amplifier – a voltage proportional to the value of the output voltage – $k.U_d$ and a voltage from an inner reference source with value about 2V – U_R. The error amplifier is gripped in the feedback between the output and the inverting input; the elements in this feedback from the desired control law and the gear function of the system for the automatic control. The instant value of the output voltage of the error amplifier is amplified by the voltage proportional to the value of the input voltage of the boost converter – $|k.U \sim|$, and as a result, the reference sinusoidal wave for the current I_R is formed. In the comparator, the instant value of the reference curve is compared to the voltage drop across the resistor R_S – U_S, which is proportional to the inductance current I_S during the interval of the conduction of the transistor VT. When the two signals become equal, a signal to turn off the transistor is passed to the driver. The signal for turning on the transistor is generated from the inductance auxiliary winding, which is used to monitor the fall of the inductance current to 0 – i_0.

It appears in the diagram shown in Figure 34 that this method is characterized by a variable control frequency of the boost converter, which increases the requirements of the EMI filter to be able to remove higher harmonics with a wider spectrum.

It is possible to modify the method and obtain constant frequency, which is used by several PFC-controllers producers. The waveforms are shown in Figure 36a. The constant operation frequency simplifies the implementation of the EMC – filter.

To be able to obtain a specified power, both methods studied up to here are characterized with a high amplitude value of the current through the inductance and switches. This is a disadvantage of the methods. This disadvantage in the same power can be eliminated if a monitoring of an effective value,

Figure 35. Inner block structure of PFC-controller

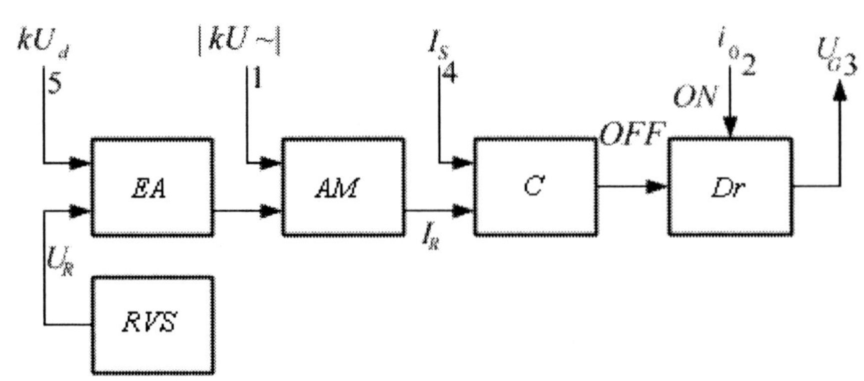

which is depicted in Figure 36.b, is applied (Unitrode, 1997). The method is also used in practice and for its realization several additional signals are required in comparison to those shown in Figure 33.

From Figure 35 is obvious that the active power factor correction allows also to stabilize the DC voltage in the output of the converter.

A significant advantage of the topology using the effective value monitoring is the wide operational range of the change of the AC mains voltage– for example, form 90V to 250V, and of AC voltage frequency – for example, 50Hz and 60Hz, i.e. the converters can operate in different countries using different standards for the mains voltage without problems.

As it has been clarified during the study of the bidirectional converter shown in Figure 27, to obtain the power factor very closed to 1 using an appropriate regulation and passing the energy into both directions are possible. If only AC/DC conversion is required, the schema can be changed as it shown in Figure 37.a. Only the switches in one of the groups are fully-controlled – in this case the upper ones of

Figure 36. a) Diagrams under operation with constant frequency and b) monitoring the effective value

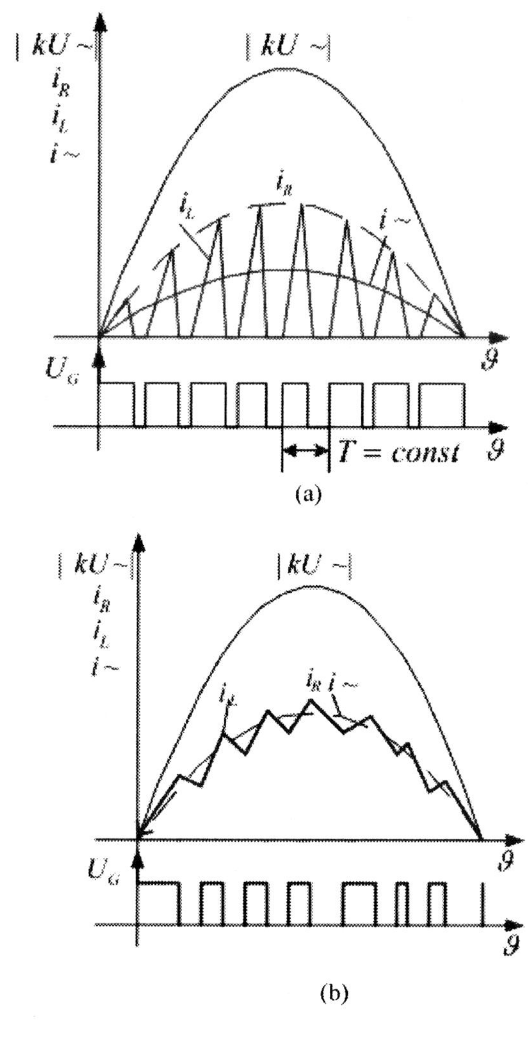

the schematic. These converters are often called rectifiers with hysteresis - current control, because of the operational principle described in Figure 38.

A sinusoidal waveform i_R for the source current is formed in phase with the source network voltage in the control system. Transistor *VT*1 turns on at the moment $\vartheta = 0$. The input of the rectifier is short-circuit connected through *VT*1 and the anti-paralel diode of *VT*2. The mains current rises and its velocity is limited only by the inductance *L*. At the moment when the current transitory value exceeds the transitory value of the reference current Δi_R, the transistor *VT*1 turns off – the moment $\vartheta 1$.

From this moment on, the diode *VD*1 and the anti-paralel diode of *VT*2 conduct. The source current *i* decreases and the capacitor C_d charges. At the moment $\vartheta 2$ the value of the current *i* becomes less than the value of the reference current with difference equal to Δi_R and the transistor *VT*1 turns on again. The process repeats during the positive half-period of the source voltage *u*. During its negative half-period, transistor *VT*2 turns on and off.

In the schematic shown in Figure 37.b, the transistor *VT* turns on and off during both half-periods.

Figure 37. Electrical schematics of rectifiers with hysteresis-current control

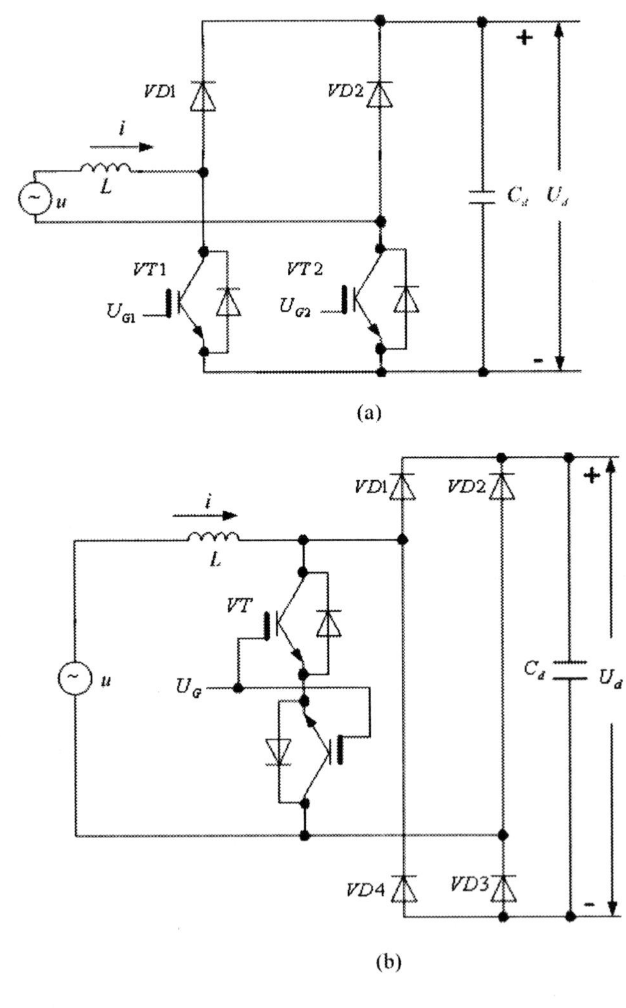

Figure 38. Waveforms of the rectifiers with hysteresis-current control

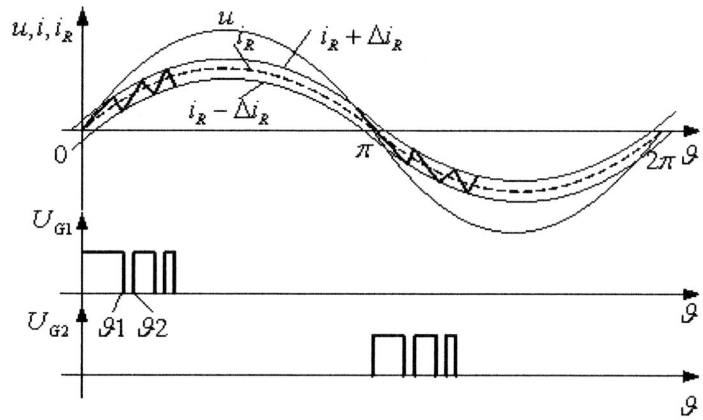

The transistor operates as a bidirectional switch that can be implemented in another than the shown way. When the transistor turns on, the diagonally connected diodes of the bridge schema conduct, and the capacitor *Cd* charges.

The operational principle of the bidirectional converters clarifies that the value of the output voltage is higher than the value of the input voltage because of the stored energy in the inductance. The value of the output voltage can be regulated by changing the reference current *i*.

Figure 39 displays a structural schema to implement a control system of the rectifier with hysteresis-current control shown in Figure 37b.

The abbreviation in the figures are:

HCCR – a hysteresis-current control rectifier; CT – a current transducer; PIC – a proportional-integral regulator; SF – a synchronizer and filter; AM – an analogue amplifier; HC – a hysteresis comparator; Dr – a driver.

The part of the output voltage of the rectifier kU_d is compared to the set value U_R. The output signal of the comparison is passed to the proportional-integral regulator and its output signal is applied to one

Figure 39. Block schema for a control system for the rectifier shown in Figure 37b

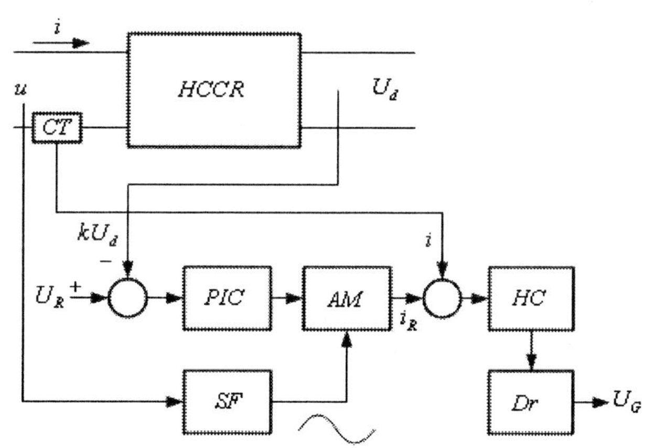

of the inputs of the analogue multiplier. A sinusoidal signal in phase with the input rectifier voltage u generated from the synchronization and filtering circuit is applied to the other input of the amplifier. The filtering is made with a purpose to get a pure sinusoidal waveform because its multiplication with the output regulator signal gives the reference current signal i_R to the output of the analogue multiplier. The reference signal value is continuously compared to the instantaneous value of the current i gained from the current transducer. The difference between the two currents is passed to the hysteresis comparator. In its circuit the value of the hysteresis Δi_R is set. The comparator output signal of the comparator has high or low levels dependent on the sign of the comparison between the two currents. The output signal is used to control the transistor via the driver.

In order to improve efficiency, switching losses and conduction losses in PFC –circuts should be reduced. The conduction losses decreases with the number of semiconductor components in the line current path – bridgeless PFC. Zero –voltage and zero-current transition are the techniques for reduction of switching losses in conventional PWM converters. The bridgeless converter with soft-switching for PFC is introduced (Mahdavi, 2009).

A new control method for PWM boost-type rectifier is described (Stankovic, 2009).

Methods to Improve Power Factor in Controlled Rectifiers

Switches that can be turn on and off using their control electrode to improve the power factor can also be applied in the controlled rectifiers with active-inductive load. According to high power use of these rectifiers, they are implemented using (gate turn off – GTO) thyristors or IGBT without a freewheeling diode.

Figure 40 shows the basic electrical schematics of the single-phase controlled rectifiers used to improve the power factor. The control is made following two methods – linear modulation (see Figure 41.a) and pulse-width modulation (see Figure 41.b). The assumption that the inductance in the load is infinitely large is made, and also, the commutation processes are neglected. In the schema shown in Figure 40.a during the positive half-period of the source voltage VT1 is turned on, while, during the negative half period – VT2 is turned on. Pauses of the output voltage and of the source current are provided from the diode VD, through which the energy of the load inductance is dissipated.

Additionally, the transistors in Figure 40.b are turned on and off in different half-periods. The diodes connected in series in one of the arms sustain the possibility to dissipate the energy stored in the load inductance during the pauses. The number of the conducting intervals in a half-period of the source voltage is marked as N.

The duration of the conducting interval in i-th switching in is marked as Δx_i, the angles of turning on and off in this interval are α_i and β_i, respectively. They are determined using different laws dependent on the modulation type.

The regulation of the value of the output voltage U_d is made by a simultaneous change of the duration of the intervals in accordance with the following equations:

-for a linear modulation

$$\Delta x_i = \Delta x = \frac{\pi}{N+1} \cdot M; \quad i = 1,2,...N \tag{4.52}$$

Figure 40. Single-phase rectifiers with width-pulse modulation -a) with a middle point of the transformer; b) bridge asymmetrical rectifier

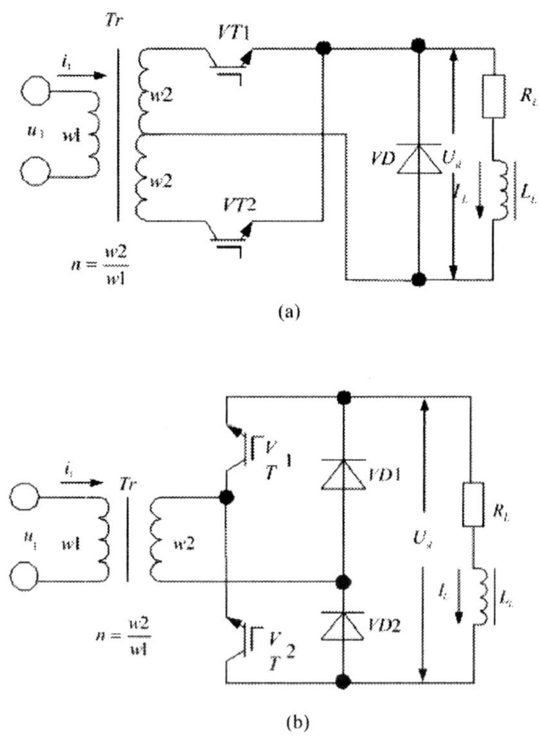

-for a sinusoidal modulation

$$\Delta x_i = \frac{\pi}{N+1} \cdot M \cdot \sin\left(i \cdot \frac{\pi}{N+1}\right); \quad i = 1, 2, \ldots N \tag{4.53}$$

Changing the modulation coefficient M between 0 and 1, the duration of the conducting intervals is changed causing a change in the value of the output voltage of the rectifier. In Figure 41 for both cases appears that the first order harmonic of the source current will remain in phase with the source voltage, thereby the displacement factor is $\cos\phi = 1$. Consequently, the main influence upon the power factor will be caused by the harmonic spectra of the source current and the distortion factor ν.

Equations for the control characteristics are obtained from the relationship:

$$U_d = \frac{1}{\pi} \cdot \sum_{i=1}^{N} \int_{\alpha_i}^{\beta_i} U_m \cdot \sin\theta d\theta \tag{4.54}$$

where in α_i and β_i are the angles to turn on and off, respectively, the devices in the i-th interval of the conductivity and they are determined from the following equations:

-for linear modulation

$$\alpha_i = \frac{\pi}{N+1} \cdot \left(i - \frac{M}{2} \right); \quad i = 1, 2, \ldots N$$
$$\beta_i = \frac{\pi}{N+1} \cdot \left(i + \frac{M}{2} \right); \quad i = 1, 2, \ldots N \tag{4.55}$$

-for sinusoidal modulation

$$\alpha_i = i \cdot \frac{\pi}{N+1} - \frac{1}{2} \cdot \frac{\pi}{N+1} \cdot M \cdot \sin\left(i \cdot \frac{\pi}{N+1} \right); \quad i = 1, 2, \ldots N$$
$$\beta_i = i \cdot \frac{\pi}{N+1} + \frac{1}{2} \cdot \frac{\pi}{N+1} \cdot M \cdot \sin\left(i \cdot \frac{\pi}{N+1} \right); \quad i = 1, 2, \ldots N \tag{4.56}$$

After substituting (4.55) in (4.54) and then of (4.56) in (4.54), it is found:
-for linear modulation:

$$S_R = \frac{U_d}{U_{do}} = M \cdot \frac{1}{2} \cdot \frac{\pi}{N+1} \cdot \sum_{i=1}^{N} \sin\left(i \cdot \frac{\pi}{N+1} \right) \approx M \tag{4.57}$$

-for sinusoidal modulation

$$S_R = \frac{U_d}{U_{do}} = M \cdot \frac{1}{2} \cdot \frac{\pi}{N+1} \cdot \sum_{i=1}^{N} \sin^2\left(i \cdot \frac{\pi}{N+1} \right) \approx 0.7854 \cdot M \tag{4.58}$$

Figure 41. Linear and sinusoidal pulse- width modulation for the rectifiers shown in Figure 37

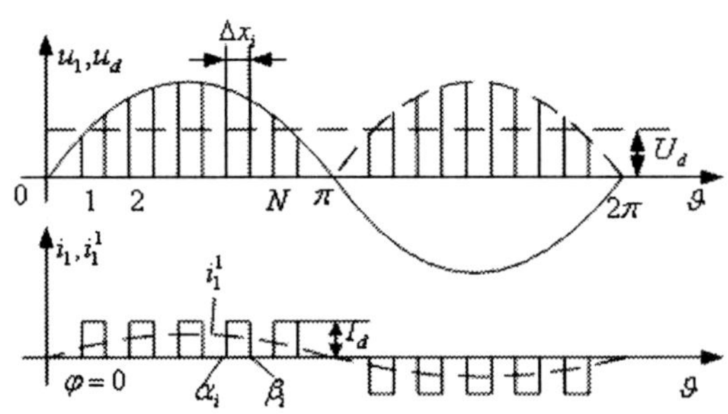

(a)

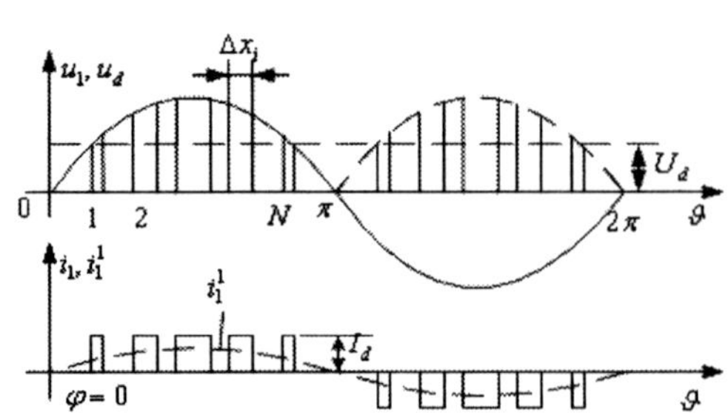

(b)

where in S_R is power of regulation and it is the ratio of the output voltage to its value in the corresponding uncontrolled rectifier.

One of the advantages of the rectifiers with pulse-width modulation is seen from (4.57) and (4.58) - linear control characteristic, which facilitates the operation in the systems of automatic regulation.

The maximum value of the n-th harmonic can be determined through a harmonic analysis of the source current from Figure 41 when the transform ratio is 1, as:

$$\frac{I_{nm}}{I_d} = \frac{2}{n \cdot \pi} \cdot \sum_{i=1}^{N} \left(\cos n\alpha_i - \cos n\beta_i \right) \tag{4.59}$$

After substituting (4.55) and (4.56) in (4.59) the relationship of the maximum value of the n-th harmonic (or its effective value) in relative units to Id and modulation coefficient M can be founded. The lack of typical for the phase regulation harmonics with numbers 3, 5, 7, etc, is an advantage of the sinusoidal modulation. Here, the main influence has the harmonics with numbers $2(N+1) \pm 1$ within the whole range of regulation. When the N is high enough, the frequencies of these harmonics are high and their filtering is facilitated.

Three-phase controlled rectifiers are also implemented using fully controlled devices. Figure 42 displays an electrical schematic of such rectifier. Pauses in the output voltage are provided by a simultaneous firing of the switches connected in one leg. Different modulation methods as linear, quasi sinusoidal, etc (Weinhold, 1991; Zargari, 1999), can be applied in this schema. As in the single-phase rectifiers shown in Figure 40, here, the control characteristic is also a straight line. Figure 43 depicts the waveforms under quasi sinusoidal width-pulse modulation. The figure also describes the firing pulses for U_{G1} to U_{G6} in the six parts during which the transistors have to be turned on. The load current flows through two transistors connected to one and the same phase of the source voltage during the pulses shown with a black color. The period of the source voltage of phase R is divided in six parts marked as 1, 2...6. The operational principles in the parts are analogues to each other. For example, in part 1, when the transistors *VT*1 and *VT*6 are turned on, the output voltage is determined by the phase-to-phase voltage u_{RS}, while when *VT*5 and *VT*6 are turned on the output voltage is determined by the phase-to-phase voltage u_{TS}. The output voltage is equal to zero when *VT*3 and *VT*6 are turned on. Namely, if in a particular part the output voltage is formed by two transistors of a group (**top** or **bottom**) in a combination with a transistor

Figure 42. Electrical schematic of a three-phase bridge rectifier with width-pulse modulation

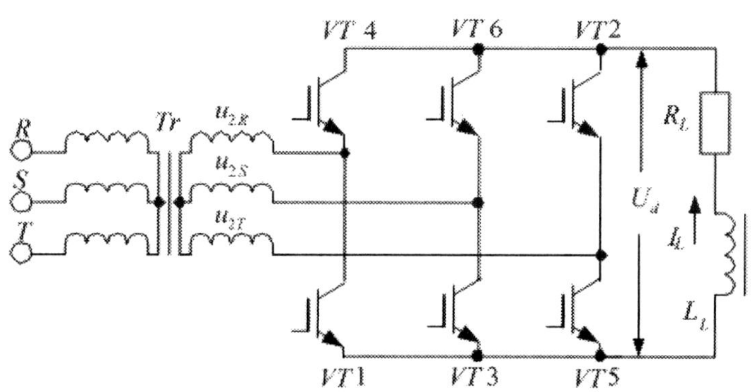

connected to the third phase but to the other group, then the firing pulses for the third transistor of the first group make the output voltage 0.

From the shown current waveforms for the three phases is seen that the first current harmonics stay in phase with the corresponding phase voltages, i.e. $\cos\phi = 1$, at a change of the duration of the firing pulses (regulation of the output voltage).

As it has been mentioned before in the text, the displacement coefficient in the examined single-phase and three-phase rectifiers with pulse-width modulation is $\cos\phi \approx 1$. Therefore, to improve the power factor, only the elimination of the higher harmonics of the mains current is required. The higher harmonics frequency is determined by the number of the switchings and it is higher than those of the conventional thyristor controlled rectifiers. The harmonic elimination is made through passive LC -filters (Katic, 2002). An example is given in Figure 44.

Besides the filtering, the capacitors also accumulate the energy stored in the inductance of each phase during the pauses of the output voltage, corresponding to pauses in the source current.

AC/DC converter shown in Figure 45, called a dual converter, derives its name from dual conversion of energy realized in it. Often it is also called "a rectifier with thyristors in the primary side of the supply transformer".

Figure 43. Waveforms for the rectifier shown in Figure 42

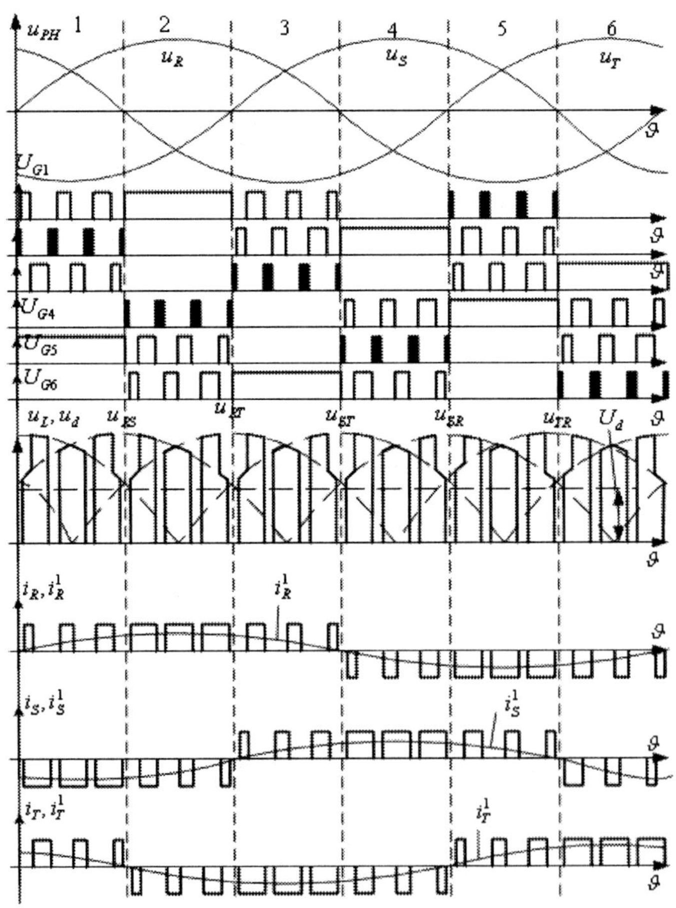

Figure 44. Electrical schematic of an input filter for rectifier with pulse- width modulation

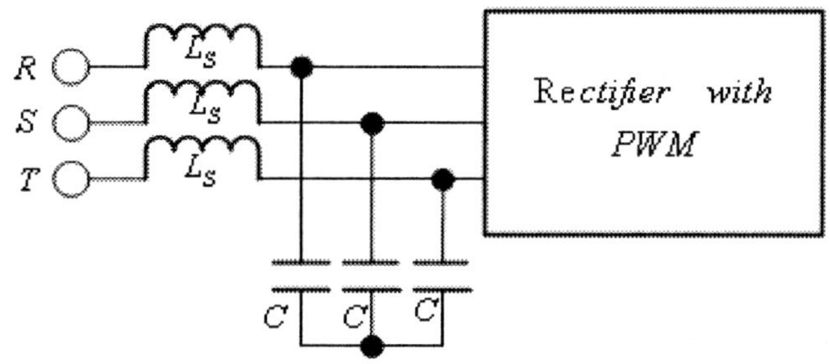

The dual converter consists of two parts – a three-phase thyristor AC regulator (VS1 – VS6) (see Chapter 5, Figure 12) and a three-phase bridge uncontrolled rectifier (VD1 – VD6) (see Figure 12). Therefore, at first an AC/AC conversion is performed, and then AC/DC conversion. The operation of the dual converter reminds of that of the three-phase bridge symmetric controlled rectifier (see Figure 19) but with a commutating diode connected in the rectifier output. The waveforms of the output volt-

Figure 45. Electrical schematic of a three-phase dual converter

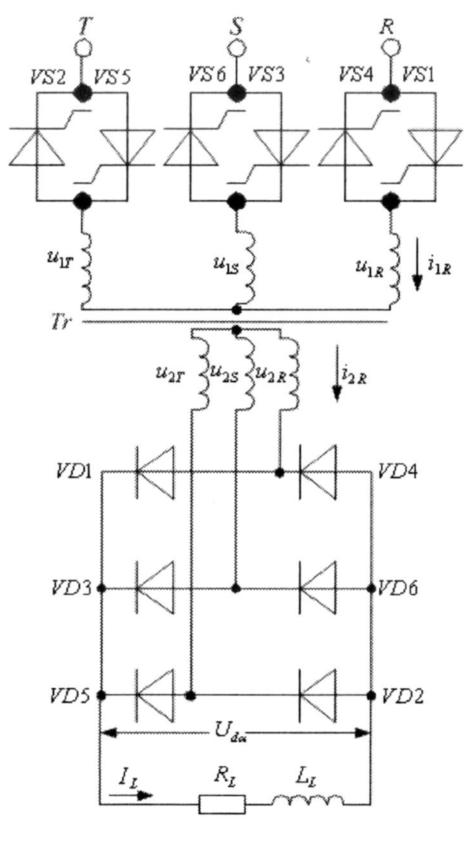

age $U_{d\alpha}$ are analogical to these depicted in Figure 20. The regulation characteristics are also analogical.

When $0 \le \alpha \le \frac{\pi}{3}$

$$U_{d\alpha} = U_{d0} \cos\alpha \tag{4.60}$$

When $\frac{\pi}{3} < \alpha \le \frac{2.\pi}{3}$

$$U_{d\alpha} = \frac{3.\sqrt{6}}{\pi} U_2 \left[1 - \sin\left(\alpha - \frac{\pi}{6}\right)\right] = 2.34.U_2.\left[1 - \sin\left(\alpha - \frac{\pi}{6}\right)\right] = U_{d0}\left[1 - \sin\left(\alpha - \frac{\pi}{6}\right)\right] \tag{4.61}$$

In the schematic shown in Figure 19, when $\frac{\pi}{3} < \alpha \le \frac{2.\pi}{3}$ the energy stored into the load inductance L_L dissipates through the commutating diode *VD* during the pauses in the waveform of the output voltage. Thus, this energy does not exchange with the source network leading to the increase of the power factor of the rectifier. In Figure 45 the whole diode bridge VD1-VD6 acts as a commutating diode. When $\frac{\pi}{3} < \alpha \le \frac{2.\pi}{3}$, the energy stored in the load inductance L_L dissipates through the bridge diodes and thus an increase in the power factor is realized also in the dual converter. Besides, into this range of change of the firing angle α, in the schematic shown in Figure 45, the diode bridge connects also in short-circuit the secondary side of the source transformer and thus it dissipates the energy stored in the windings. A feature of the converter in Figure 45 is the possibility of a contactfree switching off from the source network by turning off the firing pulses of the thyristors.

REFERENCES

Andrieu, Ch., Ferrieux, J.P., & Rocher, M. (1996). The AC-DC stage: A survey of structures and chopper control modes for power factor correction. *European power Electronics and Drives Journal, 5*(3/4), 17-22.

Garsia, O., Cobos, J. A., Prieto, R., Alou, P., & Uceda, J. (2003). Single phase power factor correction: a survey. *IEEE Transactions on Power Electronics, 18*(3), 749–755. doi:10.1109/TPEL.2003.810856

Katic, V. A., & Graovac, D. (2002). A method for PWM rectifier line side filter optimization in transient and steady states. *IEEE Transactions on Power Electronics, 17*(3), 342–352. doi:10.1109/TPEL.2002.1004242

Mahdavi, M., & Farzanehfard, H. (2009). Zero-current-transition Bridgeless PFC without extra voltage and current stress. *IEEE Transactions on Industrial Electronics, 56*(7), 2540–2547. doi:10.1109/TIE.2009.2020078

Mohan, N., Undeland, T. M., & Robbins, W. P. (1995). *Power Electronics Converters, Applications and Design.* New York: John Wiley & Sons.

Rodriquez, J. R., Dixon, J. W., Espinoza, J. R., Pontt, J., & Lezana, P. (2005). PWM regenerative rectifiers: state of the art. *IEEE Transactions on Industrial Electronics*, *52*(1), 5–22. doi:10.1109/TIE.2004.841149

SGS – Thomson. (1995). *Power Factor Corrector. Application Manual*. Italy

Skvarenina, T. (2002). *The Power Electronics Handbook*. New York: John Wiley & Sons.

Stankovic, A., & Chen, K. (2009). A New Control method for Input-Output Harmonic Elimination of the PWM Boot-Type Rectifier Under Extreme Unbalanced Operating Conditions. *IEEE Transactions on Industrial Electronics*, *56*(7), 2420–2430. doi:10.1109/TIE.2009.2017550

Unitrode. (1997). *Applications handbook*. USA.

Weinhold, M. (1991, October). Appropriate pulse width modulation for a three-phase PWM AC-to-DC converter. *European Power Electronics and Drives Journal*, 139-147.

Zargari, N. R., Xiao, Y., & Wu, B. (1999). Near unity input displacement factor for current source PWM drives. *IEEE Industry Applications Magazine*, *5*(4), 19–25. doi:10.1109/2943.771363

Chapter 5
AC/AC Conversion

BASIC INDICATORS IN RESPECT TO THE SUPPLY NETWORK

The operational principle of a classical converter of AC into AC energy is described in Section 1 Chapter 3. It includes an explanation of the basic idea on which the operation of the further down studied power electronic converters is based.

The basic indicators of a power electronic converter supplied by an AC power have been studied in Chapter 4. The conclusions made in it according the ways of the increase of the power factor K_p are also valid in AC/AC converters. The purpose is to consume current with a waveform as close as possible to a sine wave (with low contents of harmonics) and whose first harmonic to be in phase with the source voltage.

DOI: 10.4018/978-1-61520-647-6.ch005

SINGLE-PHASE AND THREE-PHASE AC REGULATORS

Single-Phase AC Regulators

Figure 1 (a) depicts a base electrical schematic of a thyristor single-phase AC regulator (TSPACR) implemented with an antiparallel connected thyristors. Figure 1.b displays the typical waveforms of the voltages of the TSPACR.

At a positive half-period of the AC input voltage, it is applied to the thyristor *VS*1; thus the output voltage is zero before the firing pulse is passed. Concerning the thyristor *VS*2, during the same interval the AC input voltage is a reverse for it and the thyristor is turned off. When the phase of the input voltage is α , a firing pulse is passed via the gate electrode circuit, and the thyristor *VS*1 turns on. The voltage applied across the thyristor *VS*2 decreases to the value of the forward voltage of a conducting switch $U_{Fcond\max}$. If an ideal V-A characteristic of a thyristor is considered, the forward voltage is zero, also, the output voltage and the input AC voltage are of the same values. The value of the current flowing through *VS*1 is determined by the input voltage and load. When the transitory value of the input voltage tends to zero, the thyristor current also decreases (in the active load case the voltage and current have

Figure 1. a) electrical schematic of a thyristor single-phase AC regulator; b) typical waveforms of the voltages of the TSPACR

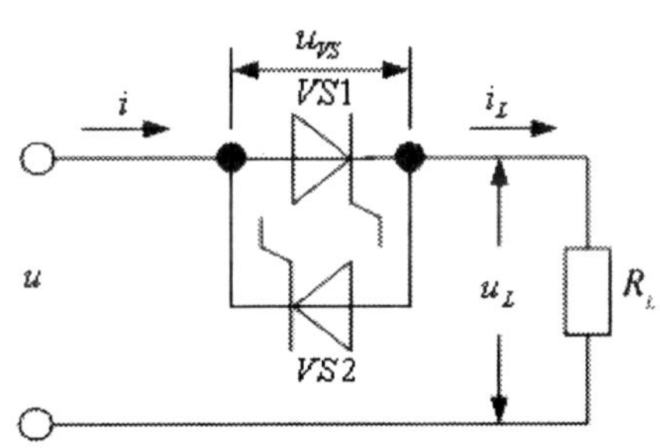

(a)

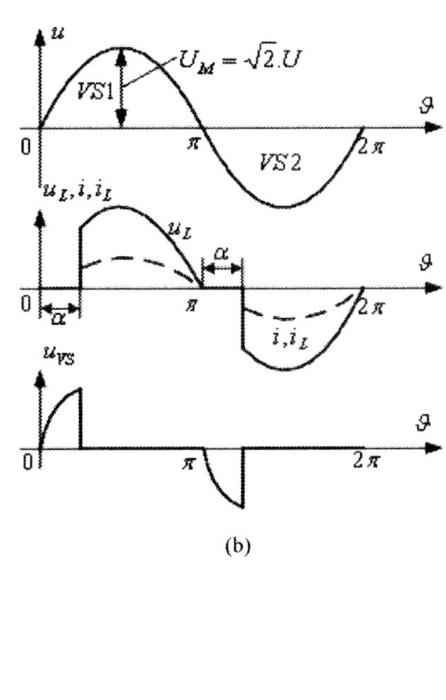

(b)

identical shapes); then in a particular moment, when the current becomes less than the holding current of the thyristor, it turns off. During the negative half-period, the voltage across *VS*1 is a reverse voltage equal to the input voltage before the firing of *VS*2 happens, after the firing till the end of the half-period the reverse voltage across *VS*1 is equal to the forward voltage of the conducting thyristor *VS*2. The processes during the negative half-period, when *VS*2 is turned on and it conducts, are analogous to those during the positive half-period.

The most interesting fact from the point of view of the power delivered to the load is the effective values of the load voltage and current.

The effective value of the load voltage is:

$$U_L = \sqrt{\frac{1}{2 \cdot \pi} \int_0^{2\pi} u^2(\omega \cdot t) d\omega t}, \tag{5.1}$$

where in

$$u(\omega t) = \sqrt{2} \cdot U \cdot \sin \omega t \tag{5.2}$$

is the transitory value of the load voltage at the intervals $\alpha \div \pi$, $(\pi + \alpha) \div 2\pi$, etc. And $\omega t = \theta$ is the running angle.

After substituting (5.2) in (5.1) the load voltage effective value is derived as:

$$U_L = U \sqrt{\frac{1}{2 \cdot \pi} \left[2 \cdot (\pi - \alpha) + \sin 2\alpha \right]} \tag{5.3}$$

The equations for the active and total power can be consistently found for determining the power factor:

$$P = \frac{1}{2 \cdot \pi} \cdot \int_0^{2\pi} u(\theta) \cdot i(\theta) \cdot d\theta = \frac{U^2}{R_L} \left[\frac{2 \cdot (\pi - \alpha) + \sin 2\alpha}{2 \cdot \pi} \right] \tag{5.4}$$

$$S = U \cdot I = U \cdot \sqrt{\frac{1}{2 \cdot \pi} \int_0^{2\pi} i^2(\theta) \cdot d\theta} = \frac{U^2}{R_L} \cdot \sqrt{\frac{1}{2 \cdot \pi} \cdot \left[2 \cdot (\pi - \alpha) + \sin 2\alpha \right]} \tag{5.5}$$

So, the power factor is derived as:

$$K_P = \frac{P}{S} = \sqrt{\frac{2 \cdot (\pi - \alpha) + \sin 2\alpha}{2 \cdot \pi}} \tag{5.6}$$

Figure 2 displays the change of the active power *P*, the power factor K_p, the displacement factor $\cos \phi$ and the distortion factor ν at the change of the firing angle α

If the two antiparallel connected thyristors in Figure 1 are replaced by a triac, the power schema and control systems are simplified, but problems in the operation in the high powers and active-inductive load occur.

Figure 2. Dependence of the change of the different variables on the firing angle

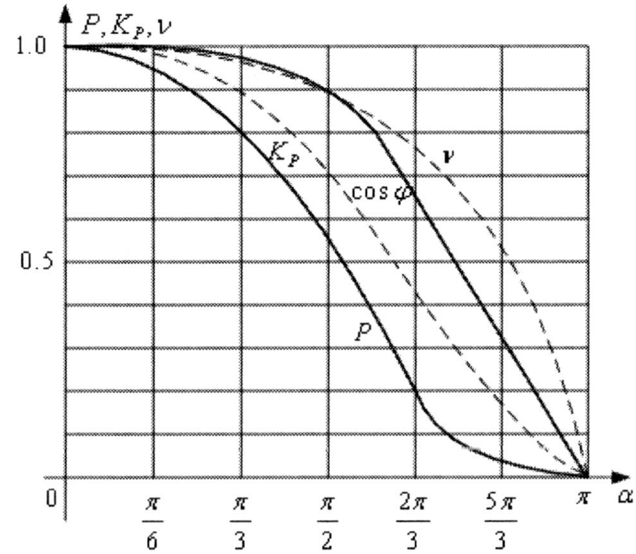

The combination of a thyristor and antiparallel diode makes asymmetrical regulation (only in one of the half-periods) of the output voltage possible. The electrical schematic and the typical waveforms of an asymmetrical TSFACR are presented in Figure 3. This regulator circuit has two main disadvantages. The first is the presence of a DC component of the source current due to the asymmetric. The second is the presence of even harmonics in the spectrum of the source current. Because in the supply systems usually passive filters for the odd harmonics are implemented, the even harmonics will not be decreased and they will cause a negative influence upon the waveform of the voltage supplying the rest of the consumers connected to the same supply network.

Figure 3. Electrical schematic and typical waveforms of an asymmetrical TSFACR

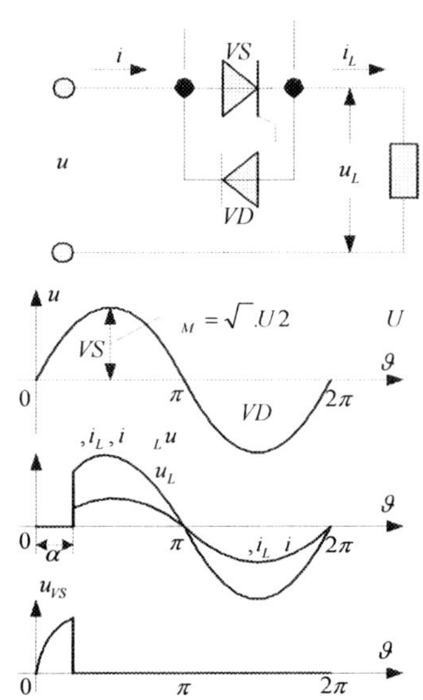

Figure 4. Circuit of TSFACR and its output voltage

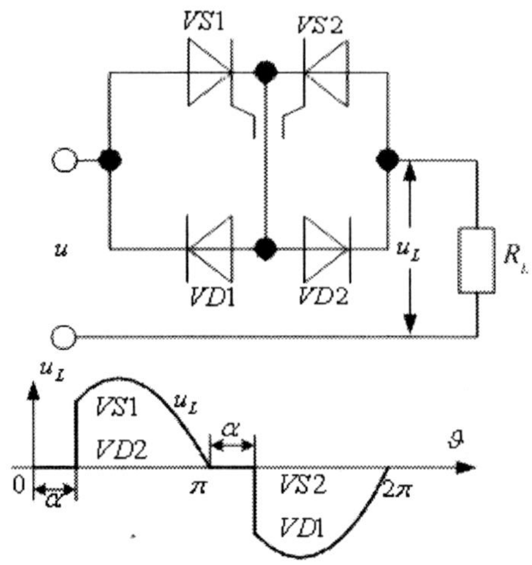

Figure 4 and Figure 5 display two more circuits of TSFACR. The regulator circuit shown in Figure 4 is implemented using thyristors and diodes connected in a bridge with a short-circuit diagonal. A thyristor and diode conduct simultaneously (at one of the periods *VS*1 and *VS*2, and at the other *VS*2 and *VS*1) which leads to increase of the losses in comparison to those of the circuit in Figure 1. The common point of the cathodes of the thyristors is a schema advantage, which simplifies the control system implementation.

The regulator circuit shown in Figure 5 is implemented using a thyristor and four diodes. In the positive half-period the devices *VD*1, *VS* and the diode *VS*3 conduct, in the negative half-period- *VD*2, *VS* and *VS*4. The regulator circuit is characterized with a lower efficiency coefficient in comparison with the others studied regulators, because three devices conduct simultaneously.

Figure 5. Circuit of TSFACR and its output voltage

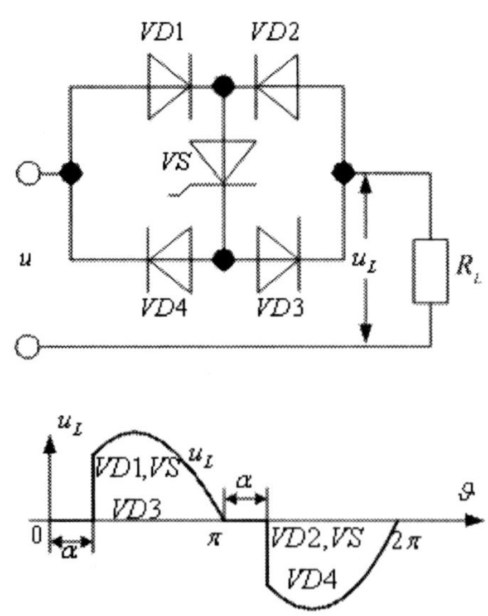

A propagation of the control method applied to the circuit in Figure 1 and described in Figure 6 is also known (Shepherd, 1975). The method is known as "pulse-width regulation with lower than the source frequency", "integral cycling switching", "cycling selection". The thyristors conduct during a particular number of whole periods (or half-periods) of the source voltage. The cycle is repeated over a constant number of whole periods (or half-periods) of the voltage. The number of the periods during which the thyristors conduct is N and the number of the repetition periods is T. The change of N regulates the load power. For waveforms in Figure 6, $N = 2$ and $T = 3$. The shapes of the load and source currents are identical to this of the voltage but their amplitude is $\sqrt{2} \cdot I$. The thyristors turn on when the source voltage crosses 0, for which specially designed optron circuits are available. A primary expectation for improved indicators in respect to the source network is available.

The main part of the source current spectrum is shown in Figure 7.

The value of the fundamental harmonic is proportional to the ratio of N to T. The displacement between the first harmonic and the source voltage is 0. It means tha $\cos\varphi = 1$. Besides the fact that during the conduction of the thyristors the source current has a sinusoidal waveform, yet the power factor K_p is not equal to 1 because there are intervals during which the current is 0.

The main disadvantages of the cycling selection switching method are the presence of sub and even harmonics of the source current. The sub-harmonics are harmonic components with a frequency lower than that of the source voltage.

The effective value of the load voltage is:

$$U_L = \sqrt{\frac{1}{2 \cdot \pi} \int_0^{\frac{2\pi}{T} \cdot N} 2 \cdot U^2 \cdot \sin^2 T\theta \cdot d\theta} = U\sqrt{\frac{N}{T}} \qquad (5.7)$$

The effective values of the load and source currents are derived as:

$$I = I_L = \frac{U_L}{R_L} = \frac{U}{R_L} \cdot \sqrt{\frac{N}{T}} \qquad (5.8)$$

Figure 6. Description of the cycling selection method

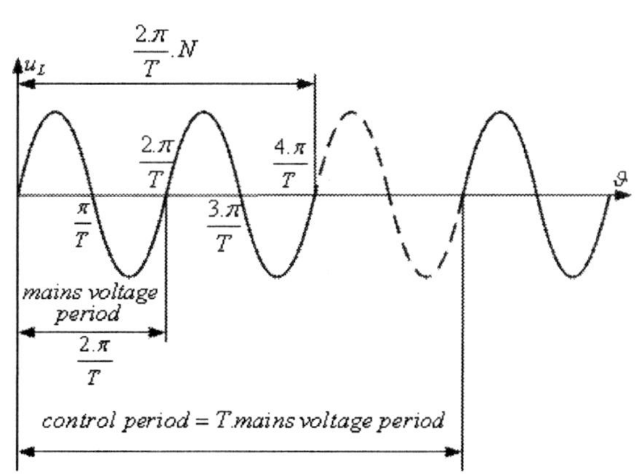

The active power consumed by the load P_L and the active power consumed from the source P are equal on the assumption that $\eta = 1$:

$$P = P_L = I_L^2 \cdot R_L = \frac{U^2}{R_L} \cdot \frac{N}{T} \tag{5.9}$$

Therefore, the power factor is:

$$K_P = \sqrt{\frac{N}{T}} \tag{5.10}$$

The distortion factor can be found as:

$$\nu = \frac{I_1}{I} = \sqrt{\frac{N}{T}} \tag{5.11}$$

The equation (5.11) proves that the displacement coefficient is equal to 1 ($\cos\phi = 1$).

The presence of series connected inductance in the load significantly changes the numerical equations and the operation of the TSFACR. The accumulated energy in the inductance causes this change. As a result of the accumulated energy an electromotive voltage of self-inductance appears at the end of a half-period. The electromotive voltage maintains the thyristor current over the holding current, because of that it continuous to conduct during a part of the next negative half-period (angle δ in Figure 8). The value of the angle δ depends on the quantity of the stored energy in the inductance. The energy is proportional to the second power of the load current and the value of the inductance itself.

Two operation modes dependent on the ratio of the firing angle α and the phase angle of the load

Figure 7. The source current spectrum for the circuit shown in Figure 1 controlled by the method described in Figure 6

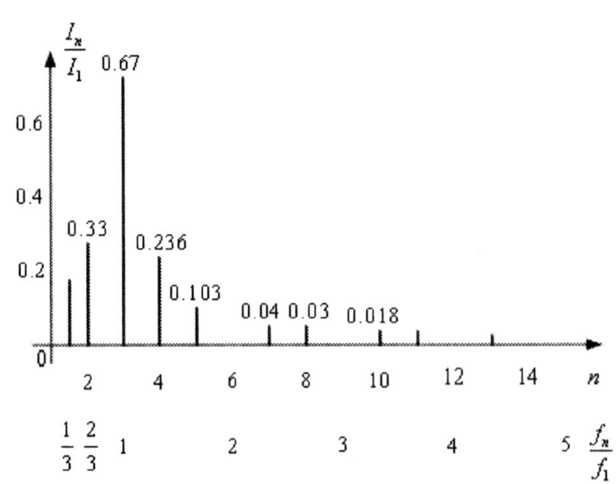

Figure 8. Electrical circuit and typical waveforms of SPACR with active-inductive load, $\alpha > \phi$

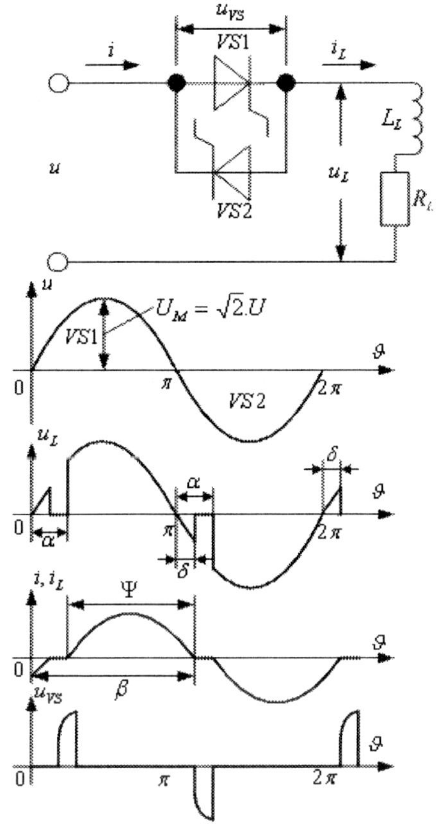

ϕ are possible. When $\alpha > \phi$, the thyristor that has been conducting turns off before the turning on of the next thyristor. The load current does not flow during particular intervals and it is a series of non-sinusoidal pulses. When $\alpha \le \phi$, the load current is discontinuous; the load voltage is equal to the input voltage. This is the mode of the discontinuous current in which the possibility to regulate the output voltage is terminated.

The equation describing the process in the regulator is:

$$L_L \cdot \frac{di(\omega t)}{d(\omega t)} + R_L \cdot i(\omega t) = \sqrt{2} \cdot U \cdot \sin(\omega t) \tag{5.12}$$

The solution of (5.12) is:

$$i(\omega t) = A \cdot e^{-\frac{\omega \cdot t}{tg\phi}} + \frac{\sqrt{2} \cdot U}{\sqrt{R_L^2 + (\omega \cdot L_L)^2}} \cdot \sin(\omega \cdot t), \tag{5.13}$$

where in $tg\phi = \dfrac{\omega \cdot L_L}{R_L}$.

After determining the constant A from the condition for discontinuous current $i(\alpha) = -i(\omega \cdot t - \phi)$ and impose the requirement for the operation of continuous mode as $i(\alpha) = 0$ at the moment $\omega \cdot t = \pi + \alpha$, the current is found as:

$$i(\omega t) = \frac{\sqrt{2} \cdot U}{\sqrt{R_L^2 + (\omega \cdot L)^2}} \cdot \left[\sin(\omega \cdot t - \phi) - \sin(\alpha - \phi) \cdot e^{-\frac{\omega \cdot t - \alpha}{tg\phi}} \right] \tag{5.14}$$

The waveform of the load current is presented in Figure 9. It appears from (5.14) that the load current has two components – sinusoidal and transient ones. At the moment $\omega \cdot t = \alpha$ the sinusoidal component has the value $\sqrt{2} \cdot I \cdot \sin(\alpha - \phi)$, where in

$$I = \frac{U}{\sqrt{R_L^2 + (\omega \cdot L_L)^2}}.$$

The transient component attenuates according to an exponential law $e^{-\frac{\omega \cdot t}{tg\phi}}$. The load current is negative when $\omega \cdot t > \beta$ (see Figure 9). This means (if a thyristor turning off at this moment presences) that the current will stop flowing. The angle β, at which the load current is terminated, is called angle of extension (this name reflects the flowing of the current after the moment π).

Based on (5.14), the angle of extension is determined as:

$$\sin(\beta - \phi) - \sin(\alpha - \phi) \cdot e^{-\frac{\beta - \alpha}{tg\phi}} = 0 \tag{5.15}$$

The solution of (5.15) is depicted in Figure 10. When $\alpha = \phi$, which corresponds to a boundary mode (current of sinusoidal waveform), the equation (5.15) becomes $\sin(\beta - \phi) = 0$. The possible result is

Figure 9. Voltage and current waveforms

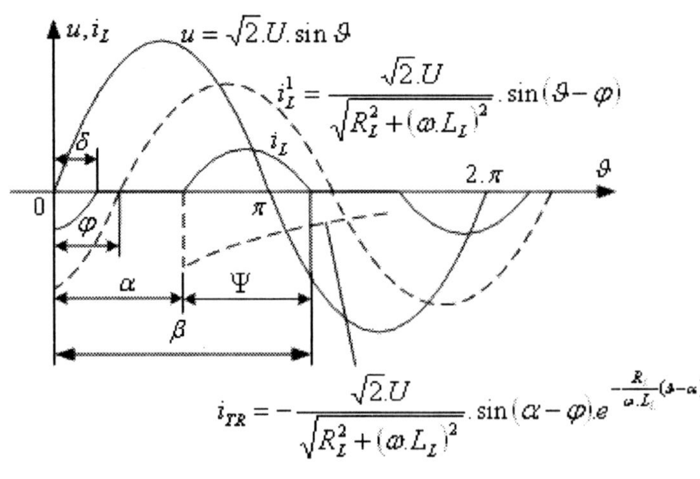

Figure 10. Dependence of values of β on the firing angle

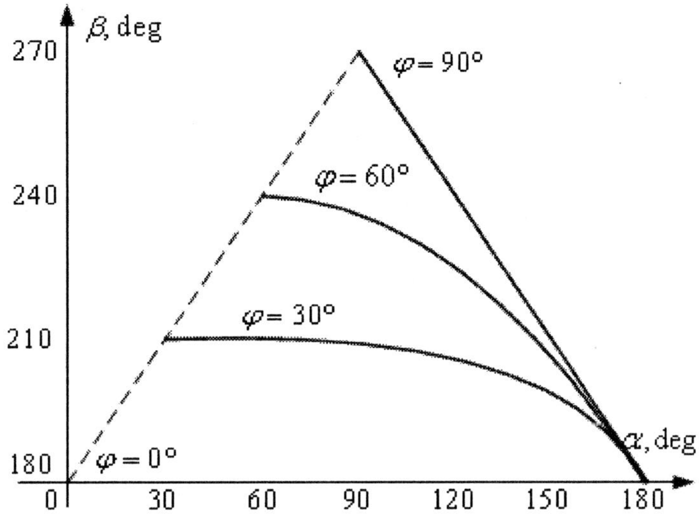

$\beta = \pi + \phi$. The values of β making the regulator to operate in accordance with a sinusoidal law are situated ont the dotted line in Figure 10.

Figure 11 displays the control characteristics of the TSPACR when the angle $\phi = \text{var}$ varies. The characteristic when ϕ corresponds to an active load and it is actual for the waveforms shown in Figure 1.b. When the firing angles $\phi = 0$ are lower than the corresponding angles α, there is no change of the effective value of the load voltage, which equals to the source voltage. Moreover, the load current has sinusoidal waveform and lags to the voltage at the corresponding angle ϕ.

Figure 11. Control characteristics of TSPACR when ϕ =var

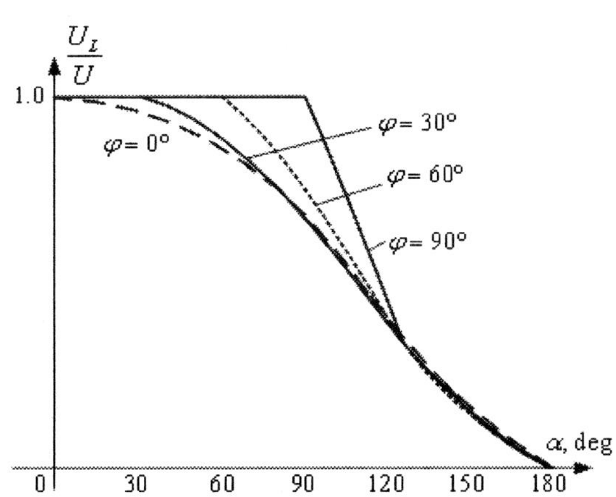

Figure 12. Electrical circuit of a TTPACR with load connected in Y configuration

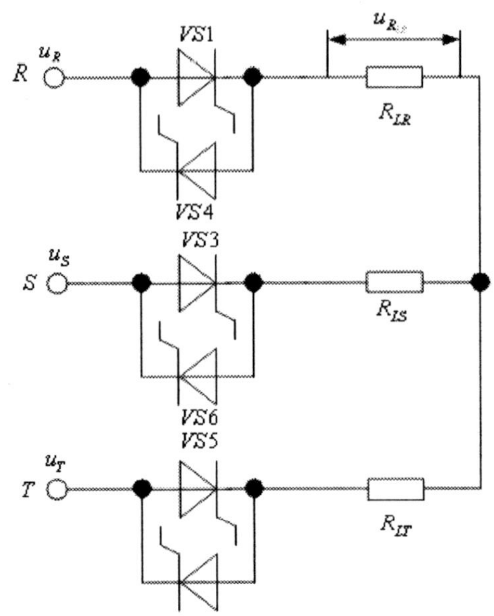

Three-Phase AC Regulators

The load of a thyristor three-phase AC regulator (TTPACR) can be connected in two ways – delta connection (Δ) and wye or star connection (Y).

Figure 12 displays a TTPACR with load connected in Y configuration. Typical waveforms for the load voltage of phase R for several values of the firing angle α are depicted in Figure 13. The load is pure active one. The moments, at which the thyristors receives the firing pulses, are shown.

The firing angle α in the three-phase regulators is measured from the moment, at which the corresponding phase voltage crosses zero.

If the regulator operates with a taken out star center and the firing angles for the three thyristors are of the same values, then the waveforms of the output voltage and the control characteristics for each of the phases are identical to those of the single-phase AC regulator.

The regulator shown in Figure 12 can operate in three modes dependent on the value of the firing angle α.

Mode 1: $0 < \alpha < \pi / 3$. Three or two thyristors conduct in the circuit. The output load voltage for a phase is equal to the phase voltage, to the half of the phase-to-phase voltage between the phase and one of the rest phases, or to zero. When the thyristor of a particular phase does not conduct, the load voltage of this phase is zero. When the three thyristors conduct, the load for this time interval appears to be connected to a three-phase symmetrical system, therefore, the voltage across each of the load resistances is equal to the corresponding phase voltage. When only two thyristors conduct, the voltage across each load connected to them is equal to the corresponding phase-to-phase voltage. For example, see Figure 13.b, after the firing pulse for the thyristor *VS*2 is passed at $\theta = \pi / 2$ for $U_{R'}$, thyristors *VS*1, *VS*2 and *VS*6 conduct. The voltage of the load in phase R is equal to $U_{R'}$. At $\theta = \dfrac{2\pi}{3}$ the thyristor *VS*6 turns off;

Figure 13. Typical waveforms for the load voltage of phase R for several values of the firing angle α

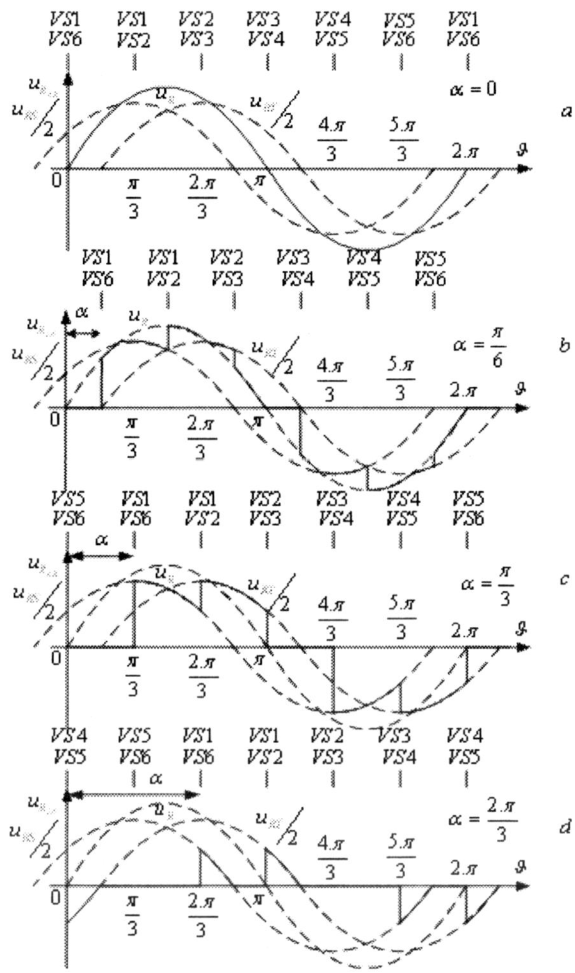

only the two thyristors $VS1$ and $VS2$ conduct and thus the voltage across the load in phase R is equal to the phase-to-phase voltage U_{RT}.

Mode 2: $\pi / 3 < \alpha < \pi / 2$. Always two thyristors conduct in the circuit. The output voltage is equal to the half of the phase-to-phase voltage between a particular phase and one of the rest phases or to zero. For example, see Figure 13.c, it appears that the supply of the firing angles to turn on $VS1$ and $VS6$ coincide with the turning off of $VS5$.

Mode 3: $\pi / 2 < \alpha < \frac{5\pi}{6}$. Two thyristors conduct in the circuit or all of them are turned off. Each thyristor is turned on by the basic and auxiliary firing pulse and the thyristor conducts twice in a period. It appears in Figure 13.d that after the turning on of $VS1$ by the basic firing pulse at $\alpha = \frac{2\pi}{3}$, the thyrsitors $VS1$ and $VS6$ conduct till the decrease of the phase-to-phase voltage U_{RS} to 0. The thyrsitor $VS1$ turns on again with $VS2$ by the auxiliary firing pulse at $\theta = \pi$ for U_R, which moment corresponds to $\alpha = \frac{2\pi}{3}$ for $VS2$.

Dependent on the operational mode, the relationship of the value of the output voltage for the load connected to one of the phases in relative units in accordance with the effective value of the source voltage is given by one of the following equations:

$$\frac{U_{R_{LR}}}{U} = \sqrt{\frac{2}{\pi}\cdot\left(\frac{\pi}{2}-\frac{3}{4}\cdot\alpha+\frac{3}{8}\cdot\sin 2\alpha\right)} \quad when\ 0\leq\alpha\leq\frac{\pi}{3}$$

$$\frac{U_{R_{LR}}}{U} = \sqrt{\frac{2}{\pi}\cdot\left(\frac{\pi}{4}+\frac{9}{16}\cdot\sin 2\alpha+\frac{3\sqrt{3}}{16}\cdot\cos 2\alpha\right)} \quad when\ \frac{\pi}{3}\leq\alpha\leq\frac{\pi}{2}$$

$$\frac{U_{R_{LR}}}{U} = \sqrt{\frac{2}{\pi}\cdot\left(\frac{5\pi}{8}-\frac{3}{4}\cdot\alpha+\frac{3\sqrt{3}}{16}\cdot\cos 2\alpha+\frac{3}{16}\cdot\sin 2\alpha\right)} \quad when\ 0\leq\alpha\leq\frac{\pi}{3} \tag{5.16}$$

The fact that the load is a pure active one can be used in the process of deriving an equation for the change of the power factor. Let mark the ratio of the effective value of the load voltage to the effective value of the source voltage as a random expression A:

$$\frac{U_{R_{LR}}}{U} = A \tag{5.17}$$

Because of the equality of the effective values of the source current for a phase and the load current, total power is found as:

$$S = U\cdot I = U\cdot\frac{U_{R_{LR}}}{R_{LR}} = \frac{U^2}{R_{LR}}\cdot A \tag{5.18}$$

If $\eta = 1$, then the active power consumed from the source network and the load active power will be of the same values:

$$P = P_{R_{LR}} = \frac{U^2_{R_{LR}}}{R_{LR}} = \frac{U^2}{R_{LR}}\cdot A \tag{5.19}$$

Then from equations (5.17) and (5.18) the power factor is found:

$$K_P = \frac{P}{S} = A \tag{5.20}$$

The above made conclusion for the power factor is proved by the already found equations for TSFACR – compare the pairs of equations (5.3), (5.6) and (5.7), (5.10) to (5.17), (5.20).

Equations (5.16) and (5.20) lead to:

Figure 14. Electrical circuit of a TTPACR when the load is connected in delta connection

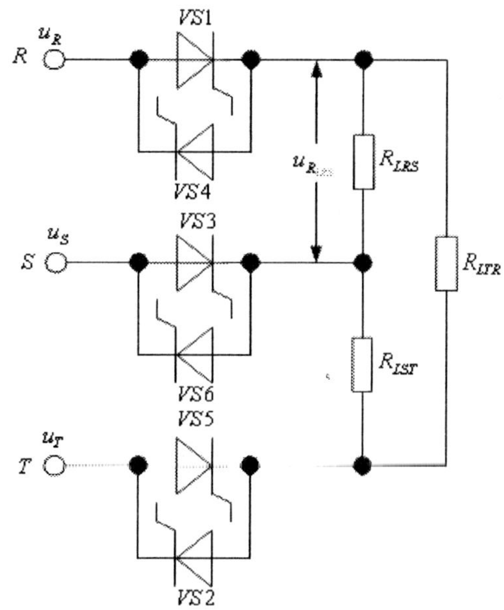

$$K_P = \sqrt{\frac{2}{\pi}\cdot\left(\frac{\pi}{2}-\frac{3}{4}\cdot\alpha+\frac{3}{8}\cdot\sin 2\alpha\right)} \quad when\ 0\le\alpha\le\frac{\pi}{3}$$

$$K_P = \sqrt{\frac{2}{\pi}\cdot\left(\frac{\pi}{4}+\frac{9}{16}\cdot\sin 2\alpha+\frac{3\sqrt{3}}{16}\cdot\cos 2\alpha\right)} \quad when\ \frac{\pi}{3}\le\alpha\le\frac{\pi}{2}$$

$$K_P = \sqrt{\frac{2}{\pi}\cdot\left(\frac{5\pi}{8}-\frac{3}{4}\cdot\alpha+\frac{3\sqrt{3}}{16}\cdot\cos 2\alpha+\frac{3}{16}\cdot\sin 2\alpha\right)} \quad when\ 0\le\alpha\le\frac{\pi}{3} \tag{5.21}$$

Figure 14 displays the TTPACR when the load is connected in delta connection.

The waveforms of the voltage across one of the loads $U_{R_{LRS}}$ are depicted in Figure 15.

In this case, three modes of operation are also typical:

Mode 1: $0 < \alpha < \pi / 3$. Three or two thyristors conduct in the circuit. The output load voltage for a phase is equal to the phase-to-phase voltage, between two phases where the load is connected to, in two cases: if three thyristors conduct or if the two thyristors connected to the same phases as the load is connected to conduct. In the rest cases, two thyristors conduct, the voltage value across the studied load is half of the one of the others two phase-to-phase voltages. For example, it appears in Figure 15.a that at $\alpha = \pi / 6$ the thyristor *VS*1 turns on and it conducts with thyristors *VS*5 and *VS*6, which has been conducting till this moment. The voltage across the load R_{LRS} is equal to U_{RS}. At $\theta = \pi / 3$, *VS*5 turns off because its current decreases and reaches 0. Only *VS*1 and *VS*6 continue to conduct. Thus, the voltage across the load R_{LRS} continues to trace the same phase-to-phase voltage. At $\theta = \pi / 2$ $VS2$ turns on, again three thyristors conduct, thus, the voltage across the load R_{LRS} remains equal to U_{RS}. Just at $\theta = 2\pi / 3$ the thyrsitor *VS*6 turns off while the thyristors *VS*1 and *VS*2 remain conducting. The loads R_{LRS} and R_{LST} appear to be in series. They are connected to U_{RT}. Thus, the half of this voltage is the voltage across the load R_{LRS}.

Figure 15. Typical waveforms for the regulator shown in Figure 14 for several value of the firing angle α

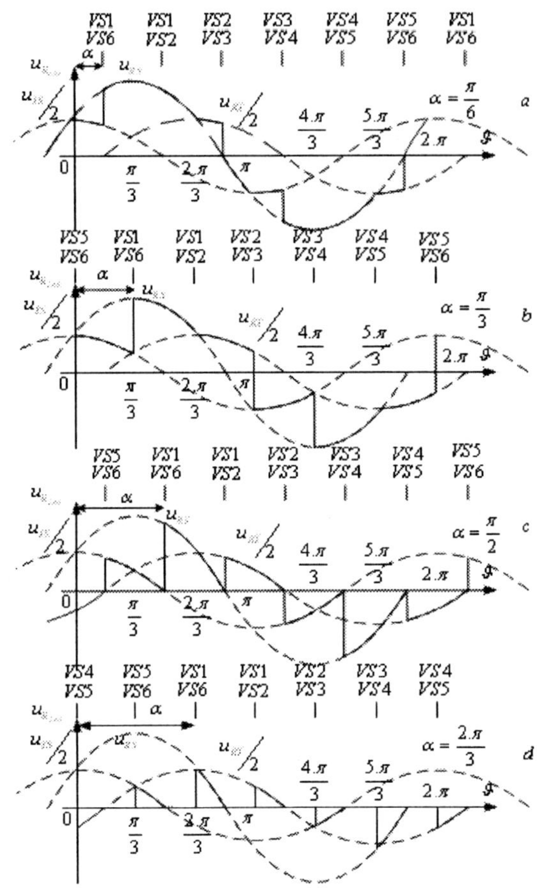

Mode 2: $\pi / 3 < \alpha < \pi / 2$. Always two thyristors conduct in the circuit. When two thyrsitors, connected to the same phases to which the studied load is connected to, conduct, then the voltage across the load is equal to the phase-to-phase voltage between these phases. In the rest cases, the voltage across the studied load is equal to the half of the one of the others two phase-to-phase voltages. It appears in Figure 15.c that at $\alpha = \pi / 2$, the thyristors *VS*1 and *VS*6turn on making the voltage across R_{LRS} equal to U_{RS}. At the moment of turning on *VS*2, the thyristor *VS*6 turns off, therefore, the voltage value across the studied load becomes half of U_{RT}.

Mode 3: $\pi / 2 < \alpha < \frac{5\pi}{6}$. The operation of the circuit is analogue to this of the mode 2. The only difference is that there are intervals in which all the thyristors in the circuit do not conduct and the voltage of all loads is equal to 0. Each thyristor turns on twice in a period - by the basic and auxiliary firing pulses. In Figure 15.d it appears that the thyristor *VS*1 turns on by the basic firing pulse at $\alpha = \frac{2\pi}{3}$ and it conducts with *VS*6 till $\theta = \frac{5\pi}{6}$, therefore, the voltage across R_{LRS} is $U_{RS} / 2$. After that moment the load voltage is 0 because all the thyristors are turned off. The thyrsitor *VS*1 turns on again with *VS*2 by the auxiliary firing pulse at $\theta = \pi$ and the voltage value across the load R_{LRS} becomes $U_{RT} / 2$.

Figure 16. Electrical circuits of TPACR

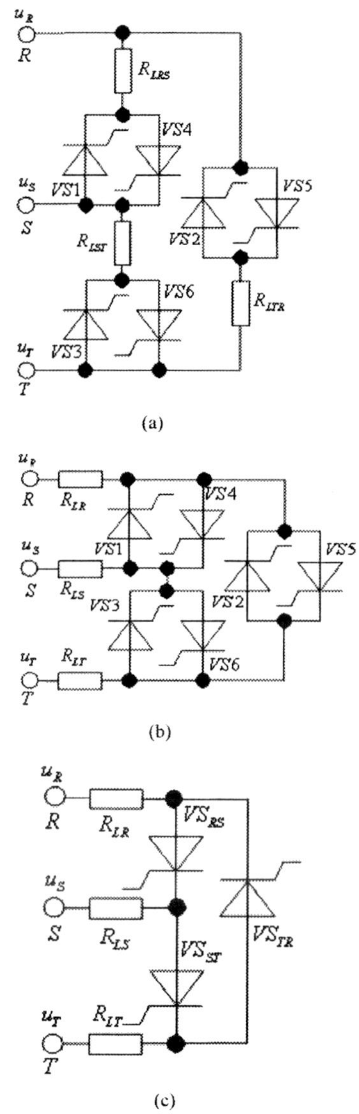

Figure 16 presents different from the above described schemas of TTPACR. They are used comparatively rarely in practice, because they require a possibility to separate the load.

The operation of the regulation circuit shown in Figure 16.a is similar to this of the schema in Figure 12, but with a taken out star center and it requires a change of the firing angle between 0 and π. Third harmonic in the load current is a typical feature.

Even harmonics in the source current are typical for the circuit shown in Figure 16.c. This is a disadvantage of the circuit. Its advantage is the reduced number of thyristors.

The use of antiparallel connected diode-thyristor is also possible in the three-phase AC regulators as in the TSPACR, but even harmonics in the source current are the main disadvantage. Moreover, the second harmonic predominates.

Figure 17. a) equivalent schema of network and load; b) and c) vector diagrams of an increase and decrease of the reactive power, respectively

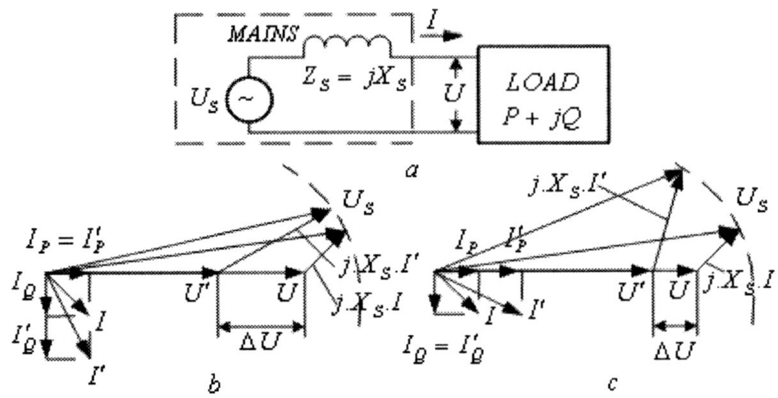

METHODS TO IMPROVE POWER EFFICIENCY IN AC/AC CONVERSION

In an AC supply network, the voltage value is necessary to be maintained in a definite range around its nominal value. In different countries the most imposed standards determine a range of from +5% to -10% around the nominal voltage value. Besides, the loads in the three-phases should be identical to escape the currents with zero and reverse sequences because they cause undesirable effects such as additional heating of the equipment. The load connected to the network varies, thus, the value of the voltage may leave the standard range. Reactive power of the load causes the major effect on the network because the inner resistance of the supply network (including transmission and distribution lines, transformers, generators, etc.) has an inductive character for source frequency of 50Hz or 60Hz.

Figure 17 displays a simplified equivalent schema for one phase in which the network is assumed to have entirely inductive resistance. Vector diagrams assume a lag of the total load current $I = I_p + j \cdot I_q$ to the voltage when the loading of the power is $P + j \cdot Q$ and the value U of the voltage supplying the load differs from the source voltage value U_S.

Figure 17.b shows an increase of the reactive power ΔQ at which the reactive component of the current ΔI_q has increased and the active component I_p has not changed. As a result, the input load voltage has decreased with ΔU, which leads to a decreasing of the load active power. Figure 17.c displays the same percentage of increase of the active component of the current I_p while the reactive component has not changed. In this case the change of the input load voltage ΔU is smaller.

First, the static var compensators of reactive power will be examined. Their operation is close to the one of the thyristors AC regulators thereby they are studied in this chapter. The static var compensators are implied to improve the power factor not only of the AC regulators but also of other loads having a worsen indicators in respect to the supply network. The static var compensators should have the following two capabilities: 1.To accumulate reactive power in case of its excess in the supply system because this may lead to the voltage increase; 2. To deliver reactive power in case of its lack in the system because this may lead to decrease of the source voltage.

The first task is solved using so called thyristor-controlled inductance, and the second one – using thyristors-switched capacitors.

Figure 18. Basic electrical schematic and typical waveform describing the operation of a thyrsitor-controlled inductance

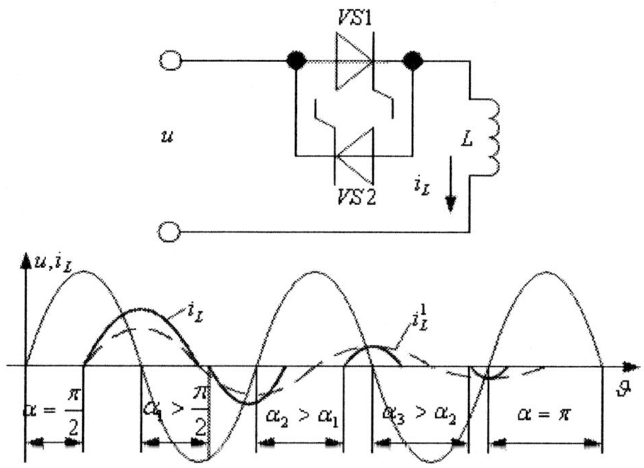

Basic electrical schematic and typical waveforms describing the operation of a thyrsitor-controlled inductance is shown in Figure 18. Using the inductance the network reactive power loading can be changed significantly fast.

The source current has a sinusoidal waveform and lags to the source voltage at angle $\pi / 2$ when the change of the firing angle is within the range of $0 \leq \alpha \leq \frac{\pi}{2}$. The current effective value is:

$$I_L = I_{L1} = \frac{U}{\omega \cdot L} \tag{5.22}$$

If the firing angle of the thyristors increases then the shape of the current differs from the sinusoidal one and effective value of its first harmonic is:

$$I_{L1} = \frac{U}{\pi \cdot \omega \cdot L} \cdot \left(2 \cdot \pi - 2 \cdot \alpha + \sin 2\alpha\right) \tag{5.23}$$

Therefore, the behavior of the circuit is as an inductance with effective value as follows:

$$L_{EFF} = \frac{U}{\omega \cdot I_{L1}} \tag{5.24}$$

Then, the additional reactive power for the phase, to which the circuit is connected, is:

$$Q = U \cdot I_{L1} = \frac{U^2}{\omega \cdot L_{EFF}} \tag{5.25}$$

Besides the first harmonic, the source current contains also higher harmonics, whose effective value for an n odd number is defined as;

Figure 19. Dependence of the change of the fundamental and higher harmonics with low numbers on the firing angle

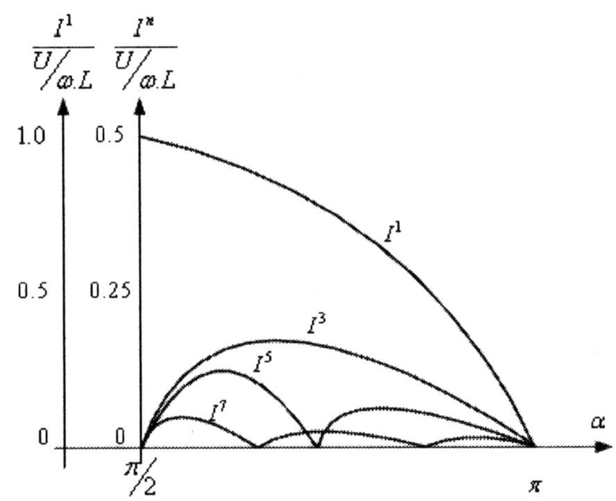

$$I_{Ln} = \frac{U}{\pi \cdot \omega \cdot L} \cdot \frac{1}{n} \cdot \left[\frac{\sin\left[\left(n-1\right)\cdot\left(\pi-\alpha\right)\right]}{n-1} - \frac{\sin\left[\left(n+1\right)\cdot\left(\pi-\alpha\right)\right]}{n+1} \right]. \qquad (5.26)$$

Figure 19 depicts the change of the fundamental and higher harmonics with low numbers when the firing angle changes (Mohan, 1995). The most important harmonics are those of No.3, 7, 9, 11 and 13 whose maximum values are reached in different firing angles. Three thyristor-controlled inductances are usually connected in delta topology to the three-phase system thus the harmonics multiple to 3 do not enter the system. The harmonics with numbers $n = 6 \cdot k \pm 1, \quad k = 1, 2, 3...$ remain. Therefore the static var compensators usually contain also filters for the lowest of those harmonics (5th, 7th, 11th and 13th). Besides, in the electrical systems assumed to be symmetrical source voltage system, in a steady mode others than the above-numbered harmonics can appear due to the different firing angles of the thyristors in the two half-periods or due to the different firing angles of the thyristors connected in the different phases.

Therefore, the maintaining of a high accuracy of the control system in dynamic and steady modes is of a great importance. As it has been shown, the thyristor-controlled inductance is a consumer of controlled reactive power because of that sometimes capacitor battery is also connected in parallel to the corresponding phase. It is possible the capacitor batteries to act as filters for higher harmonics and putting additional filter to be unnecessary.

As an additional reactive power source to secure the consumers connected to the system with quantities of reactive power bigger than those that can be and also are harmless to be supplied by the system, capacitor batteries are used. As it has been mentioned, it is possible thyristor-switched capacitor batteries to be used. They are usually connected in parallel to the load and they make so-called shunt compensation. The operational principle is explained with the schema and vector diagrams in Figure 20. A capacitor battery is connected in parallel to the load. Before the connection, the load current was I_{L1} and the displacement coefficient was determined by the angle ϕ_1. After the connection of the capacitor battery, the current decreases with $\Delta I = I_{L1} - I_{L2}$ and $\cos\phi_2 > \cos\phi_1$. Thus, the system has been unloaded

Figure 20. Electrical circuit and vector diagrams explaining the operation of a thyristor-switched capacitor

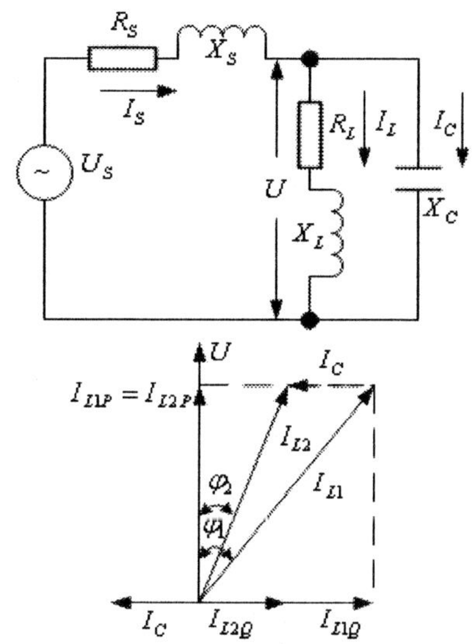

because of the generation of reactive power from the capacitor battery. The capacity of the battery C and its reactive power required to increase $\cos\phi_1$ to $\cos\phi_2$ can be determined using the vector diagram.

It appears from the vector diagrams that:

$$I_C = I_{L1Q} - I_{L2Q} = I_{LP} \cdot \left(tg\phi_1 - tg\phi_2\right) \tag{5.27}$$

Form (5.27) it is found:

$$U \cdot \omega \cdot C = \frac{P}{U} \cdot \left(tg\phi_1 - tg\phi_2\right) \tag{5.28}$$

or

$$C = \frac{P}{\omega \cdot U^2} \cdot \left(tg\phi_1 - tg\phi_2\right) \tag{5.29}$$

The battery power is determined as:

$$Q_C = U^2 \cdot \omega \cdot C = P \cdot \left(tg\phi_1 - tg\phi_2\right) \tag{5.30}$$

However, at a load change, pre-compensation is possible to obtain and the capacity of the battery may appear too big; the vector of the total current will lag to the vector of the voltage, also, the capacitive current in the system will increase as much as the capacity of the battery is bigger than the required capacity. Therefore, a possibility to assume that expandable reactive power has to exist and it can be secured through the already studied thyrsitor-controlled inductance. Besides, several capacitor batteries

Figure 21. Usual connection of several capacitor batteries

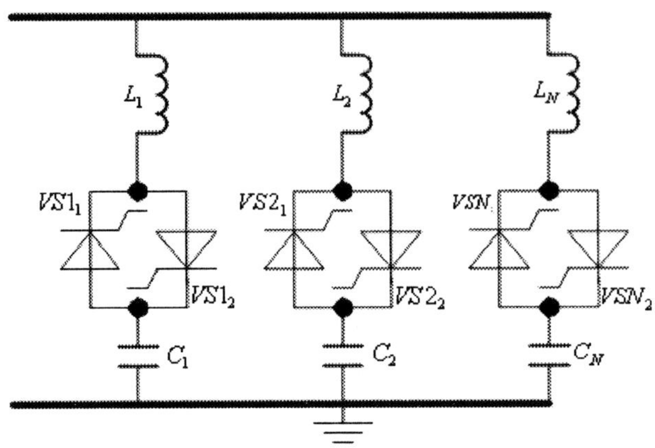

in accordance with the necessity of the system for bigger or lower value of the total capacitance C are usually incorporated - see Figure 21.

The thyrsitors of a particular capacitive battery are turned on for a definite period. The moment of turning on depends on the passing of the firing pulses. After the termination of the firing pulses to a thyristor, it turns off at the moment its current becomes lower than its holding current. As a result of the displacement between the current and voltage, the capacitor battery will remain charged after the turning off of the thyristors. The current through the thyristor which turns on at the next switching will be developed in different way dependent on the value of the remaining battery voltage. Two examples of current development are shown in Figure 22. To limit the current in the transient process, the inductances L are connected in series in Figure 21. Also, the value of the inductances has to be correctly chosen so the frequency of the series circuit containing one of them and the corresponding capacitor to be lower than the lowest possible harmonic frequency of the AC system. Otherwise, appearance of a resonance at the frequency of one of the higher harmonics in the system is risky.

The compensators of a combinational type usually contain two types of controlled reactive elements – thyristor-controlled inductances and thyristor-switched capacitors. To secure a constant control

Figure 22. Examples for the different current waverforms

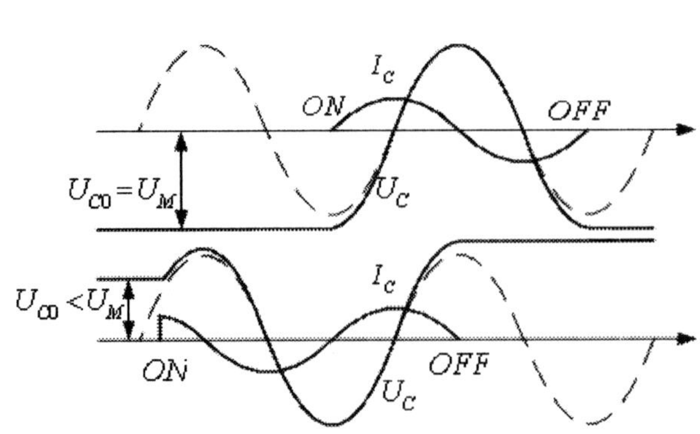

Figure 23. Compensator of a combinational type

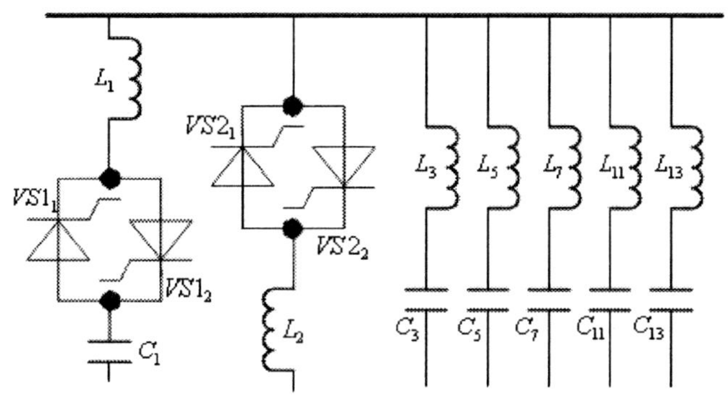

characteristic one inductance commensurable with the power of the capacitors is used for two to four capacitor batteries. Figure 23 displays a compensator of a combinational type for a phase of the three-phase system. Also, series resonant filters for 3th, 5th, 11th and 13th harmonics are put.

Special attention is paid to the system of monitoring and controlling the compensators. The contemporary control systems are of microprocessor type. The purpose is to maintain a firm operation not only in the steady mode but also in the dynamic modes. The embedded regulators in the automatic control system are obliged to secure a set quality of the dynamic processes without overloading of the elements to be possible. For this purpose different control strategies of different companies, as Siemens, Brown-Boveri, Wesringhouse, etc, are used.

The so-called method of phase-leveled regulation is a specific one to improve the indicators of the AC regulator in respect to the supply network. An electrical schematic of a single-phase AC regulator with phase-leveled regulation and its typical waveforms are shown in Figure 24.

The firing pulses is supplied to *VS*1 (at positive half-period) and to *VS*2 (at negative half-period) at the moments of zero crossing of the source voltage. A firing pulse is passed to *VS*3 at $\theta = \alpha$, it turns on while the thyristor *VS*1 turns off. Analogue to the positive half-period in negative half-period of the source voltage when the thyristor *VS*4 turns on, the thyristor *VS*2 that has been conducting till this moment turns off. It is clear, that the value of the output voltage will be minimum if the thyristors *VS*2 and *VS*4 do not conduct, and it will be with maximum value if the thyristors fire at $\alpha = 0$. The relationship between the effective value of the output voltage and the firing angle can be derived using the following equation, assuming that the turn ratio of the transformer is 1 ($n = 1$):

$$U_L = \sqrt{\frac{1}{\pi} \cdot \left[\left(s \cdot \sqrt{2} \cdot U\right)^2 \cdot \sin^2\theta \cdot d\theta + \int_{\alpha}^{\pi} \left(\sqrt{2} \cdot U\right)^2 \cdot \sin^2\theta \cdot d\theta \right]} \tag{5.31}$$

After several transformations the ratio of the effective values of the output and source voltages is gained as:

$$\frac{U_L}{U} = \sqrt{\frac{1}{\pi} \cdot \left[\left(\alpha - \frac{\sin 2\alpha}{2}\right) \cdot \left(s^2 - 1\right) + \pi \right]} \tag{5.32}$$

Figure 24. Electrical schematic of a single-phase AC regulator with phase-leveled regulation and its typical waveforms

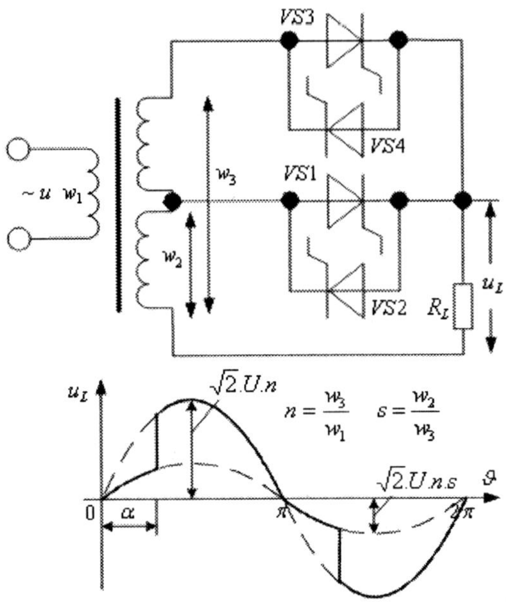

The equation (5.32) is the control characteristic of the single-phase AC regulator with control phase-leveled method.

To find the equation for the change of the power factor at regulation, total and active powers from the source have to be found:

$$S = U \cdot I = U \cdot \frac{U_L}{R_L} = \frac{U^2}{R_L}\sqrt{\frac{1}{\pi} \cdot \left[\left(\alpha - \frac{\sin 2\alpha}{2}\right) \cdot \left(s^2 - 1\right) + \pi\right]}$$

$$P = \frac{1}{\pi} \cdot \int_0^{\pi} u(\theta) \cdot i(\theta) \cdot d\theta = \frac{U^2}{R_L} \cdot \frac{1}{\pi} \cdot \left[\left(\alpha - \frac{\sin 2\alpha}{2}\right) \cdot \left(s^2 - 1\right) + \pi\right] \qquad (5.33)$$

Figure 25. Dependence of the relative effective value of the output voltage and the power factor on firing angle

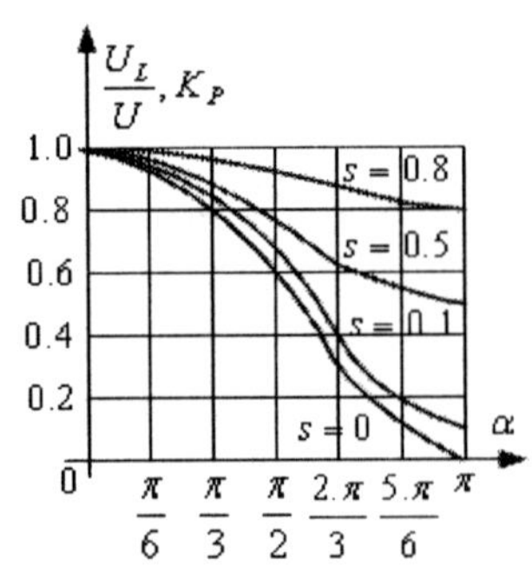

Then for the power factor it is valid:

$$K_P = \frac{P}{S} = \sqrt{\frac{1}{\pi} \cdot \left[\left(\alpha - \frac{\sin 2\alpha}{2} \right) \cdot \left(s^2 - 1 \right) + \pi \right]} \tag{5.34}$$

From the above stated it makes an impression that the equations (5.32) and (5.34) coincidence as the corresponding equations for the single-phase AC regulator with phase control and of the three-phase regulators. This also proves the correctness of (5.20). From (5.34) when $s = 0$ the equation (5.6) for the SPACR with phase control is obtained.

The change of the relative effective value of the output voltage and the power factor is depicted in Figure 25. The characteristic when $s = 0$ is the same as this of the TSPACR with phase control.

As it is clarified by the description of the operation and by the control characteristic, using this method the change of the load voltage can be between two values – the minimum value is when $\alpha = \pi$ and the maximum valuc - $\alpha = 0$.

There is a possibility to improve the power factor of a system with two AC regulators operating to a common network but with different loads using integrator cycling switching (Shepherd, 1975). The currents consumed by each regulator, as well as the total current are shown in Figure 26.

As it is appears, the common control system should secure the operation of the one of the regulators during the pause of the other one, furthermore, a change of N_1 and N_2 is possible, as well as it is possible the load currents not to be equal. If the equations for the current is written down, the total power factor might be derived:

$$i = \sqrt{2} \cdot I_1 \cdot \sin\theta \,\Big|_0^{2\cdot\pi\cdot N_1} + \sqrt{2} \cdot I_2 \cdot \sin\theta \,\Big|_{2\cdot\pi\cdot(N_1+N_{01})}^{2\cdot\pi\cdot(N_1+N_{01}+N_2)} \tag{5.35}$$

The equation for the effective value of the source current is derived on the basis of (5.35) and taking also the following:

Figure 26. Consumed currents from the system of two regulators using integral cycling method control

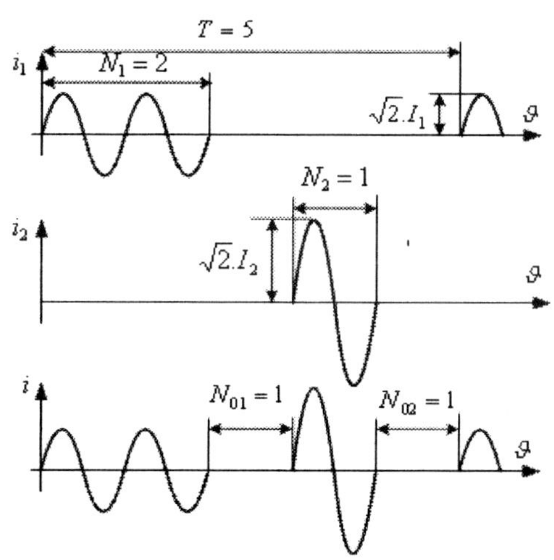

$$I^2 = \frac{I_1^2 \cdot N_1 + I_2^2 \cdot N_2}{T} \tag{5.36}$$

Total power is:

$$S = U \cdot \sqrt{\frac{I_1^2 \cdot N_1 + I_2^2 \cdot N_2}{T}} \tag{5.37}$$

The active power is the sum of the active power of the two regulators:

$$P = P_1 + P_2 = U \cdot \frac{I_1 \cdot N_1 + I_2 \cdot N_2}{T} \tag{5.38}$$

Dividing (5.38) to (5.37) leads to:

$$K_P = \frac{I_1 \cdot N_1 + I_2 \cdot N_2}{\sqrt{T(I_1^2 \cdot N_1 + I_2^2 \cdot N_2)} \cdot} \tag{5.39}$$

The total power factor calculated using (5.39) is bigger than the power factors of each separate regulator calculated using (5.6).

In the case of equality of the loads of the regulators then the currents are also equal $I_1 = I_2$ and from (5.39) it is found:

Figure 27. Electrical circuits of AC regulators using fully-controlled switches

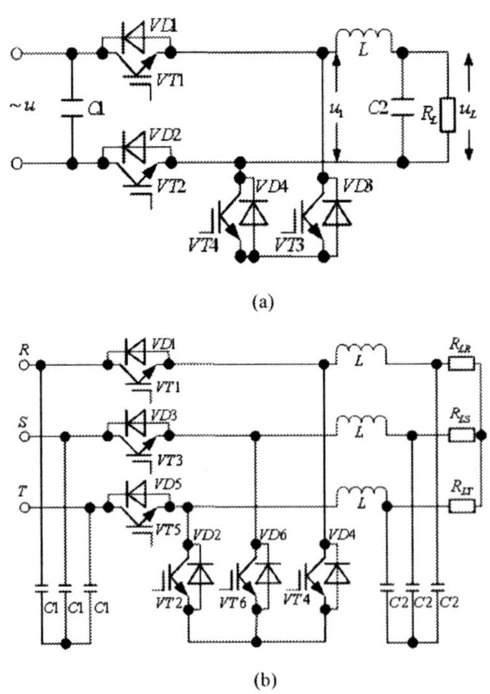

$$K_P = \sqrt{\frac{N_1 + N_2}{T}} \tag{5.40}$$

The equation (5.6) is derived from (5.40) for a regulator if the number N for the other regulator is 0. If the sum of N_1 and N_2 is T, then the power factor is 1.

If (5.39) is analyzed, it is worthy to mention that nevertheless the sum of N_1 and N_2 is T, the power factor will be less than 1 dependent on the values of I_1 and I_2. The reason is that the consumed current will be of sinusoidal waveform but it will contain higher harmonics due to periodical differences in its values.

A possibility to use fully-controlled power devices combined with the control method of pulse width modulation exists also in the AC regulators (Kazerani, 2003; Glinka, 2005; Chen, 2008). Electrical schematics of such single-phase and three-phase regulators are shown in Figure 27.

Power modules can be used in the schemas, for example pairs $VT1, VT3$ and $VT2, VT4$ from Figure 27.a. More often, power isolated gate bipolar transistors (IGBT) are used. The operation of the circuit shown in Figure 27.a is as follows: the transistors $VT1, VT2$ turn on simultaneously, and after them, transistors $VT3, VT4$ conduct simultaneously. Thus the voltage of the input filter $L - C_2$ will be equal either to the input voltage or to zero, and the current direction will be determined by the inductance current direction. For example, if the input voltage is positive and transistors $VT1$ and $VT2$ are turned on, the current flows left to right trough the inductance, then $VT1$ and $VD1$ will conduct. If the current flows right to left through the inductance then $VT2$ and $VD1$ will conduct. During the pauses the input filter voltage is zero, and $VT4, VD3$ or $VT3, VD4$ conduct. This is a constant frequency control. The duty cycle of the firing pulses for the transistors $VT1, VT2$ is d and with its change the effective value of the first harmonic of the output voltage across R_L changes. The explaining diagrams of the operation are depicted in Figure 28.

Let the input voltage of the circuit in Figure 27.a be:

$$u = U_M \cdot \cos\left(\omega t\right) \tag{5.41}$$

The progress of the control signal, which is a series of square pulses with a constant frequency ω_S and duty cycle d, in Fourier series is:

Figure 28. Explanation of the constant frequency control used in schema shown in Figure 27

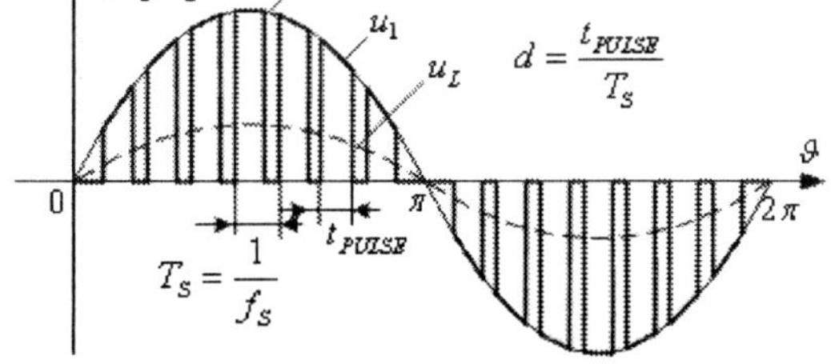

$$S = d + \sum_{n=1}^{\infty}\left(\frac{2 \cdot \sin\left(n \cdot d \cdot \pi\right)}{n \cdot \pi} \cdot \cos\left(n \cdot \omega_s \cdot t\right)\right) \tag{5.42}$$

The analytical equation for the input voltage of the filter $L - C_2$ is found by multiplication of (5.41) and (5.42):

$$u_1 = u \cdot S = d \cdot U_M \cdot \cos\left(\omega \cdot t\right) + \sum_{n=1}^{\infty}\left(\frac{2 \cdot \sin\left(n \cdot d \cdot \pi\right)}{n \cdot \pi} \cdot \cos\left(\left(n \cdot \omega_s \pm \omega\right) \cdot t\right)\right) \tag{5.43}$$

The purpose of the filter placed is to separate only the first component of the right side of (5.43). From the addend left, the change of the output voltage at the change of the duty cycle d is seen.

If the effective value of the first harmonic of the input filter voltage is marked as U_1, then the first harmonic of the input current which is also the current through the inductance L in a complex form will be:

$$I_{L1} = \frac{\left(R_L - j \cdot X_{C_2}\right) \cdot U_1}{X_{C_2} \cdot X_L + j \cdot R_L \cdot \left(X_L - X_{C_2}\right)} \tag{5.44}$$

where in:

$$U_1 = d \cdot U \tag{5.45}$$

The equivalent circuit for the first harmonics in respect to the supply network is shown in Figure 29. The first harmonic of the source current in a complex form will be:

$$I_1 = d \cdot I_{L1} + I_{C1} \tag{5.46}$$

After substituting from (5.44) it is derived:

$$I_1 = \frac{R_L \cdot \left(d^2 \cdot X_{C_1} - X_L + X_{C_2}\right) + j \cdot X_{C_2} \cdot \left(X_L - d^2 \cdot X_{C_1}\right)}{X_{C_1}\left[X_L \cdot X_{C_2} + j \cdot R_L \cdot \left(X_L - X_{C_2}\right)\right]} \cdot U \tag{5.47}$$

The equation for the phase angle between the source voltage and the first harmonic of the current is found from (5.47):

$$\phi = arctg\frac{X_{C_2} \cdot \left(X_L - d^2 \cdot X_{C_1}\right)}{R_L \cdot \left(d^2 \cdot X_{C_1} - X_L + X_{C_2}\right)} - arcg\frac{R_L \cdot \left(X_L - X_{C_2}\right)}{X_L \cdot X_{C_2}} \tag{5.48}$$

If the condition $d^2 \cdot X_{C_1} > X_{C_2} > X_L$ is true, then using (5.48) leads:

$$\phi = -arctg\frac{X_{C_2}}{R_L} + arcg\frac{R_L}{X_L} \tag{5.49}$$

Figure 29. The equivalent circuit for the first harmonics in respect to the supply network

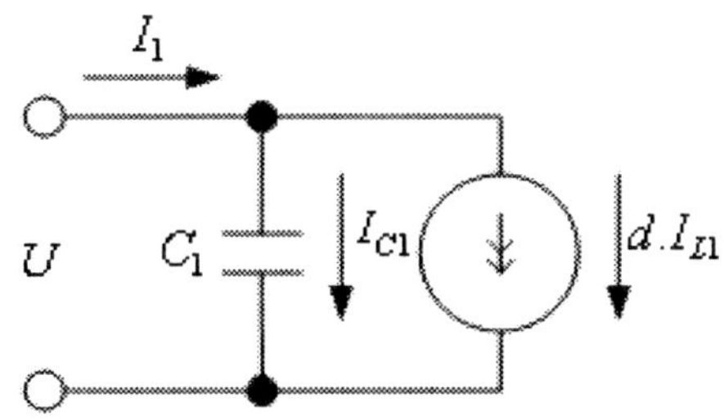

Because the input filter capacitor eliminates the higher harmonics of the inductance current, the power factor is assumed to be determined mainly from the displacement coefficient:

$$K_P \approx \cos\phi, \tag{5.50}$$

where in ϕ is determined using (5.49) and the power factor will depend very lightly on *d*.

There is a category of converters from AC into AC energy which are with a possibility to change the frequency or the number of phases of the output voltage. Also, the conversion is made with improve indicators in respect to the supply network.

Matrix converters are included in this category and they are direct converters from AC into AC energies with a possibility to change the frequency or phases without an intermediate conversion. In general, the conversion is from m-input to n-output phases (Clare, 2005). The converters name comes from the fact that there are bidirectional fully-controlled switches, which are usually implemented in the way shown in Figure 30 at each crossing point of the input and output phase wires.

Conductivity in the one direction is secured by VT_1 and VD_2, and in the other – by VT_2 and VD_1. The shown combination is frequently symbolized in the power schematics as switch *SW*. The advantages of the matrix converters are:

- Increased efficiency coefficient because the AC power of the supply network is not necessary to be converted into DC power and then again into AC one with the possibility to regulate the frequency and the output value of the voltage.
- Improved power factor in respect to the supply network and improved harmonic spectrum of the source current also using an appropriate control the consumed current will be of sinusoidal

Figure 30. Implementation of a bidirectional fully-controlled switch and its schematic symbol

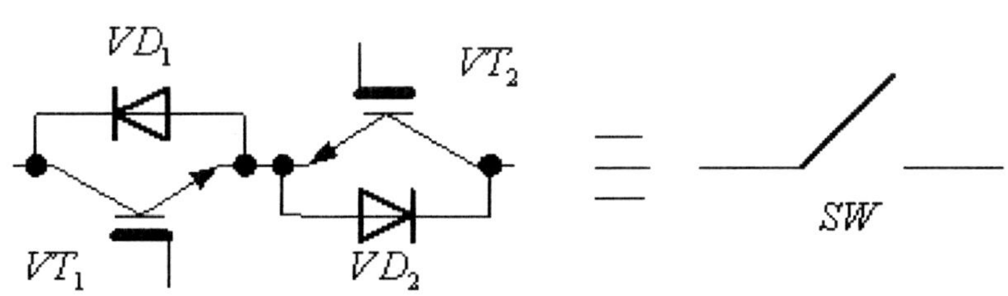

waveform and in phase to the corresponding voltage of the supply network, nevertheless the regulation of the converter.

The matrix converters require complex control and regulator methods realized with control systems based on the programming logic or digital signal processors.

Figure 31 displays a single-phase matrix converter circuit, i.e. m=n=1. The possible combinations to control power devices besides the half-period of the input voltage u_i are as follows:

- **Simultaneously turning on of** SW_1, SW_3 **:** The polarity and the transition value of the voltage between the points P and N are determined from the polarity and the transition value of the input voltage.
- **Simultaneously turning on of** SW_2, SW_4 **:** The polarity and the transitory value of the voltage between the points P and N are determined from the polarity and the transitory value of the input voltage.
- **Simultaneously turning on of** SW_1, SW_4 **or/and** SW_2, SW_3 **:** Nevertheless the polarity and the value of the input voltage, the value of the output voltage is 0.

The possibility of simultaneously turning on of SW_1, SW_2 or SW_3, SW_4 is eliminated in the matrix converters. Actually it is possible in other types of converters, but one should have in mind that in this case the input circuit of the bridge is short-circuited and the input current increases. Dependent on the load type, it is possible after that switching to turn on a pair of diagonally connected devices and the stored energy in the inductance to be assumed by the load. The filter L_F, C_F eliminates the higher disturbances produced from the switching of the power devices regarding the supply network. The state of the power device – turned on or off, is depicted in matrix:

$$[SW] = \begin{bmatrix} SW_1 & SW_2 \\ SW_4 & SW_3 \end{bmatrix} \tag{5.51}$$

Figure 31. An electrical schematic of a single-phase matrix converter

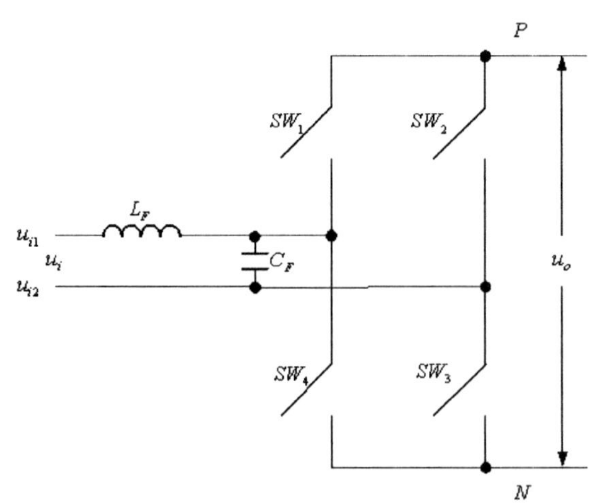

The elements of the matrix can have values 1 or 0. The input voltage u_i is presented as:

$$u_i = u_{i1} - u_{i2} \tag{5.52}$$

where in u_{i1} and u_{i2} are voltages measured in respect to a conventional point with zero potential and they are described as:

$$u_{i1} = \frac{U_M}{2}.\sin\vartheta; \quad u_{i2} = -\frac{U_M}{2}.\sin\vartheta \tag{5.53}$$

Two additional voltages u_P and u_N are introduced in respect to the output and also to the same conventional point, therefore, the output voltage is determined as:

$$u_O = u_P - u_N \tag{5.54}$$

The circuit shown in Figure 31 can be presented as shown in Figure 32.
Vectors of the input and the output voltages can be used:

$$\left[u_i\right] = \begin{bmatrix} u_{i1} \\ u_{i2} \end{bmatrix}; \quad \left[u_O\right] = \begin{bmatrix} u_P \\ u_N \end{bmatrix} \tag{5.55}$$

The connection between them with the control matrix (5.51) is:

$$\left[u_O\right] = \left[SW\right].\left[u_i\right] \tag{5.56}$$

Or after substitution of (5.51.) and (5.55.):

$$\begin{bmatrix} u_P \\ u_N \end{bmatrix} = \begin{bmatrix} SW_1 & SW_2 \\ SW_4 & SW_3 \end{bmatrix} . \begin{bmatrix} u_{i1} \\ u_{i2} \end{bmatrix} \tag{5.57}$$

Figure 32. Presentation used to analyze the circuit in Figure 31

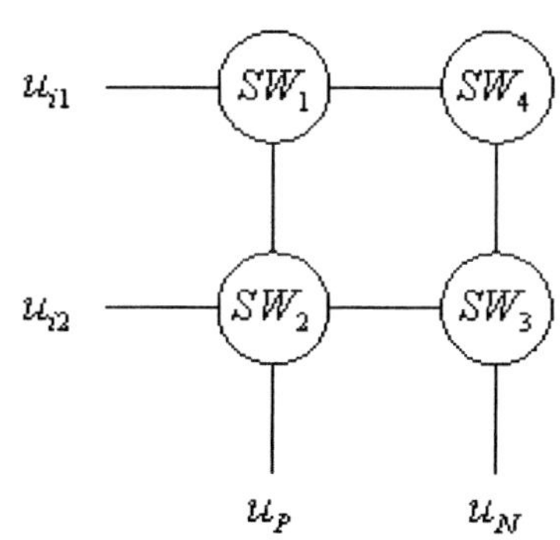

Based on (5.54) and (5.57), it is found:

$$u_O = (SW_1.u_{i1} + SW_2.u_{i2}) - \left(SW_4.u_{i1} + SW_3.u_{i2}\right) \tag{5.58}$$

$$u_O = \left(SW_1 - SW_4\right).u_{i1} + \left(SW_2 - SW_3\right).u_{i2} \tag{5.59}$$

It is assumed a simultaneous turning on only of diagonal connected devices. The control signals are a series of square pulses with amplitude of 1 and duty cycle of 0.5 with frequency ω_S. Thus, the control signals in Fourier series are:

$$\begin{aligned} \left(SW_1 - SW_4\right) &= \frac{4}{\pi}\sin\omega_S.t + \frac{4}{\pi}\sum_{n=3,5,7...}^{\infty}\frac{1}{n}.\sin\left(n\omega_S t\right) \\ \left(SW_2 - SW_3\right) &= -\frac{4}{\pi}\sin\omega_S.t - \frac{4}{\pi}\sum_{n=3,5,7...}^{\infty}\frac{1}{n}.\sin\left(n\omega_S t\right) \end{aligned} \tag{5.60}$$

Substituting of (5.53) and (5.60) in (5.59), and after transformations, it is derived:

$$u_O = \frac{4}{\pi}.U_M.\sin\omega.t.\sin\omega_S.t + \frac{4}{\pi}U_M\sin\omega.t.\sum_{n=3,5,7...}^{\infty}\frac{1}{n}\sin\left(n\omega_S.t\right) \tag{5.61}$$

Figure 33 displays the waveforms of the input u_i and output u_O voltages of a matrix converter. The output voltage waveform is determined by (5.61).

Output voltage with the shown shape can be used in electrical engineering technological uses (Antchev, 2007); it can be transformed through a high frequency transformer and converted into a DC one with a desired value, etc .The frequency of the fundamental voltage harmonic alters using ω_S.

Multiplying both sides of (5.56) with the reverse matrix of the control matrix - $\left[SW\right]^{-1}$ leads to the relationship between the input and output currents:

Figure 33. Waveforms of the voltages of the single-phase matrix converter

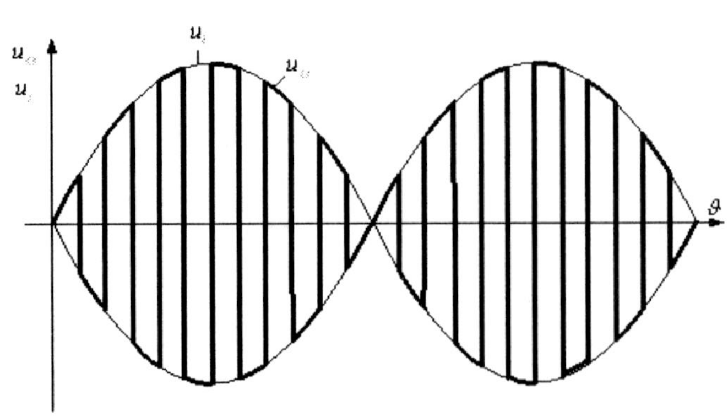

$$[I_i] = [SW]^{-1} \cdot [I_O] \tag{5.62}$$

Therefore, while the shape of the output voltage is determined in accordance with (5.56), the shape of the input current is determined by the output one in accordance with (5.62).

It is assumed that a resonant mode between the output voltage and output current is maintained, for example, in induction heating. Then the input current of the bridge circuit will contain uni-polar pulses with the polarity of the half-period and with increasing amplitude towards the maximum of the sinusoidal voltage waveform of the half-period. Using the input filter L_F, C_F, only the current first harmonic with a shape close to the sinusoidal one and in phase to the input voltage can be separated. Thus the power factor in respect to the supply network can be maintained close to 1 without additional circuit stage for power factor correction.

The operation of a three-phase matrix converter when m=n=3 is explained using Figure 34. The input filters are shown and also all voltages and currents are marked. The converter consists of nine bidirectional fully-controlled power devices, marked as switches with subscripts of the input and output three-phase systems. This version is most frequently used in control and regulation of three-phase AC motors. A simplified situation of the devices in respect to the input and output lines is illustrated in Figure 35.

The possible combinations among the switches states are 27 because each of the output lines can be plugged in only one of the input lines. Using the 27 combinations the output voltages can be synthesized in shape, changed their values and frequencies.

The following vectors for the input and output voltages and currents in the circuit are presented:

$$[U_i] = \begin{bmatrix} U_R \\ U_S \\ U_T \end{bmatrix}; \quad [I_i] = \begin{bmatrix} I_R \\ I_S \\ I_T \end{bmatrix}; \quad [U_O] = \begin{bmatrix} U_A \\ U_B \\ U_C \end{bmatrix}; \quad [I_O] = \begin{bmatrix} I_A \\ I_B \\ I_C \end{bmatrix} \tag{5.63}$$

A control matrix reflecting the states of the bidirectional devices is:

Figure 34. Schematic used to explain the operation of a three-phase matrix converter

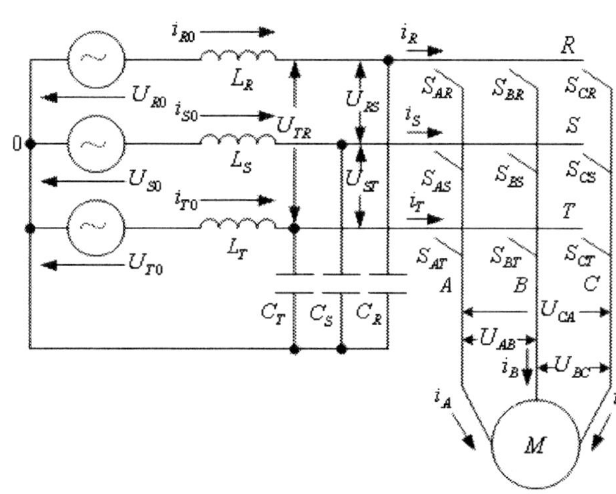

Figure 35. Presentation used to analyze the circuit in Figure 34 .

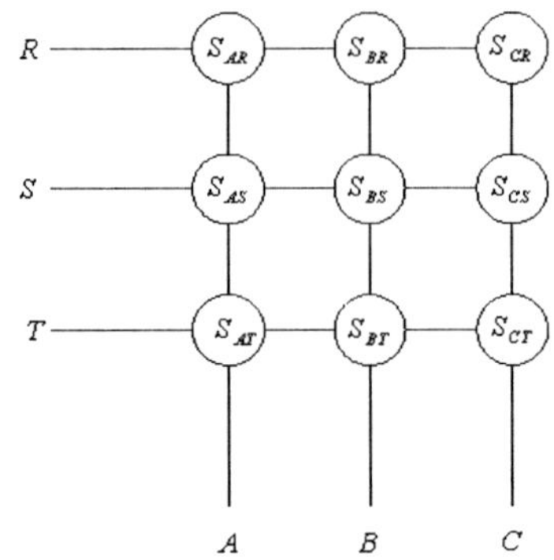

$$[S] = \begin{bmatrix} S_{AR} & S_{AS} & S_{AT} \\ S_{BR} & S_{BS} & S_{BT} \\ S_{CR} & S_{CS} & S_{CT} \end{bmatrix} \tag{5.64}$$

Then the relationship of the output voltages to the input ones is:

$$[U_O] = [S].[U_I] \text{ or } \begin{bmatrix} U_A \\ U_B \\ U_C \end{bmatrix} = \begin{bmatrix} S_{AR} & S_{AS} & S_{AT} \\ S_{BR} & S_{BS} & S_{BT} \\ S_{CS} & S_{CR} & S_{CT} \end{bmatrix} . \begin{bmatrix} U_R \\ U_S \\ U_T \end{bmatrix} \tag{5.65}$$

Thus, the relationship of the output currents to the input ones is:

$$[I_i] = [S]^{-1} \cdot [I_O] \tag{5.66}$$

Figure 36 displays the shape of the three output phase currents (the upper waveforms) and the system of the output phase voltages (the lower waveforms) of the three-phase matrix converter supplying asynchronous motor (Clare, 2005). The control systems of those converters are implemented based on a three-dimensional-vector width-pulse modulation. Other implementation of the three-phase matrix converters is in the systems of electrical source in use of wind generators.

At the end of this chapter I am going to discuss several special features of realization of the control systems for AC/AC converters. Similarly to the controlled thyristor rectifiers, the AC/AC converters are supplied from the AC network and they demand the change of the firing angle α in respect to the zero crossing of the source voltage. This change of the phase of the firing angles leads to the change of the regulator output voltage. Namely, the requirements for this are analogues to those of the controlled thyristor rectifiers. The control system for the thyristor AC regulators can be implemented on the basis of the already studied principles in Chapter 4 to implement the control systems of the controlled thyristor rectifiers. The possibilities to realize the phase-controlled circuits – horizontal and vertical methods, are also analogical.

Figure 36. Typical waveforms of the fundamental variables for the three-phase matrix converter supplying an asynchronous motor . Source:(Clare, 2005)

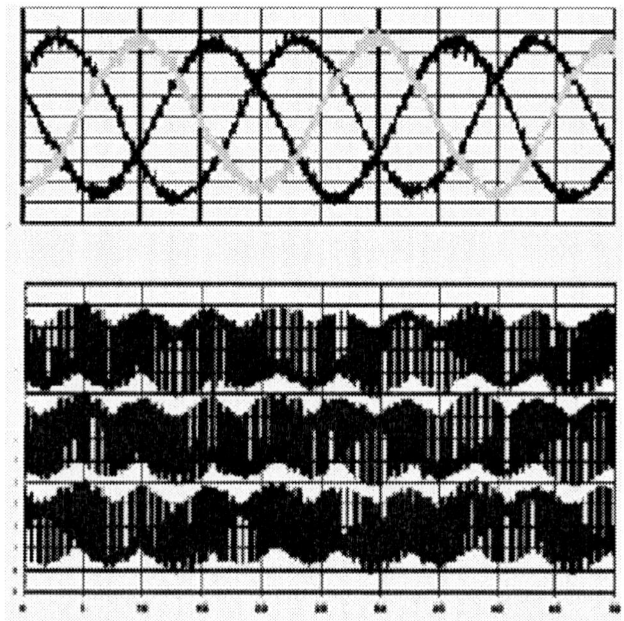

There are several special features in implementation the control system for the thyristor AC regulators operating in accordance with the integral cycling control method – Figure 6 and Figure 26. Since the change of the firing pulse phase is not required, the phase-regulation circuit from the control system block schema is missing, as it is shown in Figure 37.

The schema is for a three-phase AC regulator and it consists of three identical channels, on which inputs a three-phase system of a synchronizing voltages u_R, u_S, u_T is applied. Each of the voltages is in phase with the corresponding source voltage of the regulator circuit. The synchronizing voltages are galvanic separated from the power schema. By the use of the input synchronizing circuit (ISC), the moments of zero crossing of each of the phase voltages are determined. At these moments, the firing pulses with a desired duration are formed by the pulse former (PF). Whether these pulses will be delivered to the drivers or not is decided in the control logic by the signal for turning on or off the regulator - On/Off.

Figure 37. Block schema of a control system of the AC regulators operating in accordance with the integral cycling control method

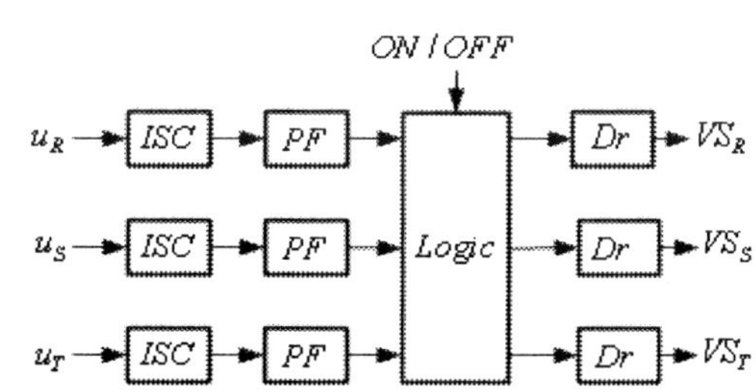

Figure 38. Control of a triac using specialized optoelectronic integral circuit

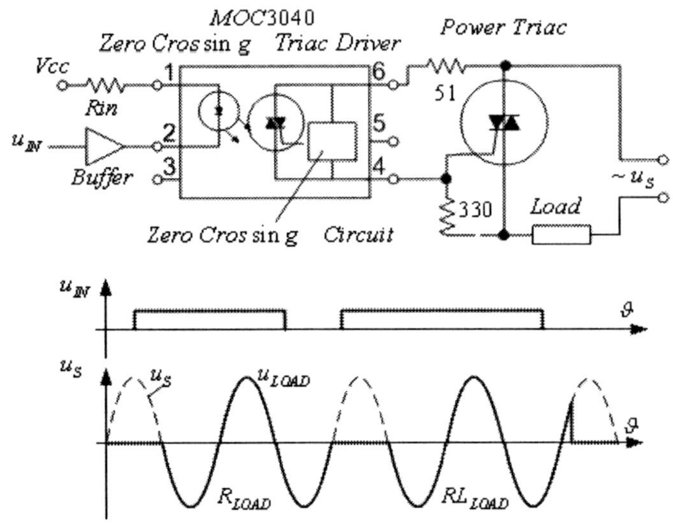

Optoelectronic integral circuits designed especially for the triac control and voltage zero crossing are very convenient to be used in implementation of the control in the integral cycling control methods. Figure 38 displays an example. The integral circuit secures also a galvanic isolation from the control signal and monitoring of the zero by the zero crossing circuit. Thus it combines several of the required blocks. It also contains and inner triac, but in the figure a connection of the outer triac for a higher power loads is shown. Waveforms of the control signals and the load voltage in the case of active and active-inductive loads are shown. As it appears, in both cases the turning on happens at a high level of the control signal but only just the source voltage zero crossing reaches. At a low level of the control signal, the turning off happens when the current through the power device decreases under the holding current. In the case of active load, it happens approximately at the moment of the source voltage zero crossing, and in case of active-inductive load – after this moment dependent on the stored energy in the inductance.

Figure 39 shows a block schema to implement the control of an AC/AC converter operating in accordance with the pulse -width modulation method (see Figure 27).

Pulses are generated at the output of the input synchronizing circuits with a frequency equal to this of the source AC voltage - f_S. Using the phase locked loop (PLL) and the Counter as a devisor of N, pulses with Nth times higher frequency than the source one are generated. Each of these pulses is a start pulse

Figure 39. A block schema to implement the control of an AC/AC converter operating in accordance with the width-pulse modulation method

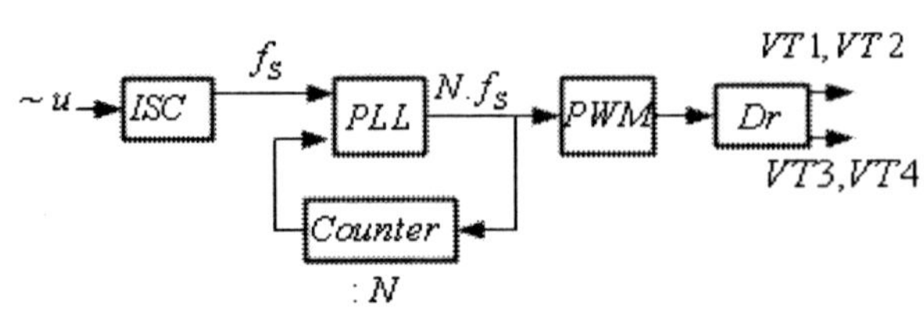

for the pulse-width modulator (PWM). Subsequently, using drivers (Dr), the firing pulses are passed to the gates of the power transistors.

The described block schema is also implemented as a basic one in the control of the matrix converters but there are some changes connected with their operation characteristics.

REFERENCES

Antchev, M. H., & Kanov, G. Tz. (2007). *Study of single-phase high frequency matrix converter for induction heating applications.* Paper presented at the 14-th Symposium Power Electronics Ee2007, Novi Sad, Serbia.

Chen, D. (2008). Novel current-mode AC/AC converters with high-frequency AC link. *IEEE Transactions on Industrial Electronics*, *55*(1), 30–37. doi:10.1109/TIE.2007.896135

Clare, J., & Wheeler, P. (2005). New technology: Matrix Converters. *Industrial Electronics Newsletter*, *52*(1), 10–12.

Glinka, M., & Marquardt, R. (2005). A new AC/AC multilevel converter family. *IEEE Transactions on Industrial Electronics*, *52*(3), 662–669. doi:10.1109/TIE.2005.843973

Kazerani, M. (2003). A direct AC/AC converter based on current – source converter nodules. *IEEE Transactions on Power Electronics*, *18*(5), 1168–1175. doi:10.1109/TPEL.2003.816184

Mohan, N., Undeland, T. M., & Robbins, W. P. (1995). *Power electronics converters, applications and design.* New York: John Wiley & Sons.

Shepherd, W. (1975). *Thyristor control of AC circuits.* UK: Bradford University Press.

Chapter 6
DC/DC Conversion

BASIC INDICATORS

The basic idea of converting DC into DC voltages has been studied in Section 1 Chapter 3. There also has been mentioned that to separate DC component a smoothing filter is connected between the output of the converter and the load. The filter consists of passive elements – inductances and capacitors, and in several cases it is included in the converters circuit (Williams, 2008). Figure 1 displays the waveform of the output voltage as the block of the converter contains and the smoothing filter. Using the figure the basic indicators of the converter will be defined.

Efficiency coefficient of the converter is the ratio of the output to the input powers:

$$\eta = \frac{P_{OUT}}{P_{IN}} = \frac{U_{OUT}.I_{OUT}}{U_{IN}.I_{IN}} \tag{6.1}$$

This coefficient should be as high as possible.

Coefficient of the ripples of the output voltage is defined analogue to those in the rectifiers (see Chapter 4):

DOI: 10.4018/978-1-61520-647-6.ch006

Figure 1. Schema used to define the indicators of DC/DC converters

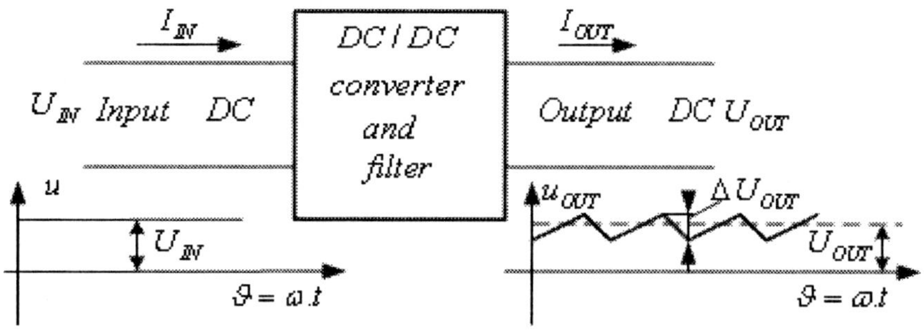

$$S_1 = \frac{U^1_{OUTM}}{U_{OUT}} \approx \frac{\Delta U_{OUT}/2}{U_{OUT}}, \tag{6.2}$$

where in U^1_{OUTM} is the maximum value of the first harmonic of the output voltage.

The right part of the equation is obtained on the assumption that the first harmonic is a sinusoidal function which positive amplitude coincides with the maximum value of the ripples shown in Figure 1, and which negative amplitude coincides with the minimum value of the ripples also shown in Figure 1. This coefficient should be as low as possible.

Output characteristics also exist in DC/DC converters like those in the rectifiers (see Chapter 4). These characteristics give the change of the output voltage U_{OUT} at the change of the output current I_{OUT}.

CONVERSION WITHOUT GALVANIC ISOLATION

The following converters without galvanic isolation are known – buck, boost, buck/boost, Cuk converter, Single-Ended Primary Inductance Converter (SEPIC) (Billings, 1999), (Kazimierczuk, 2008).

Figure 2 shows the electrical schematic of and the waveforms explaining the operation of a buck converter. It may operate in two modes – discontinuous-conduction and continuous-conduction current modes through the inductance. A fully-controlled power device able to turn on and off by a control electrode is shown using the circuit symbol for a switch in the figure. The switch can be a power BJT, power MOSFET or IGBT. Ratio of the pulse time (time during which the switch conducts) to the operation period defines duty ratio. It is important for the converter operation and it is always less than 1. When the transistor is on, the current through the inductance increases and recharges the capacitor. This is so-called forward motion; since during it energy is passed to the output capacitor, the converter is frequently called forward converter.

When the transistor is off, the current through the inductance decreases flowing through the diode VD. This is so-called reverse motion, the diode is called a commutation diode. In the case of continuous-conduction mode, the relationship between the output and input voltages is:

$$U_{OUT} = \delta.U_{IN} \text{ and } 0 \leq \delta \leq 1 \tag{6.3}$$

Figure 2. Electrical schematic and waveforms of a buck converter

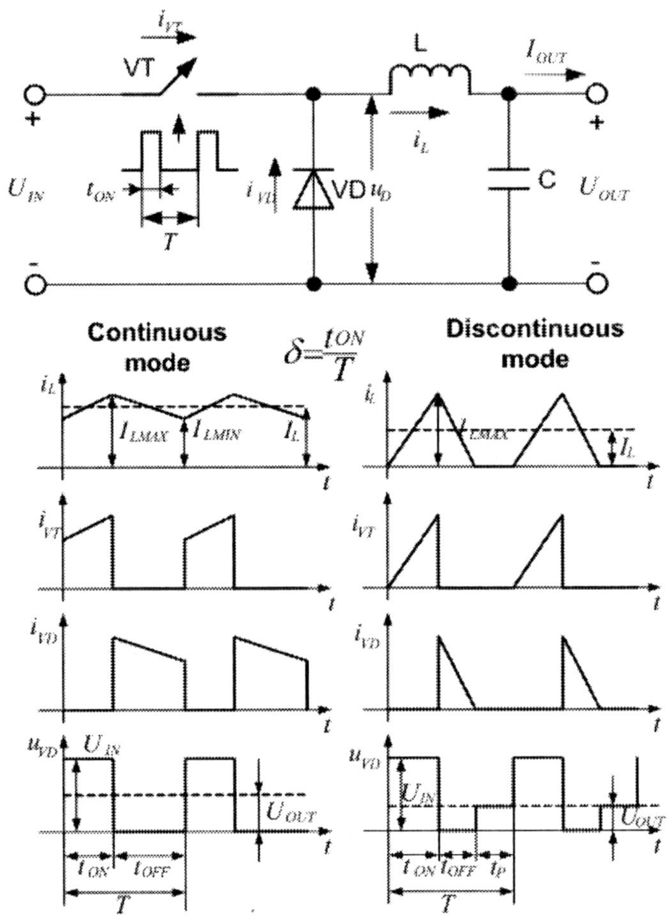

In the case of discontinuous-conduction current mode through the inductance, the relationship is:

$$U_{OUT} = \frac{U_{IN}}{1 - \frac{t_P}{T}} \tag{6.4}$$

As the time of the pause is not preliminarily known, through several transformations, a more convenient equation can be found:

$$U_{OUT} = \frac{U_{IN}}{1 + \frac{2.L.I_{OUT}}{\delta^2.T.U_{IN}}} \tag{6.5}$$

The maximum value of the transistor and diode voltages is equal to the value of the input voltage, the maximum value of the current through them is determined by this of the inductance current:

$$I_{VTMAX} = I_{VDMAX} = I_{LMAX} = I_L + \frac{U_{OUT}}{2.L}.T\left(1-\delta\right) = I_{OUT} + \frac{U_{OUT}}{2.L}.T\left(1-\delta\right) \quad (6.6)$$

As it can be seen from Figure 2, the ripples Δi_L (from I_L upto I_{LMAX}) of the current through the inductance L are calculated by the second member of the right side of equation (6.6). This is so because when the transistor is off, the output voltage is applied across the inductance for time $T\left(1-\delta\right)$. The frequency of the ripples is equal to the operating frequency of the converter.

The ripples of the output voltage ΔU_{OUT} (from the average value U_{OUT} up to the maximum value) may be determined taking in consideration the following: In steady operational mode the average values of the currents through the inductance and the load are equal. Therefore, the average value flowing through the capacitor is 0. Thus, the ripples of the voltage across the capacitor are determined only by the ripples of the inductor current Δi_L. If the balance of the electrical charges in the circuit of the inductance and capacitor for a half-period $T / 2$ is studied, the following is clear: the charge into the inductance is determined by $\frac{\Delta i_L}{2}.\frac{T}{2}$, while in the capacitor it is $2.\Delta U_{OUT}.C$.

After the charges are made equal, an equation for the ripples of the output voltage is found:

$$\Delta U_{OUT} = \frac{\Delta i_L.T}{8.C} \quad (6.7)$$

Due to the non-ideality of the capacitor – presence of series parasitic resistance and inductance, in practice the ripples are higher than the above determined ones.

The electrical schematic and waveforms for both operational modes of a boost converter are shown in Figure 3. When the transistor is on, the current through the inductance increases flowing through the transistor - a forward motion. When the transistor is off, the current through the inductance decreases flowing through the diode and energy is passed to the output capacitor – a reverse motion. This converter is frequently called flyback converter.

In the case of continuous-conduction current mode through the inductance, the relationship between the output and input voltages is:

$$U_{OUT} = \frac{1}{1-\delta}.U_{IN} \quad (6.8)$$

In the case of discontinuous-conduction current mode through the inductance, the relationship is:

$$U_{OUT} = \frac{1-\frac{t_P}{T}}{1-\frac{t_P}{T}-\delta}.U_{IN} \quad (6.9)$$

As the time of the pause is not preliminarily known, the following relationship for the discontinuous-conduction mode can be derived:

Figure 3. Electrical schematic and waveforms of a boost converter

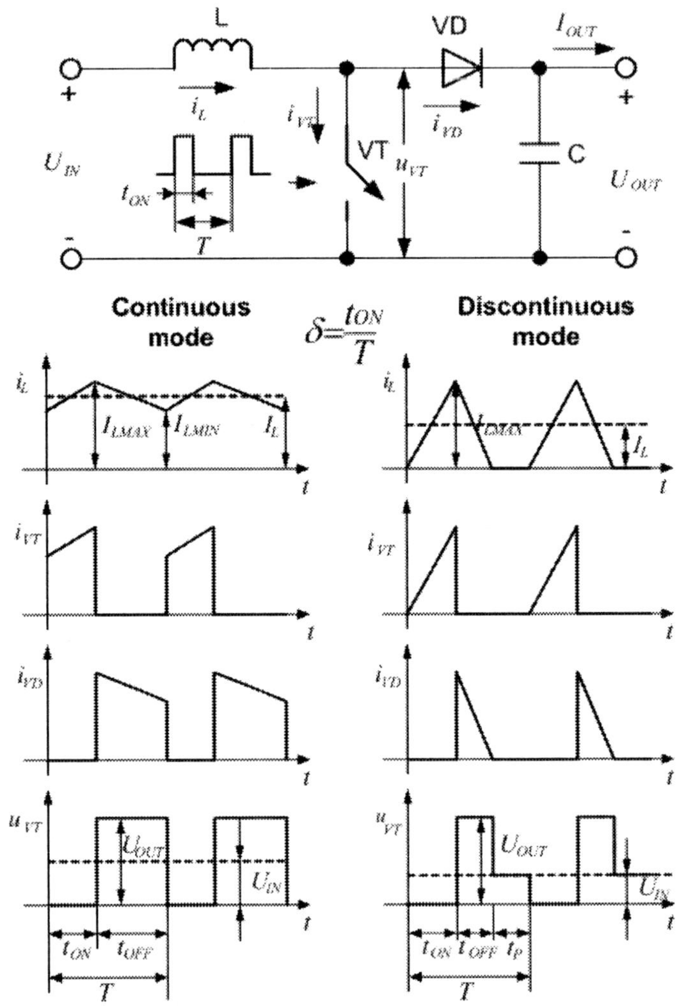

$$U_{OUT} = \left(1 + \frac{U_{IN}.T.\delta^2}{2.L.I_{OUT}}\right).U_{IN} \tag{6.10}$$

The maximum value of the transistor and diode voltages is equal to the value of the output voltage; the maximum value of the current through them is determined by this of the inductance current:

$$I_{VTMAX} = I_{VDMAX} = I_{LMAX} = I_L + \frac{U_{IN}.t_{ON}}{2L} = I_L + \frac{U_{IN}.\delta.T}{2L} \tag{6.11}$$

Here, it is also seen from Figure 3, that the ripples Δi_L (from I_L upto I_{LMAX}) of the current through the inductance L are calculated by the second member of the right side of equation (6.10). This is so because when the transistor is on, the input voltage is applied across the inductance for time $\delta.T$. The frequency of the ripples is equal to the operating frequency of the converter.

Figure 4. Electrical schematic and waveforms of a buck-boost converter

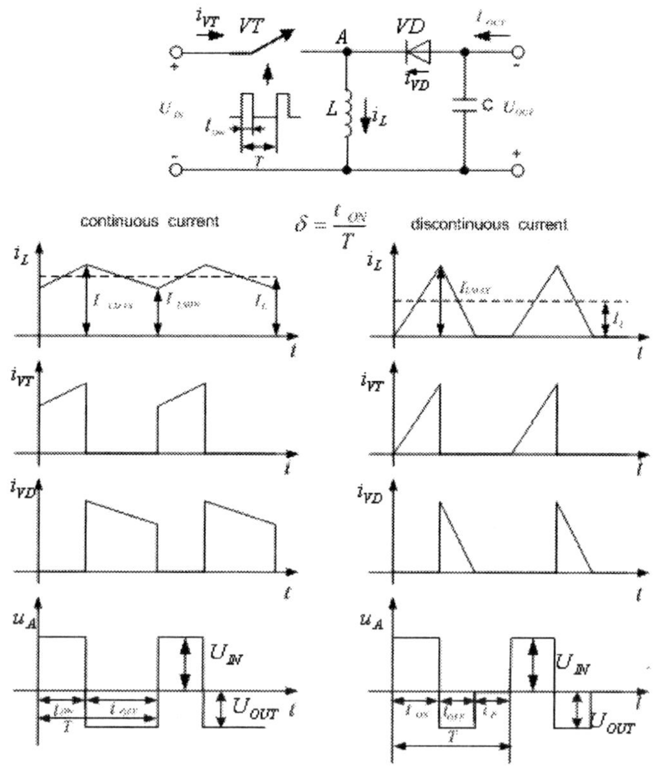

The ripples of the output voltage are determined in the following way: the change of the capacitor charge $2.\Delta U_{OUT}.C$ at turned on transistor for time $\delta.T$ happens at the influence of the output current I_{OUT} flowing through the capacitor. Therefore, the ripples ΔU_{OUT} are equal to $\frac{\delta.T.I_{OUT}}{2.C}$.

Figure 4 displays the electrical schematic and waveforms of a buck-boost converter in both operational modes – continuous-conduction and discontinuous-conduction current through the inductance. As it is seen, the converter contains the same elements as already studied ones – transistor *VT*, diode *VD* and inductance *L*, but their place in the schematic is different causing new characteristics of the converter. It has two special features. The first is that the polarity of the output voltage is the opposite to this of the input voltage. And the second is that this converter is able to operate as buck and as boost converters.

In the continuous-conduction current mode, at the turning on of the transistor, the current through it and through the inductance increases reaching its maximum value I_{LMAX} at time t_{ON}. If the voltage drop across the turned-on transistor is neglected during this interval, the whole input voltage is applied to the inductance, i.e. point A. During this time the capacitor *C* discharges through the load. At the turning off of the transistor *VT*, the diode *VD* begins to conduct, the current through the inductance decreases charging the capacitor and maintaining the load current. At the end of the period *T* the inductance current reaches its minimum value I_{LMIN}. During this interval the voltage in point A is equal to the output voltage (the voltage drop across the conducting diode is neglected). In the continuous-conducting current mode, the following relationship between the output and input voltages is derived:

$$U_{OUT} = \frac{\delta}{1-\delta}.U_{IN} \qquad (6.12)$$

As it has already been mentioned the signs of the input and output voltages are contrary to each other. Also, it is obvious from the (6.10) that when $\delta < 0.5$ the converter is a buck one and when $\delta > 0.5$ it is a boost converter. In the discontinuous-conducting current mode, after the decrease of the inductance current to 0 and dissipating of the stored energy in it, the voltage of point A becomes 0.

Because the input voltage is applied across the inductance when the transistor is on, the ripples of the current through the inductance Δi_L (from I_L upto I_{LMAX}) may be calculated as it is in the boost converter.

Because the change of the capacitor charge is a result of the output current flowing through it when the transistor is on, the ripples of the output voltage is determined as it is the boost converter.

The waveforms studied in the figures are presented at neglecting the parasitic elements in the schematic such as capacity of the inductance *L*, inductivity of the output capacitor, as well as considering ideal switching characteristics of the transistor and diode.

Similar to the buck-boost converter, the Cuk converter, named after its inventor, provides a negative polarity output voltage (Middlebrook, 1983). The electrical schematic and waveforms of Cuk converter are shown in Figure 5.

The capacitor C_1 acts as the primary means of storing and transferring the energy from the input to the output. In steady state, the average inductors voltages U_{L1} and U_{L2} are zero and the voltage across the first capacitor U_{C1} is larger than both U_{IN} and U_{OUT}. Assuming C_1 to be sufficiently large, in steady state the variation in u_{C1} from its average value U_{C1} can be assumed to be negligibly low.

When the transistor is off, the inductor currents i_{L1} and i_{L2} flow through the diode. The capacitor C_1 is charged through the diode by energy from both the input and inductance L_1. Since U_{C1} is larger than U_{IN}, the current i_{L1} decreases. Energy stored in L_2 feeds the output. Therefore, i_{L2} also decreases.

When the transistor is turned on, the voltage across the first capacitor reverse biased the diode. The inductor currents i_{L1} and i_{L2} flow through the transistor. The capacitor C_1 discharges through the switch, transferring energy to the output and to the inductance L_2, because $U_{C1} > U_{OUT}$. Also, i_{L2} increases. The input feeds energy to L_1 causing an increase of i_{L1}.

Both inductance currents i_{L1} and i_{L2} are assumed to be continuous.

The relationship between the output and input voltages is determined by (6.12).

Assuming $\eta \approx 1$ gives the following relationship between the input and output currents:

$$I_{OUT} = \frac{1-\delta}{\delta}.I_{IN} \qquad (6.13)$$

The advantage of Cuk converter is that both the input current and the current feeding the output stage are reasonably ripple free. It is possible to simultaneously eliminate the ripples in i_{L1} and i_{L2} completely, leading to lower external filtering requirements. The main disadvantage is the requirement of the capacitor C_1 with a large ripple current carrying capability.

Figure 5. Electrical schematic and waveforms of Cuk converter

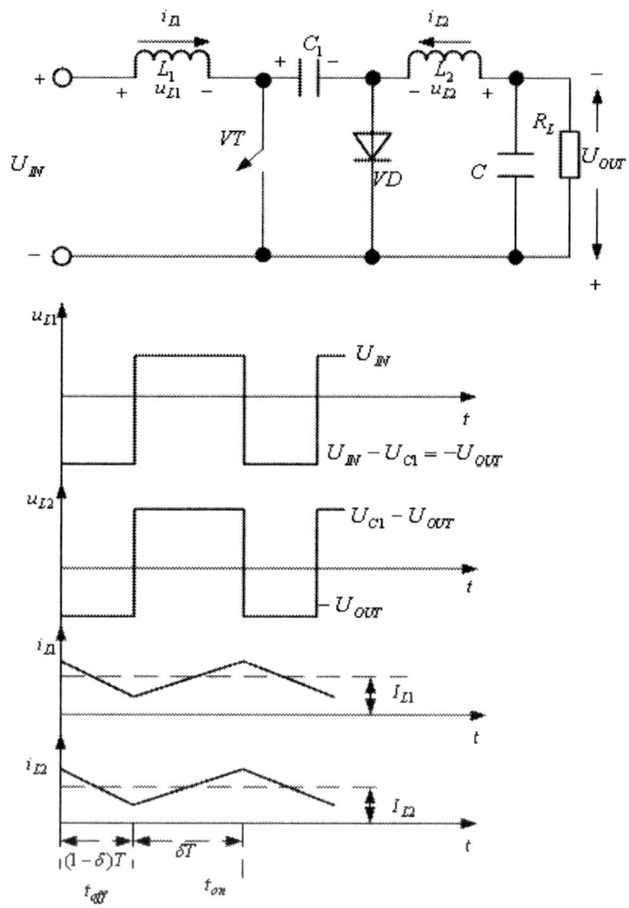

Single - Ended Primary Inductance Converter (SEPIC) combines the abilities of two converters–boost and buck-boost converters (Unitrode, 1993). It is convenient to be used in power factor correction in the uncontrolled rectifiers (see Chapter 4).

Special characteristic of SEPIC operation is that its input voltage may have the waveform of a full-wave rectification voltage without filtering. The electrical schematic of a such converter is shown in Figure 6.

Figure 6. Electrical schematic of a SEPIC converter

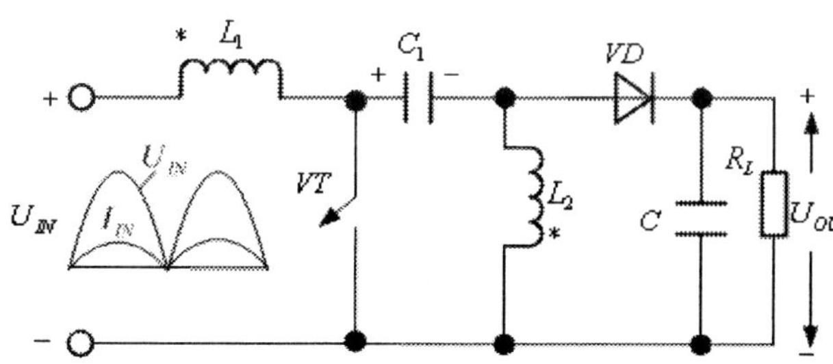

Two inductances are used. The inductance L_1 and the transistor switch are in a boost converter topology; the connection place of the inductance L_2 is as one of a buck-boost converter. Using the capacitor C, remaining of the Cuk converter, is a special feature. In the conventional schematic the inductances are different, wound up at different cores. As we will see further down during description of the operational principle of the converter, it is possible to make the inductance currents to change in the same way, i.e. it is possible the inductances to be wind up in a common core. Furthermore, the leakage inductance appears in series to the inductance L_1 and it can be use as a part of the input EMI filter. The basic advantages of SEPIC used as a power factor correction schema in comparison with the already studied use of a boost converter in Chapter 3 are:

- It is possible to obtain a desire value of the output voltage even less than that of the input voltage;
- Easier protection against current overloading and short-circuit of the output.

Disadvantages of the SEPIC are:

- Higher maximum values of the voltages and currents of the elements;
- The presence of an additional capacitor increases the price and size of the converter.

The operation of the schema will be examined taking in consideration that the switching frequency of the transistor is higher than the mains one. Thus at the period of switching the input voltage can be consider as a DC one. Actually, its real waveform contains ripples with a frequency doubled the frequency of the network voltage compared to which the switching frequency is repeatedly higher. The capacitor C is a basic element of the electrical schema. In a steady state, its voltage is equal in value to the input voltage. Since the capacitor value is little – usually its value is about $0.5\,\mu F$, its voltage traces the waveform of the rectified network voltage, i.e. it follows approximately the input voltage ripples. To be assured in this, the circuit formed from the input voltage, L_1, C and L_2 can be studied. As in the steady state, the average value of the voltage of each of the inductances for a period of a switching with a high frequency has to be 0, then the average value of the capacitor of the same period has to be equal to the input voltage. Besides, the capacitor voltage differs from the input voltage with the high ripples put over its waveform of a low value.

If the transistor VT is turned on, the transient value of the rectified network voltage V_{IN} is applied to the inductance L_1. Simultaneously, voltage of the same value as the voltage of the capacitor is applied to L_2. Therefore, the currents through both inductances increases with a speed of $\frac{V_{IN}}{L}$, if $L_1 = L_2 = L$. The current through the transistor is the sum of the two inductance currents; the diode reverse voltage is equal to the sum of the values of the output and input voltages. When the transistor turns off, the voltages of the self-inductance of the two inductances change the signs of the voltages across each of them causing turning on of the diode *VD* which conducts the sum of the two inductance currents. The output voltage U_{OUT} is applied to the inductance L_2. Studying the circuit - the input voltage, L_1, capacitor, conducting diode, output voltage, it is seen that the output voltage and the capacitor voltage are equal in values but with opposite signs, will be mutually compensated and thus the voltage across the inductance will be equal in its value to the output voltage. Therefore, the current through both inductance will decreases with $\frac{U_{OUT}}{L}$. If the duty ration is again marked as the ration of the turned-on transistor time to the operational period, the following equation is written:

$$U_{IN}.\delta = U_{OUT}.(1 - \delta), \qquad (6.14)$$

and the output to input voltage ratio is founded to be (6.12).

Therefore, the output voltage can be of higher or lower values than the input voltage and they are will the same polarity. Thus, regarding the value of the output voltage the characteristics of SEPIC are as those of buck-boost converter. Regarding the input current, the SEPIC characteristics are as those of the boost converter. It appears from the operational principle of SEPIC that monitoring the input current value makes possible for the average value of the current to trace a sine wave in phase to the source voltage – see Chapter 4.

CONVERSION WITH GALVANIC ISOLATION

The flyback converter with galvanic isolation shown in Figure 7 contains a high frequency transformer with a core of ferromagnetic material. When the transistor VT is on, energy is stored in the primary winding. After the turning off of the transistor, the voltage across the secondary winding changes its sign leading to turning on of the diode VD through which the stored energy transferred to the output capacitor C. Therefore, the duty ratio has to be less then 0.5 for the transformer to be able to demagnetize itself. Additionally, if the turning on and off times of the transistor are considered then the duty ratio has to be less than 0.45. The waveform of the voltage across the transistor has a special shape. When the transistor is turned off and the diode VD conducts, the transformed output voltage as a reverse one is summed to the input voltage across the transistor. After the secondary current decreases to 0, only the input voltage remains across the transistor.

Because of the leakage inductance of the windings, at first a voltage peak dependent on the value of this inductance (determined by the transformer constriction) is applied across the transistor. Incorporating additional elements, such as zener diode and diode parallel to the primary winding, decreases this voltage. The relationship between the output and input voltages is:

$$U_{OUT} = n\frac{\delta}{1-\delta}U_{IN} \qquad (6.15)$$

Advantages of the flyback converter: During the time of transferring the energy towards the output, the inductance of the secondary winding is connected in series with the diode VD and it acts as a current source. Namely, the smoothing inductance in the secondary side is not necessary. Therefore, each output consists only of a diode and filter capacitor. This makes the converter convenient for cheap implementations requiring multiple output voltages. The lack of the output inductance decreases the mutual influence between the outputs – cross regulation (influence of the load change in one of the outputs).

Disadvantages: From the operation principle it is obvious that the magnetizing of the transformer is one-directional, which makes its use more inefficient and requires a bigger size. The stored energy in the primary inductance is:

Figure 7. Electrical schematic and waveforms for a flyback converter

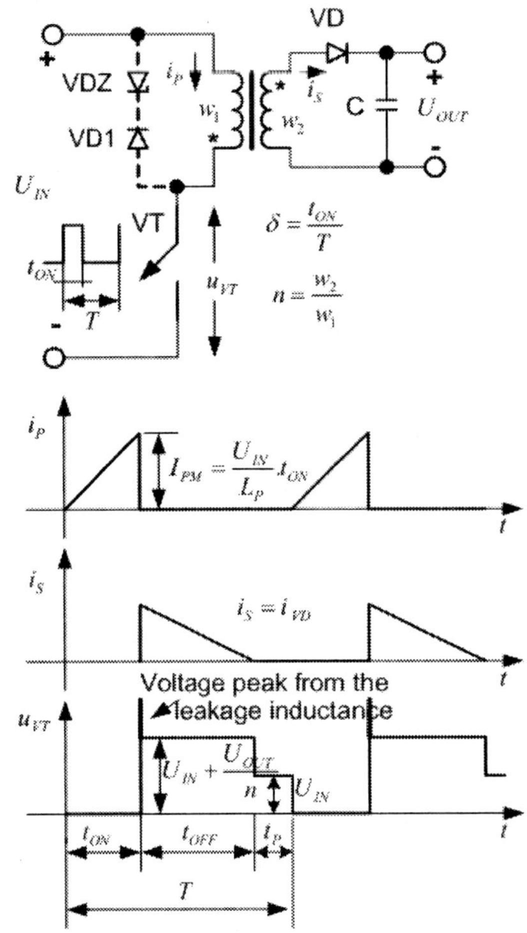

$$W_P = \frac{1}{2}.L_P I_{PM}^2 \tag{6.16}$$

A higher maximum current, which loads the transistor, is required for higher energy to be stored. Not to saturate the transformer on this high current, an air chink which leads to the decrease of the inductance is made in the core. To decrease the maximum value of the voltage across the transistor, a very good mutual connection between the windings is required to be able to decrease the leakage inductance. Different methods for winding up the transformer are known. To decrease the ripples of the output voltage at the lack of the smoothing inductance, a high capacity of the output capacitor with a little equivalent series resistance ESR is required. The flyback converter is characterized by higher ripples of the output voltage than the other converters. Figure 7 displays discontinuous-conduction operational mode. The operation in the continuous-conduction mode is characterized by a lower maximum value of the current through the transistor and the primary winding, which requires higher inductance, i.e. bigger size of the core transformer. If the value of the voltage of the leakage inductance is neglected then the maximum value of the voltage across the transistor in flyback converter is doubled the input voltage. The big size of the transformer with its design characteristics, the high output capacitor and the high values of the currents in discontinuous-conduction mode define the implementation of the flyback converter for low output voltages, multiple outputs and comparatively low powers (5 to 300W).

Figure 8 displays the electrical schematic and waveforms of a forward converter with galvanic isolation. In the secondary side it contains a filter inductance L and an additional diode VD2. The transformer has an additional demagnetizing winding, which number of turns is the same as those of the primary winding. The diode VD3 is connected in series with the demagnetizing winding. The connection of the beginnings of the three windings is representative. At the turning on of the transistor, the diode VD1 also turns on and energy is transferred to the output, also, the current through the inductance L increases – a forward motion. Into the core of the transformer very little energy is stored in contrast to the flyback converter. This is energy for magnetizing connected with low magnetizing currents shown in the forth diagram. At the turning off of the transistor, the voltages of the three windings alter their signs, the diode VD1 turns off, also the diode VD2 turns on and through it the decreasing inductance current flows – a reverse motion. Figure 8 depicts a continuous-conduction mode current regarding the inductance L. This mode is most often used because of the inductance large value. Simultaneously, the diode VD3 also turns on. Through it and the demagnetizing winding, the stored magnetizing energy in the core of the transformer is returned to the source. Moreover, at the same time the transformer demagnetizes. No high maximum values of the currents through the devices and low ripples of the output voltage characterize the continuous-conduction current mode.

The relationship between the input and output voltages is:

$$U_{OUT} = n.\delta.U_{IN} \tag{6.17}$$

Advantages: The ripples of the output voltage are low because of the smoothing inductance and the operation in continuous-conduction current mode. Since little energy is stored in the transformer core, the losses in it and its size are reduced. The low magnetizing current supposes a large inductance of the primary winding, thus, the performance of an air chink is not necessary. The standard ferromagnetic cores without air chinks are convenient here because they have a high flux density – 0.2 to 0.3 T, magnetic permeability 2000 to 3000. Yet it has to be mentioned that the transformer is used inefficiently because it magnetizes only in one-direction when the transistor is on. It is obvious that to be able to demagnetize the transformer, here, also the maximum duty ratio with the switching times considered has to be less than 0.45 too. The low ripples of the current through the inductance require much lower output capacitor in comparison with the flyback converter. The maximum voltage across the transistor in forward converter is also doubled the input one. A rising of the voltage after the turning off of the transistor can appear only if the connection between the primary and demagnetizing windings is not good. The bad connection is avoided constructively by their bifilar winding up.

Disadvantages: The forward converter requires more elements for its implementation, which makes it expansive. The transformer requires an additional winding. If there are multiple output voltages, the influence among the outputs – cross regulation, is increased compared to the flyback converter. It can be reduced by winding up the inductances of all outputs on a common core.

The smaller size of the transformer, no high maximum values of the currents through the elements, the lower output capacitor and the cross regulation make the forward converter suitable for use more frequently with one output voltage and higher output power compared to the flyback converter – from 100W to 500W.

Figure 8. Electrical schematic and waveforms of a forward converter

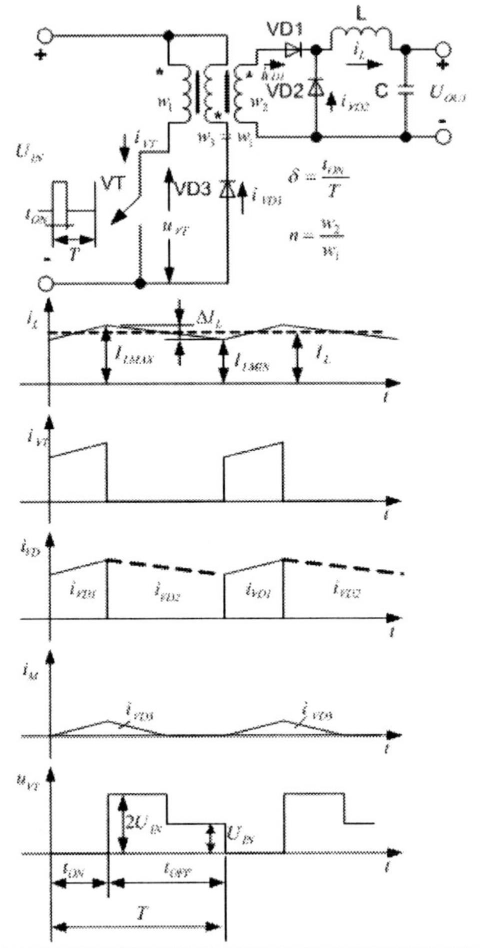

Two transistor modifications of the forward and flyback converters are known. In them connection of two transistors in series with the primary winding is performed, also during the turning off of the transistors across each of them the maximum voltage applied is the input voltage(no its doubled value as it is in the semi transistor modifications). Their disadvantage is a requirement of a galvanic isolated stage of the control system for one of the transistors, because the switches do not have a common electrode.

For better use of the transformer, the core should operate in a symmetrical regime – to magnetize in both directions. This would decrease the size of the transformer and also for a particular core it would increase the power of the converter in comparison with the studied forward and flyback converters. Bidirectional magnetizing demands the number of power devices in the electrical schematic of the converter to be even. One of the probable decisions of such a converter is push-pull converter shown in Figure 9. The primary and secondary windings are with middle points. The transistors turn on in sequence in time. Moreover, the currents in the primary winding are with opposite directions making the magnetizing of the transformer bidirectional. Thus, the necessity of a demagnetizing winding is eliminated. Energy is transferred to the output when each of the transistors is turned on and in the same time the corresponding diode of the secondary winding is turned on – a forward motion. At the turned-off transistors the current through the inductance L decreases via the turned-on diodes VD1 and VD2 – a reverse motion. Each of the transistors can operate with a maximum duty ratio of 0.45.

The relationship between the input and output voltages is:

$$U_{OUT} = 2.n.\delta.U_{IN} \tag{6.18}$$

Advantages: Power towards the output is transferred during up to 90% of the operational period allowing more fully use of the transformer and reduction of its size. The continuous current of the output leads to low ripples and to a requirement of lower output capacitor. In this operational mode the stability of the closed system of regulation of the output voltage is easier to be obtained as it is in the forward converter. Both transistors have common electrodes (emitter or source) relative to which their control is ensured without a requirement of a galvanic isolation between one another (for example, in the further down stated half-bridge and bridge converters).

Disadvantages: Commutating diodes connected in antiparallel to each transistor are necessary. Thus the energy stored in the leakage inductances is returned through the diodes to the supply network at first

Figure 9. Electrical schematic and waveforms of a push-pull converter

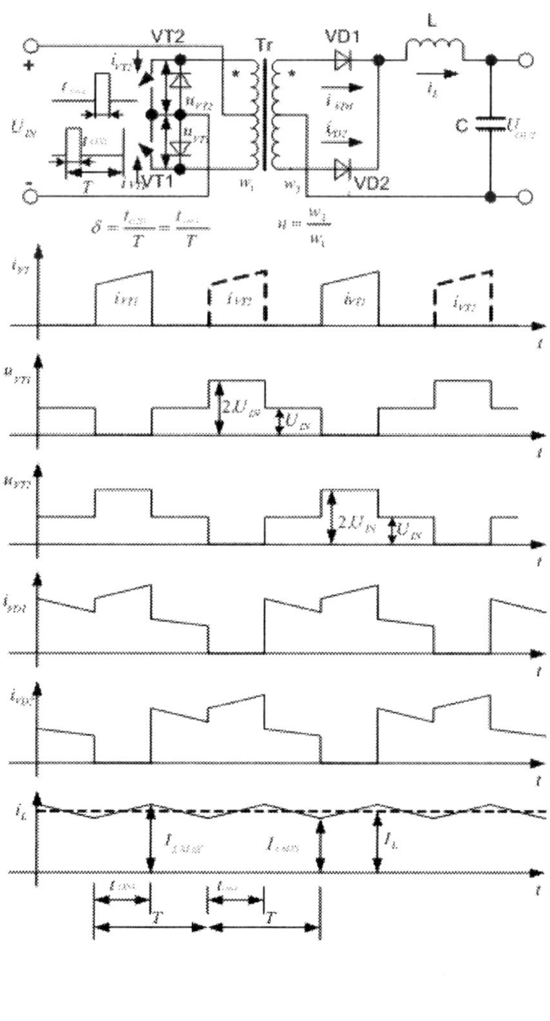

moment of the turning off of a transistor. This energy is a reason for an increase of the voltage across the transistor at the moment of turning off. A very good constructive implementation of the transformer is required to be able to maintain the symmetry and to ensure a very good connection between the two semi-windings of the primary and secondary sides. The winding up is usually bifilar.

The maximum value of the transistor voltages is equal to the doubled input voltage because when one of the transistors is turned on, the sum of the voltage across its corresponding primary winding and the input voltage is applied to the other transistor.

All this makes the converter suitable to operate in low input voltages. An unpleasant probability of lack of symmetry in the two directions of magnetizing exists in this converter. This can be a result of asymmetry in the transformer, of differences in the times of switching of the transistors or of asymmetry in their control. Moreover, saturation can be reached in one of the directions especially at high input voltages.

To avoid this effect, special measures, which in some cases are only schema-technical ones, are taken. So-called current control is very convenient in this case. At this control, during every cycle of turning on of a transistor, the maximum current value through it is monitored and at its reach the transistor turns off. Therefore through both transistors the current will reach the same value and the above described undesired effect will be eliminated by the control system.

The push-pull converter is suitable to operate in lower input voltage – 12V, 24V or 48V and in power from 200W to 800W.

The half-bridge converter shown in Figure 10 is very popular to be used in high powers. The two capacitors C1 and C2 connected in series form a middle point of the supply source and each of them charges to the half of the input voltage. The transistors turn on consequently applying the voltage across one of those capacitors each time in the opposite to direction to the preceding time to the primary winding of the transformer. Thus the transformer magnetizes bidirectional and its use is better. In this way the converter is similar to the push-pull topology but in the last one the whole supply voltage is applied to the primary winding. It appears from the waveforms that they are identical in shapes to those of the push-pull converter but the voltages across the transistors are twice lower. The maximum voltage value is equal to the input voltage value and it is applied to the turned-off transistor when the other turns on. The maximum duty ratio of each transistor is 0.45 taking in consideration the switching times of the transistors and the dead time that has to be secured between the moments of the turning off of a transistor and the moment of turning on of the other one. The dead time is required to be refrained from simultaneous conduction of both transistors which would short-circuit the supply source.

The relationship between the output and the input voltages is:

$$U_{OUT} = n.\delta.U_{IN} \tag{6.19}$$

Advantages: They are connected to the good use of the transformer, which magnetizes bidirectional and to the missing of the demagnetizing. Besides, in comparison with the push-pull schema, this one has not got a middle point. The operation in the mode of continuous-conduction current through the inductance L leads to decreased ripples of the output voltage and to necessity of lower capacitor C. Other advantage compare to the push-pull converter is the easier elimination of the danger of asymmetry in the magnetizing of the transformer in both directions. This is made by connection of a blocking capacitor Cb (usually less than 10 μF), avoiding a DC component of the current through the primary winding. A serious advantage is the twice reduced voltages across the devices.

Figure 10. Electrical schematic and waveforms of a half-bridge converter

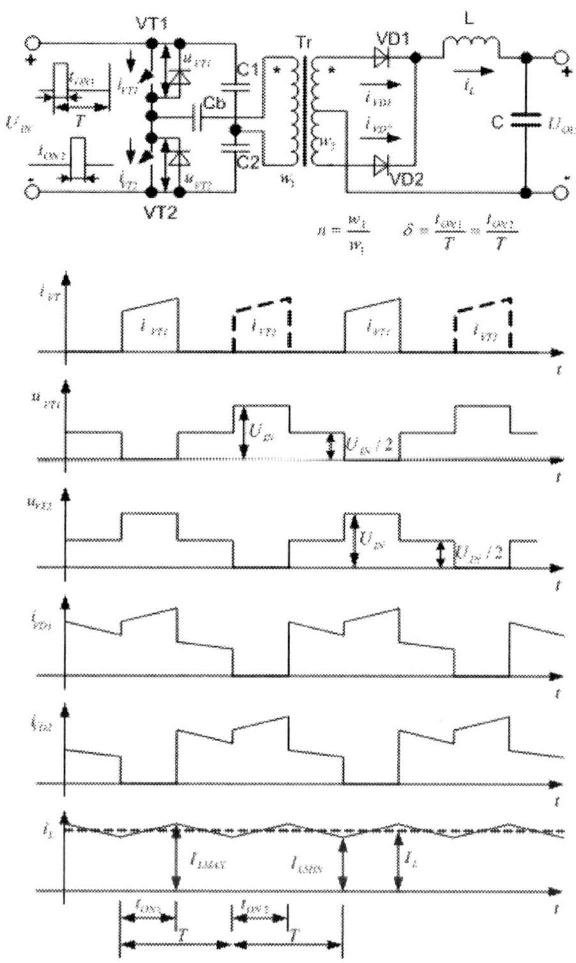

Disadvantages: The necessity of the capacitors C1 and C2, which are with a high capacity, increases the size of the converter. A commutating diode, through which the stored energy in the leakage inductance is returned to the supply source, is required for each transistor. The stored leakage inductance energy is a reason for an increase of the voltage across the transistor at the moment of its turning off. The upper transistor in the schematic requires a control stage galvanic isolated from this of the lower one because the transistors lower electrodes (emitters or sources) have not got a common point as it is in the push-pull converter. The additional capacitors and diodes increase the price of the converter.

The half-bridge converter is suitable for high input voltages because the voltage across the transistors does not exceed the source one and for power from 500W to 1500W.

The bridge converter shown in Figure 11 contains the highest number of elements compared to all converters studied till this moment determining its highest price. The transistors turn on in series by pairs conducting simultaneously two by two diagonally connected devices – VT1 with VT3 and VT2 with VT4. The whole input voltage is applied to the primary winding of the transformer, not the half of it as it is in the half-bridge converter. Therefore, using the same transistors chosen by their currents using the bridge converter approximately twice higher power can be achieved. As it is seen from the waveforms they are the same as those of the half-bridge converter. The use of the blocking capacitor Cb is also identical to this in the half-bridge converter – to eliminate the DC component of the current in the primary winding and thus to prevent eventual asymmetry in the magnetizing of the transformer core.

Because of the fact that the whole input voltage is applied to the primary winding, the relationship between it and the output voltage is:

$$U_{OUT} = 2.n.\delta.U_{IN} \tag{6.20}$$

Advantages: Using this converter the highest output power can be obtained. The two capacitors creating a middle point miss. Regarding the secondary side the mode is as those in the half-bridge or push-pull converters – continuous-conduction current mode, low ripples and low output capacitor. The maximum voltage across the transistor, if the primary increase of the voltage due to the leakage inductance is neglected, is equal to the input voltage. The value of the current through the transistors is limiting the achievement of higher power.

Figure 11. Electrical schematic and waveforms of a bridge converter

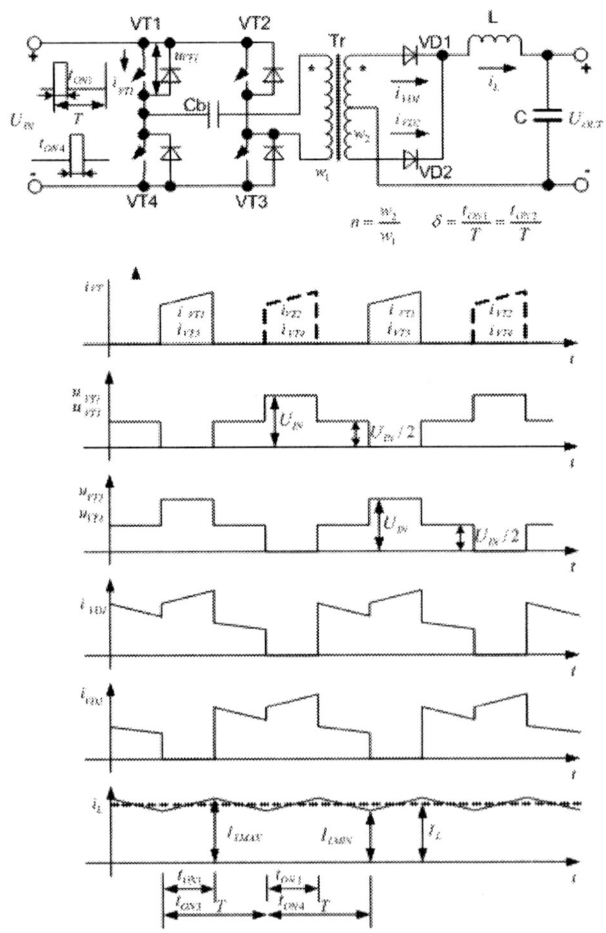

Disadvantages: The converter power schematic has the highest number of elements which makes its implementation more complicated and increases its price. The upper transistors in the schematic require galvanic isolated control stages.

The bridge converter is used in high input voltages and in power from 1000W to 5000W.

Based on the boost converter shown in Figure 3, converters with galvanic isolation may be implemented. Figure 12 depicts principle electrical schematic and waveforms illustrating the operation of a full-bridge transformer isolated boost-derived converter. The indexes of the switches that conduct in each interval are marked. When all the transistors are on, energy is accumulated in the inductance L. When a pair of diagonally connected transistors is on, the energy accumulated is transferred towards the output through a corresponding diode connected to the secondary winding of the transformer.

In a steady operation mode and continuous current through the inductance, the increase of the current through the inductance in the case of all four transistors on has to be equal to its decrease during the next interval. Therefore:

$$\delta.U_{IN} = -\left(1-\delta\right).\left(U_{IN} - \frac{U_{OUT}}{n}\right) \tag{6.21}$$

After transformation, a relationship between the input and output voltages is found:

Figure 12. Electrical schematic and waveforms of a full-bridge transformer isolated boost-derived converter

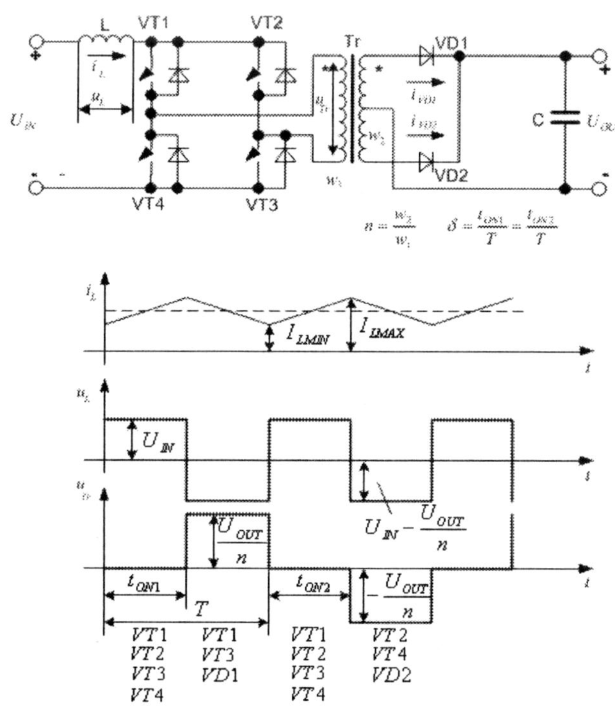

$$U_{OUT} = n.\frac{1}{1-\delta}.U_{IN} \tag{6.22}$$

A version of push-pull boost-derived converter (Ericson, 2001) is also available.

Figure 13 shows the way of obtaining Cuk converter with galvanic isolation starting from the schema without galvanic isolation Figure 5. The beginning of the secondary winding is at its lower end; also, the connection direction of the diode *VD* is altered. Thus, the polarity of the output voltage is as it is shown in the figure. When the transistor *VT* turns on, the input voltage U_{IN} is applied across the inductance L_1 and the current through it increases. The voltage of the capacitor $C_1^{'}$ is applied across the primary winding. This voltage transformed in the secondary winding leads to a turning off of the *VD*. The current through the inductance L_2 increases and it flows through the capacitor $C_1^{'}$.

At a turning off of the transistor *VT*, the current through the inductance L_1 decreases, flowing through the capacitor $C_1^{'}$. The diode *VD* turns on and secures a circuit for the decreasing current through the inductance L_2. The relationship of the input and output voltage considering the turn ratio of the transformer is:

Figure 13. Electrical schematic of a transformer isolated Cuk converter

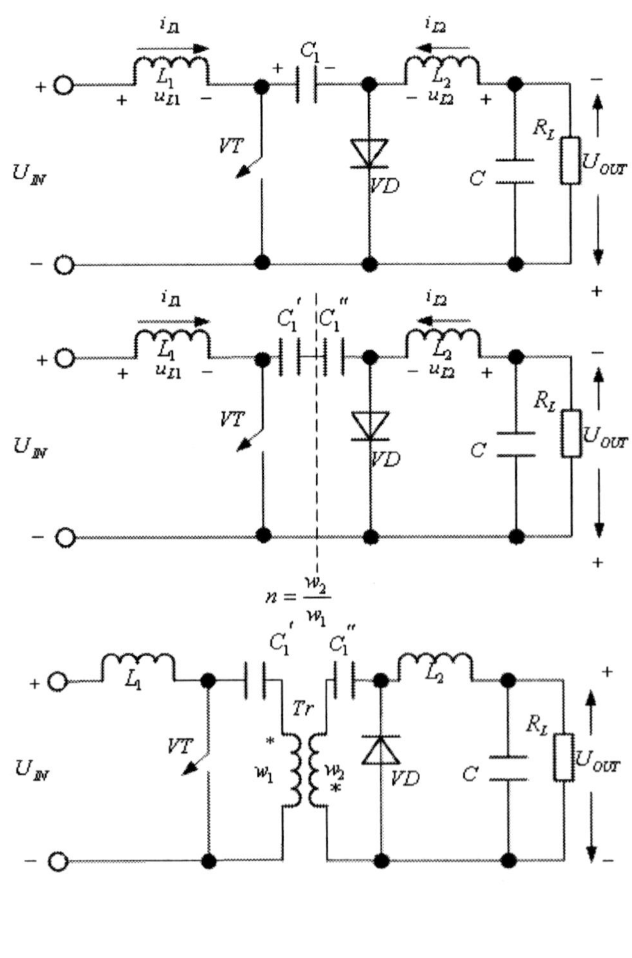

$$U_{OUT} = n.\frac{\delta}{1-\delta}.U_{IN} \tag{6.23}$$

Figure 14 shows the electrical schematic of SEPIC converter with galvanic isolation. Its operation is analogical to the schema shown in Figure 6 having in mind that the role of L_2 is played by the magnetic inductance L_μ of the transformer. The voltage across this inductance is transformed with a coefficient n in the secondary side. The relationship between the input and output voltages is defined by the equation (6.23).

So-called inverse SEPIC converter in variations with and without galvanic isolation is also known (Erickson, 2001).

Specially designed integrated circuits are usually used to control DC/DC converters. Many producers offer such circuits having different special features. Their operational principle however is based on two control methods - voltage mode control and current mode control (Kislowski,1985). Figure 15. depicts a schema and waveforms describing the voltage mode control.

The schema contains the following basic blocks:

Figure 14. Electrical schematic of a transformer isolated SEPIC converter

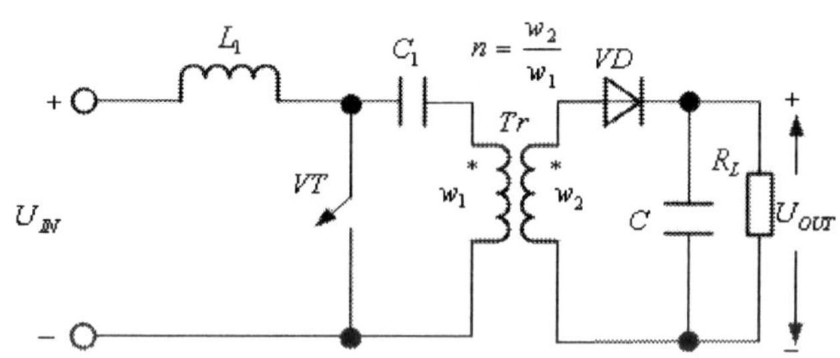

SLVG: Generator of a linear changing voltage U_{SLV}, determining the operational period T_S of the DC/DC converter;

EA: Error amplifier. Usually, inner reference voltage U_r from a stabilized source is applied to the non-inverting input, and the inverting input and output are taken out as terminals of the integrated circuit. In this way, passive circuits consisting of resistors and capacitors marked as Z_1 and Z_2 assisting in the formation of the gear characteristic of the closed system for automatic control of the output voltage are connected to these terminals. Part of the output voltage $k.U_{OUT}$ is passed to the inverting input.

PWM: Pulse-width modulator. It compares the passed voltages to its inputs and in its output pulse signals with a changeable duty ratio δ are generated.

Dr: Driver. It forms the firing pulses for power device (power devices).

The fundamental feature of the voltage mode control is determination of the duty ratio of the firing pulses directly from the voltage of the error amplifier. So-called soft start circuit is provided for an initial turning on of the converter. The circuit usually contains inner generator of current I_G, supplied from an inner source U_{IS}. The current I_G charges an outer connected capacitor C_{SS} and its voltage U_{SS} decreases

Figure 15. Schema and waveforms describing the voltage mode control

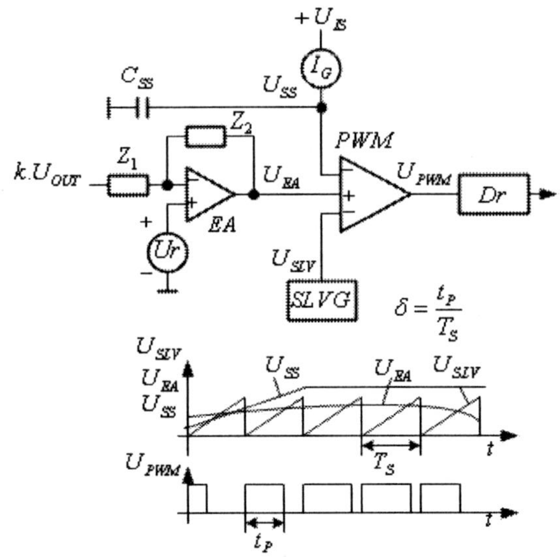

according to a linear law. PWM is designed in such a way that the duration of the pulse t_p is determined by the lower level of the voltage of its both inverting inputs. Therefore, at the beginning the duty ratio will lightly increase; the soft increase of the output voltage of the converter is also gained. After the charge of the capacitor C_{SS} to its maximum value, the duty ratio is determined directly from the output voltage of the error amplifier. For example, if a tendency of increasing the output voltage of the converter U_{OUT} above the reference voltage occurs, the output voltage of the amplifier U_{EA} will decrease. Also, the duration of the pulses for turning on t_p will decrease and this will cause a tendency of decreasing the output voltage and of its returning to the set value.

Figure 16. Schema and waveforms describing the current mode control

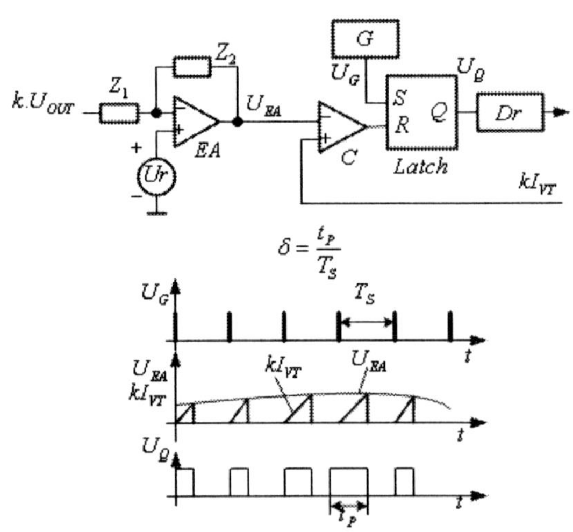

The current mode control is described in Figure 16. The schema contains the following blocks:

G: Pulse generator, determining the start of each operational cycle, and thus determining also the operational period T_S of the converter.

C: Comparator, comparing the signal of the output of the error amplifier to the signal proportional to the current through the converter transistor I_{VT}. The last mentioned signal can be obtained using current transformer or resistor.

Latch: Its inputs are controlled from the generator and comparator, and in its output control pulses for the driver are generated.

The rest two blocks – the error amplifier and driver have the same functions as those in the voltage mode control. Pulse-width modulator is missing here.

The fundament of the operational principle of the current mode control is that the output signal of the error amplifier determines the maximum value of the current through the transistor of the converter and thence indirectly the duty ratio δ. For example, if a tendency of increasing the output voltage of the converter U_{OUT} above the reference voltage occurs, then the output voltage of the amplifier U_{EA} will decrease. After the start of the next cycle, the transistor will be turned off earlier when the current through it has reached a lower value. This means a decrease of the duty ratio and a reverse to the initial tendency – a tendency for decreasing the output voltage, i.e. its return to the reference value.

Figure 16 does not show the soft start circuits which are also provided.

The control integrated circuits for both methods also include additional elements for protection of over-voltages and increasing of the converter output current, as well as additional supply circuit.

The advantage of the current mode control is the faster reaction of the automatic control system at a disturbing influence – the input voltage of the converter. For example, if the input voltage increases the gradient of the transistor current will increase, the faster the reach of the value of the output signal of the error amplifier will happen and the transistor will turn off earlier. Besides, in this control method, the protection of current exceed of the converter is easier implemented. The implementation of the current mode control is especially useful in the push-pull converters where possibilities for asymmetrical magnetizing of the transformer core and for DC magnetizing exist at different maximum values of the currents through the two transistors.

Control methods for DC/DC converters based on hysteresis control, as well as on sliding mode control exist but they have found a comparatively limited implementation.

BIDIRECTIONAL DC/DC CONVERSION

Special feature of bidirectional DC/DC converters is their ability to transfer DC energy into two directions (Tolbert, 2002: Chiu, 2006). Dependent on the voltage and current directions the converter can operate in different quadrants. The operational principle is described in Figure 17.

The current and voltage directions are assumed to be as those shown in Figure 17. According to the shown directions the source is u_1 and the consumer is u_2, i.e. energy is transfer from left to right. If the polarity of the voltages stays the same but the current directions are changed, then the energy will be transferred from right to left. Therefore, the converter is two-quadrant regarding both right and left sides of the schema.

If there is a possibility to change also the direction of the voltage in one of the sides, then this converter will be four-quadrant regarding this side.

Figure 17. Principle of bidirectional converters

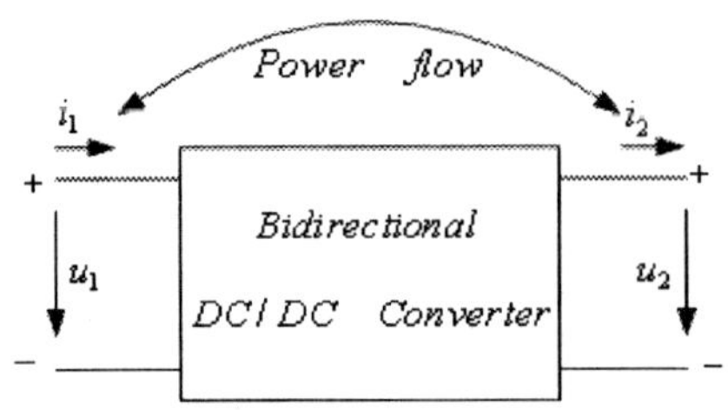

Figure 18 shows an example of a two-quadrant converter regarding both sides without galvanic isolation.

It appears from the figure that $VT1$, L, $VD2$ are connected in a buck converter circuit. Also, $VT2$, L, $VD1$ are connected in a boost converter circuit. Therefore, voltage U_1 is higher than voltage U_2. Figure 18 displays also the waveforms describing the operation of the converter. In the first case, the buck converter operates and energy is transferred from left to right sides. In the second case, the boost converter operates and energy is transferred from right to left sides. The converter is two-quadrant one regarding both sides, because the current directions change and the voltage directions stay unchangeable.

Circuits of bidirectional DC/DC converters with galvanic isolation exist (Zhang, 2004). An example is given in Figure 19 based on a bridge schema of the converter. Usually, one of the two voltages – U_1 or U_2 is higher than the other. Typical examples are 250-350V and 24-28V. Thus, the converter is boost in one direction and it is buck in the other direction. Since the polarity of the two voltages can not be changed, the converter is two-quadrant regarding both sides.

It is obvious, that if it is necessary to transfer energy form $U1$ towards U_2, the transistors from $VT1$ to VT4 will be turned on and transistors form VT5 to VT8 will be maintained off. The rectification at the right side is made using the diode bridge VD5 to VD8.

Figure 18. Bidirectional two-quadrant converter

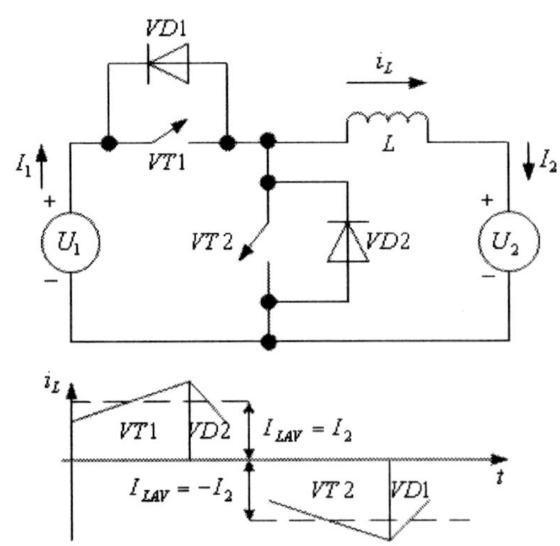

Analogically, when energy is necessary to be transferred from right to left sides, the transistors from VT5 to VT8 have to be on, and the transistors from VT1 to VT4 will be maintained off. A special feature of the control of this converter is traversing form one mode to the other which is connected to turning on the appropriate pair of diagonally connected devices in one of the sides depending on which pair of diagonally connected devices in the other side have been last on.

There are also other topologies of bidirectional converters with galvanic isolation, for example one described in (Gunningham, 2007). In this bidirectional converter topology, an energy exchange is realized among three DC lines – two low-voltage lines 14V/42V in one side and a high-voltage line 200-500V in the other side.

METHODS TO IMPROVE POWER EFFICIENCY IN DC/DC CONVERSION

The weight and the size of the DC/DC converters are of a great importance because very often their implementations are connected with limitations in this respect (Luo,2003). A possibility to obtain good results for the weight and size is the use of a current-doubler rectifier instead of the full – wave rectifier in push-pull and bridge converters (Balosh, 1997). Figure 20 shows the electrical schematic and the waveforms of a full-wave rectifier.

Because of the similarity of the operation of this rectifier to the operation of the single-phase uncontrolled rectifier with a middle point of the transformer, studied in Chapter 4, here the operation will not be described in details. The current through the L increases when the voltage of a half-winding of the transformer differs form 0 and it is flowing through one of the two diodes. The current through the L decreases when the voltage of a half-winding of the transformer is 0, during this interval both diodes conduct. The most important conclusion which will be used further down for comparison among converter topologies is that the average value of the inductance current in the full-wave rectifier is equal to the output current I_{OUT}, and the frequency of the ripples of the inductance current is the same as the converter operational frequency.

Figure 21 shows the electrical schematic and waveforms of a current-doubler rectifier.

Its operation has several special features and will be examined more detailed. At the positive half-period of the voltage of the secondary winding u_2, the diode $VD1$ conduct, while the diode $VD2$ is turned off. The current through the inductance L_1 decreases flowing through the diode $VD1$ and the output capacitor. The voltage across this inductance according to its chosen direction in Figure 21 is positive and equal to the output voltage. This current does not flow through the secondary winding of the transformer. At the

Figure 19. Bidirectional DC/DC converter with galvanic isolation

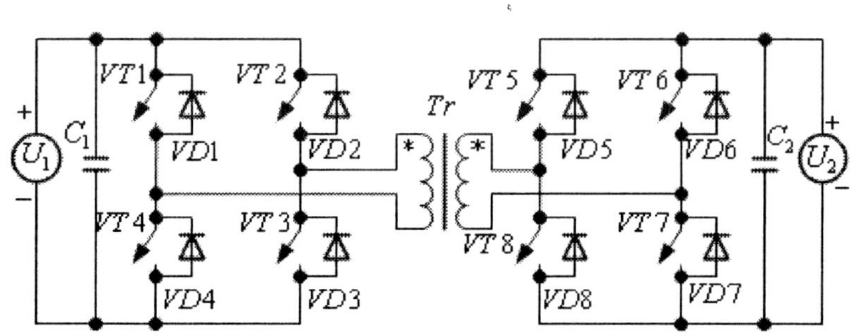

Figure 20. Electrical schematic and the waveforms of a full-wave rectifier

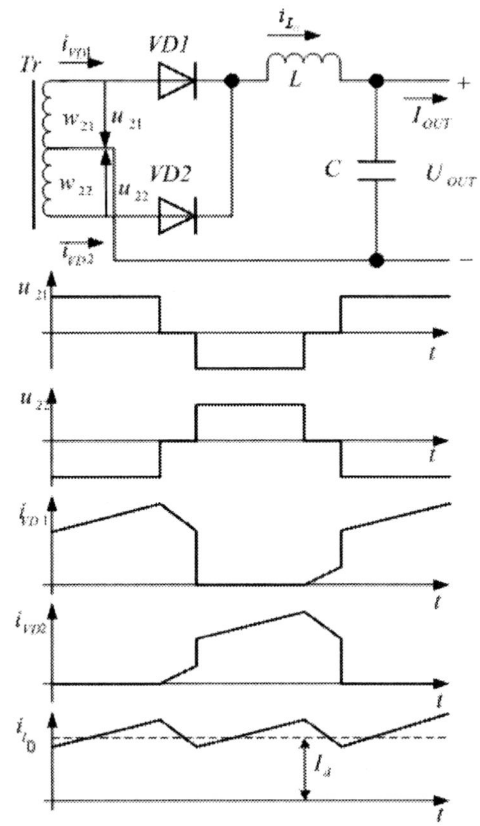

same positive half-period of the voltage of the secondary winding, the current through the inductance L_2 increases flowing through the secondary winding, diode $VD1$ and output capacitor. The voltage across this inductance is positive and equal to the difference between the secondary winding and output voltage. Then an interval during which the voltage of the secondary winding is zero follows. Therefore the voltage u_{L_2} across the inductance L_2 changes its sign and becomes equal to the output voltage causing the decrease of the current through this inductance. The state of the inductance L_1 stays unchanged and its current continues to decrease. The operation at the negative half-period of the secondary winding of the transformer is analogical, during it the diode $VD2$ conducts. The current through the inductance L_2 continues to decrease; the output voltage is applied to it through the diode $VD2$. The current through the inductance L_1 begins to increase and the voltage across it changes its sign. A new pause follows. Then the process begins again. The most important feature of the current-doubler rectifier, which arises during the description of the operation and from the waveforms in Figure 21, is that the average value of the current flowing through each inductance is equal to the half of the value of the output current. This permits to reduce the total size and weight of the inductances compared to the full-wave rectifier nevertheless the increased number of the inductances – two in comparison to 1 in the full-wave rectifier. Also, one should have in mind that the frequency of the ripples of the current through each of the two inductances is twice lower than the converter operational frequency. This reduces the requirements of the magnetic material of the core of the inductance and also reduces the losses in it.

Figure 21. Electrical schematic and waveforms of a current-doubler rectifier

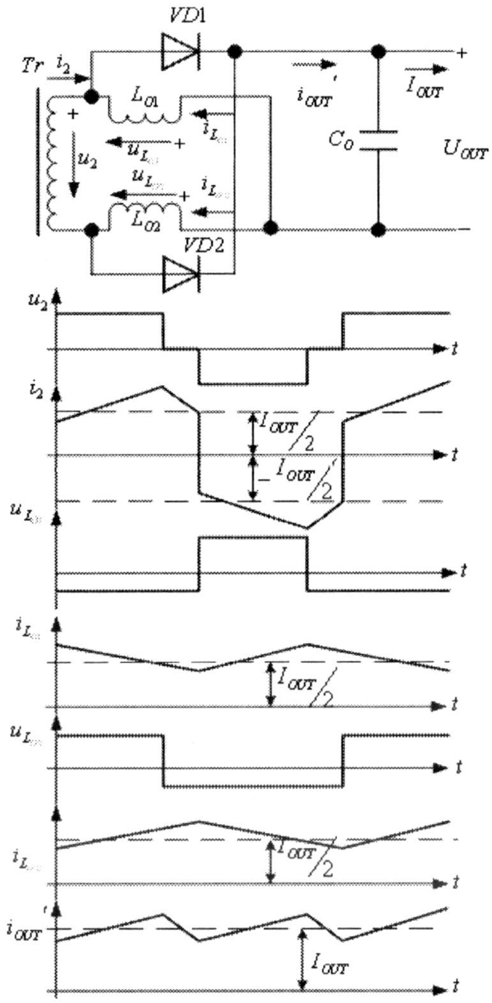

The next step of decreasing the size and weight of the DC/DC converters is connected with an increase of their operational frequency. This will cause the possibility to reduce the size of the transformer. The increase of the operational frequency when the PWM method combined with the above studied schemas is used, however, leads to increase of the commutation losses of the devices. In these electrical schematics, the switching is called hard-switching because at the intervals of turning on and off the current through and the voltage across the switch are simultaneously changed. Also, an increase of power exactly at these intervals happens. That is the reason why the schemas of DC/DC converters with resonant switching, in which a soft-switching of the devices is used, are developed. More often these are schemas with zero voltage switching (ZVS) or/and zero current switching (ZCS). In general, under ZVS the turning on of the transistor is at a zero voltage across it, and under ZCS the turning off of the transistor is when the current through it decreases to 0.

At first, the operational features of load –resonant converters will be examined. A special feature in them is the use of series oscillating circuit, as a result the voltage across the capacitor of the circuit and the current through it are changed in accordance with harmonious curves. Thus, ZCS and/or ZCS of the power devices is/are obtained. For simplicity, the electrical schematics are examined by a direct connection of an output rectifier but it may be implemented using high-frequency transformer with the purpose to obtain an appropriate value of the output voltage and also to have a galvanic isolation between the input and output of the converter. The load may be connected in series with the resonant circuit, in parallel to capacitor of the circuit or as a hybrid connection. Therefore, three topologies of load-resonant converters exist.

Figure 22 shows the electrical schematic and waveforms of a series-loaded resonant converter, which can operate in discontinuous-conduction or continuous-conduction current modes through the inductance. When current flows through the inductance, a pair of diagonally connected diodes of the output rectifier always conducts. The load appears in series with the resonant tank. The discontinuous-conduction current mode is obtained when the switching is made with a frequency twice as lower as the resonant frequency. The mode of the rest two cases, dependent on the ratio between the switching and resonant frequencies, is continuous-conduction one.

In discontinuous-conduction current mode, the transistors and diodes turn on and off at a zero current. In the second case shown, only the turning off of the transistors and the turning on of the diodes are at zero voltage, while the turning on of the transistors and the turning off of the diodes is a hard-switching. The

Figure 22. Electrical schematic and circuit waveforms of a series-loaded resonant converter

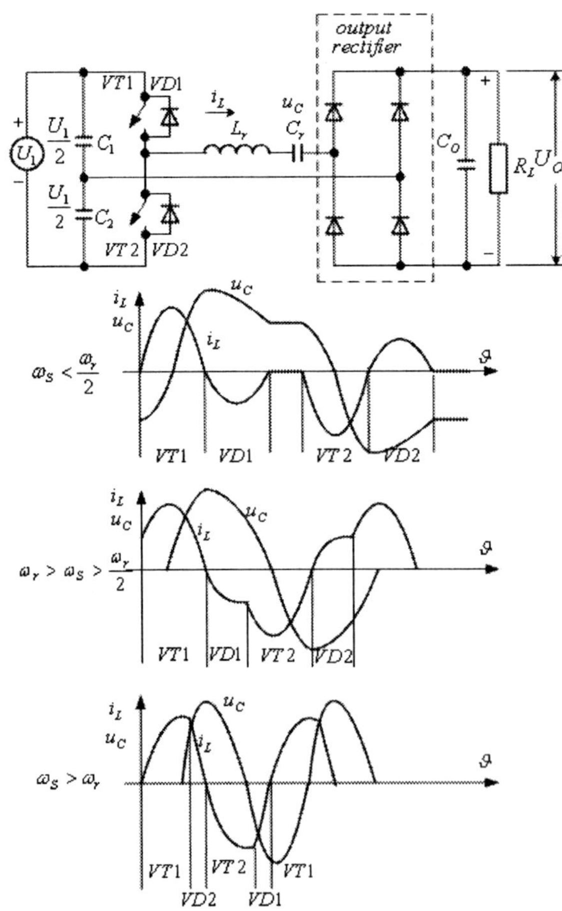

third case, which is at a switching frequency higher than the resonant one, is of interest. As it appears the transistors turn on at a zero current through them and at a zero voltage across them. This happens owing to the fact that before the conducting interval of a transistor, its antiparallel diode has been conducing.

A special characteristic of this converter is that in the discontinuous-conduction mode it operates as a current source in respect to the load, i.e. the converter has inherent overload protection capability. Without a transformer, the converter can operate only as a buck one.

Figure 23 displays the electrical schematic and waveforms of a parallel-loaded resonant converter. As it is seen, the load is in series with capacitor C_r of the resonant tank. The connection can also be made via a transformer. The output inductance L_o is typical, therefore, the whole output circuit can be replaced by a current source equal to the load current I_O during the analysis. The voltage across the capacitor C_r is rectified by the output rectifier. In the case when this voltage is equal to 0, continuous current through the inductance L is secured by simultaneously turning on of the phase-leg diodes of the output rectifier.

The converter can also operate in two modes in respect to the current i_L of the resonant inductance L_r - discontinuous-conduction current mode (first waveforms) and continuous-conduction current mode (second and third waveforms).

In discontinuous-conduction current mode when the turning on of the next transistor happens, the capacitor charge starts with a delay just when the inductance current increases the load current. Since the capacitor is discharged by the load current, its voltage decreases according to a linear law after the turning off of the diode which has been conducting. From the operational principle and waveforms, it is clear that in this mode all devices turn on and off at ZCS.

If $0.5 \cdot \omega_0 < \omega_S < \omega_0$, then the inductance current and the capacitor voltage change following continuous curves. As it is seen in the second diagrams, the turning off of the transistors and turning on of the diodes is at zero current. The turning on of the transistors and turning off of the diodes however is a hard-switching.

When $\omega_S > \omega_0$ - the third diagrams in Figure 23, the turning off of the transistors is hard, while becuase of their antiparallel diodes have been conducting till their turning on, it is at ZCS and ZVS.

A special characteristic of this converter is the possibility to operate as a buck-boost converter without presence of a transformer. In discontinuous-conduction mode the converter operates as a voltage source undependable on the consumed load. In contrast to the series-loaded resonant converter, the parallel-loaded resonant converter does not have inherent overload protection capability in no one of the modes.

A hybrid variant also exists where in the connection of the load is in series with the part of the capacitor of the resonant tank. Thus, the advantages of the previous two topologies can be combined.

In resonant-switch converters, elements of the resonant tank are connected in parallel to or in series with a transistor which leads to ZCS or ZVS (Andreycak, 1997).

Let us examine buck DC/DC resonant–switch converter without galvanic isolation, which is shown in Figure 24. The schematic is similar to this of the buck converter with hard-switching presented in Figure 2, only, here elements of the resonant tank circuit L_r, C_r are connected to the transistor. Output inductance L_o is more higher than the resonant one thus the current through L_o may be assumed to be unchangeable during the time of resonant processes and the circuit right from diode VD can be replaced by a current generator with value I_O. At initial state $\vartheta = 0$, the resonance capacitor is charged to the input voltage, and the diode is on conducting the output current. At $\vartheta = 0$ the transistor *VT* turns on, the current through it increases according to a linear law and at $\vartheta = \vartheta_1$ the current reaches I_O, the diode turns off. From this moment on, a resonant process starts in the parallel circuit containing C_r, VT, L_r.

Figure 23. Electrical schematic and waveforms of a parallel-loaded resonant converter

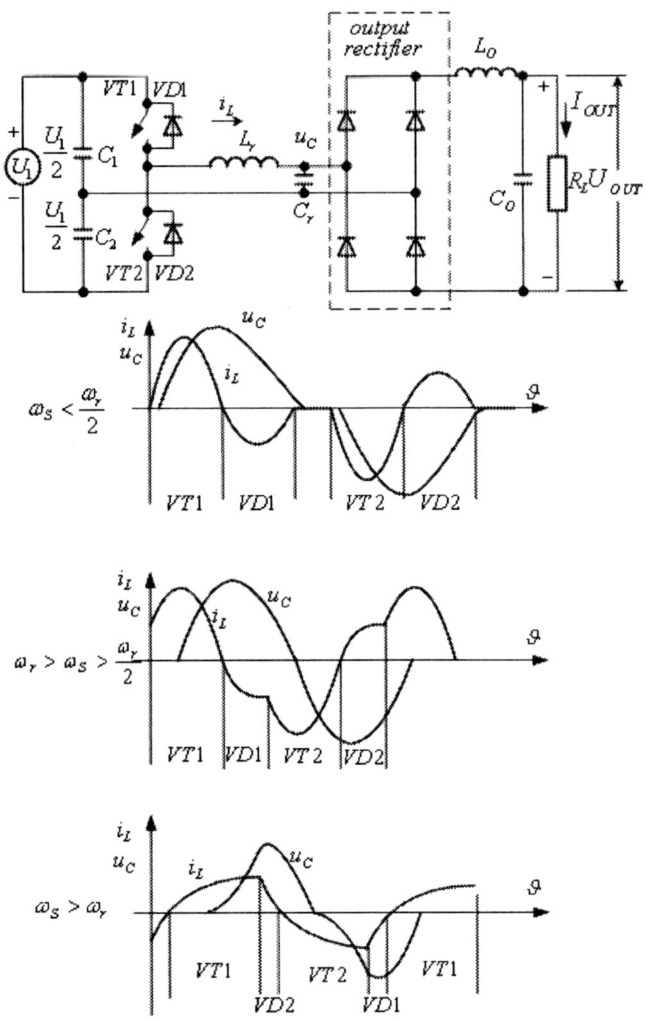

The current through the resonant circuit reaches its maximum value higher than the output current value at $\vartheta = \vartheta_2$ then it decreases and at $\vartheta = \vartheta_3$ the current reaches again the value of I_O. At this moment the voltage across the capacitor C_r has its maximum negative value, while the voltage across the diode has its maximum positive value $2.U_{IN}$. The current through the transistor continue to decrease and at $\vartheta = \vartheta_4$ it equals to 0, i.e. this is the moment of ZCS. As it is seen, the transistor turns on and off at zero current. After the turning off of the transistor, the output current of a constant value I_O recharges the capacitor and during the interval $\vartheta_4 - \vartheta_5$ its voltage increases according to a linear law reaching the value of the input voltage at $\vartheta = \vartheta_5$. At this moment the diode *VD* turns on and a new turning on of the transistor may happen – for example at $\vartheta = \vartheta_6$. Output inductance L_o and capacitor C_O separate across the load the average value of the diode voltage – the last diagram. From it and operational principle, it is clear that a change of the output voltage can be achieved by a change of the switching frequency but only if the duration of the interval $\vartheta_5 - \vartheta_6$ changes.

Figure 24. Electrical schematic and waveforms of a ZCS resonant-switch buck converter

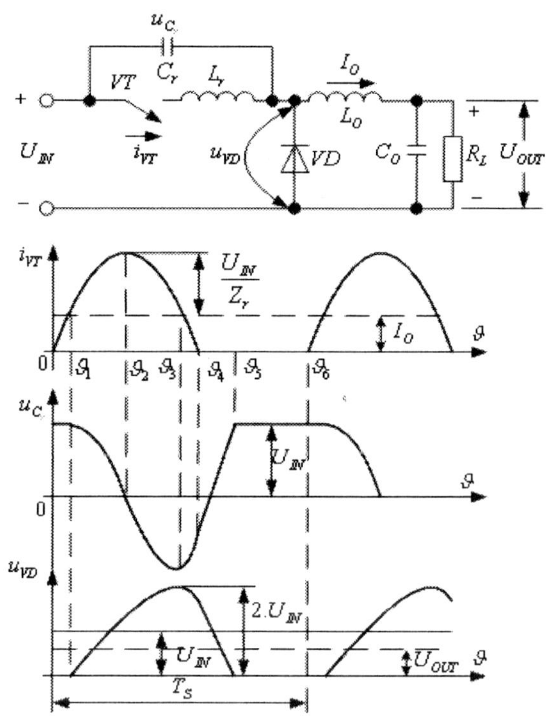

A disadvantage of this converter is the high maximum value of the transistor current increasing the output current, as well as the high value of the voltage across the diode increasing twice the input voltage value.

Figure 25 depicts the electrical schematic and waveforms of a resonant-switch DC/DC buck converter and ZVS.

Again, according to the abovementioned reasons, the whole circuit right from diode VD may be replaced with a current generator of value I_O, which stays constant during the commutation process.

The transistor VT is on before moment $\vartheta = 0$ conducting output current I_O, and the voltage across the diode VD is equal to the input voltage. At $\vartheta = 0$ the transistor turns off and capacitor C_r starts its charging by current I_O according to a linear law. The charge is rather slower than the time of the turning off of the transistor. Therefore, it may be assumed that the transistor turns off at zero voltage. At $\vartheta = \vartheta_1$ the voltage of the resonant capacitor reaches the value of the input voltage and the diode VD turns on. From this moment on the current I_O, as well as the resonant current of the series resonant tank C_r,L_r, flows through the diode. At $\vartheta = \vartheta_2$ the resonant current reaches zero, the voltage of the resonant capacitor – its maximum value increasing the value of the input voltage. Then the current through the resonant inductance changes its direction, the voltage across C_r reduces and at $\vartheta = \vartheta_3$ it reaches again the value of the input voltage. Stored energy in the inductance L_r continues to maintain the current through it, the voltage of the resonant capacitor decreases and at $\vartheta = \vartheta_4$ reaches zero. At this moment, the antiparallel diode VD_r of the transistor VT turns on and a pulse for turning on the transistor may be passed, nevertheless, it will start to conduct just at $\vartheta = \vartheta_5$ when the current through its antiparallel diode reduces to zero. Thus the transistor turns on at ZVS and ZCS. At the interval $\vartheta_5 - \vartheta_6$ the current through the inductance increases according to a linear law as it reaches the value of the output current at $\vartheta = \vartheta_6$ and it is the cause of the diode VD turning off. Then new turning off of the transistor may happens for example at $\vartheta = \vartheta_7$.

The average value of the voltage across the diode VD is separated as an output voltage across the load using the output filter L_O, C_O. The output voltage value can be changed by changing the switching frequency thence the interval during which the transistor is turn on changes.

As it is seen in the waveforms, the voltage across the diode VD does not increase the input one but the transistor voltage is higher than the input voltage with a value dependent on the output current.

There are others schematics of resonant DC/DC converters (Philips, 1991; Skvarenina, 2002) used to avoid the disadvantage of the studied topologies such as high maximum value of currents and voltages. The schemas and principles which have been studied till now give orientation in the field to increase efficiency coefficient of the DC/DC converters together with reducing their sizes and volume.

From the operational principle of resonant converters a conclusion is made that to regulate the output voltage it is not necessary to change the duty ratio as it is in the hard-switching converters with PWM but to change the frequency. This is also seen from a block schema shown in Figure 26. It contains the basic blocks of a controller family used to control such converters with ZVC or ZCS.

The purpose of the error amplifier EA is the same as already described in the converters with a hard-switching. A new element is VCO – voltage controlled oscillator, whose frequency depends on the output voltage of the error amplifier. The connected capacitor and resistors to the VCO determine

Figure 25. Electrical schematic and waveforms of a resonant-switch DC/DC buck converter and ZVS

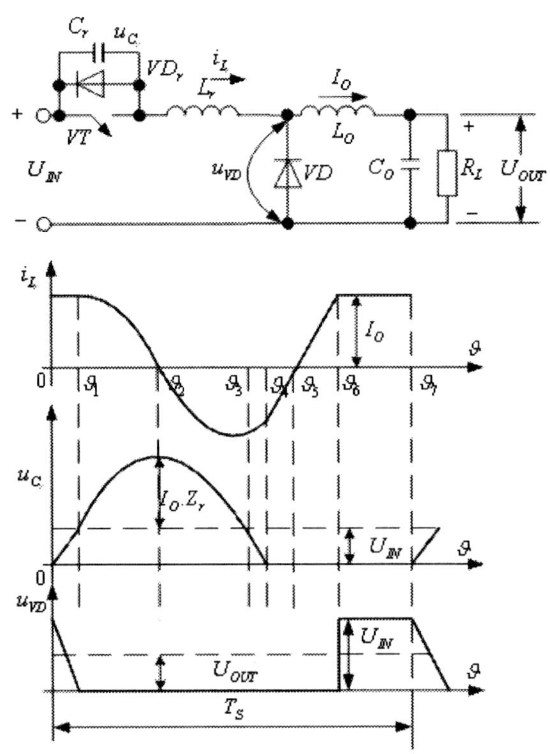

Figure 26. Block schema used to control resonant converters

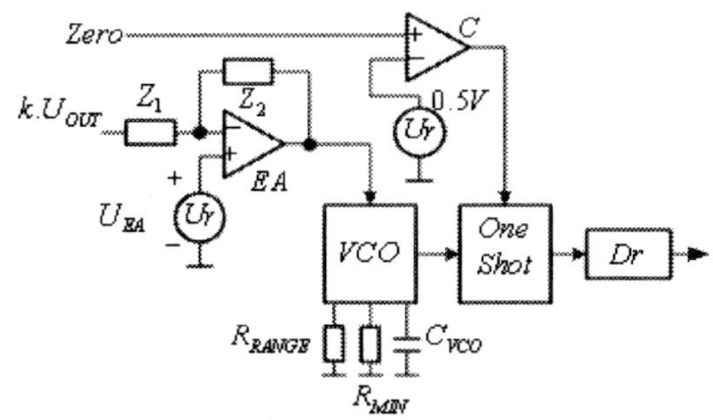

the minimum frequency of generation and the range of the change of the frequency. A special feature is the separated input Zero, used to monitor the current through or the voltage across the transistor of the converter. For example, when a mode of zero voltage switching is used, the one shot generator One Shot is programmed for a maximum time for the transistor turning off. This time can be changed by the output signal of the comparator C. The change of the output comparator signal depends on the voltage value across the switching device monitored by special circuits. The voltage across the transistor is compared to a threshold voltage usually equal to 0.5V in such a way that a delay between the moments of turning the comparator and of turning on the transistor to be guarantee.

The block schema in Figure 26 is also expanded with the characterizing blocks for the other control methods of the DC/DC converters such as: soft-start circuit, protections, generators of inner supplying voltages, etc.

A new method to increase energy efficiency in DC/DC converters is described from (Huang, 2007). This method is modification of the buck regulator and it is called Hyper Light Load[1]. “Hyper” refers to a very short time of a transient process under a load change, while “Light Load” refers to a high efficiency coefficient at low currents – even lower than 10mA.

REFERENCES

Andreycak, B. (1997). *Zero voltage switching resonant power conversion.* USA: Unitrode Application Note U-138.

Balosh, L. (1997). *The current - doubler rectifier: an alternative rectification technique for push-pull and bridge converters*. USA: Unitrode Design Note-63.

Billings, K. (1999). *Switchmode power supply handbook* (2nd ed.). New York: McGraw-Hill.

Chiu, H. J., & Lin, L. W. (2006). A bidirectional DC-DC converter for fuel cell electric vehicle driving system. *IEEE Transactions on Power Electronics*, *21*(4), 950–958. doi:10.1109/TPEL.2006.876863

Erickson, R., & Maksimovic, D. (2001). *Fundamentals of Power Electronics* (2nd Ed.). Norwell, MA: Kluwer Academic.

Gunningham, G. S., & Lixin Tang, J. P. (2007, June). *A reduced-part triple-voltage DC-DC converter for electric vehicle power management.* Paper presented at the Power Electronics Specialists Conference PESC 2007. USA/Orlando, Florida.

Huang, B. (2007, October). Beyond LDOs and switching regulators. *Bodo's Power Systems*, pp. 44-46.

Kazimierczuk, M. (2008). *Pulse-width modulated DC-DC power converters*. Chichester, UK: Wiley.

Kislovski, A. (1985). *Introduction to dynamic analysis of switching DC-DC converters.* Bern, Switzerland: EWV Engineering.

Luo, F., & Hong, Y. (2003). *Advanced DC/DC converters*. Boca Raton, FL: CRC Press.

Middlebrook, R. D., & Cuk, S. (1983). *Advanced in switched – Mode power conversion. Volumes I and II.* Irvine, CA: TESLAco.

Philips (1991). *Power semiconductor applications.* The Netherlands: Philips Export B.V.

Skvarenina, T. (2002). *The power electronics handbook.* New York: John Wiley & Sons.

Tolbert, L. M., Peterson, W. A., White, C. P., Theiss, T. J., & Scudiere, M. B. (2002). *Bi-directional DC-DC converter with minimum energy storage elements.* Paper presented at the Industry Applications Conference, 2002. 37th IAS Meeting. USA/Pittsburgh, PA.

Unitrode (1993). *High power factor preregulator using SEPIC converter.* Unitrode Power Supply Design Seminar.

Williams, B. W. (2008). Basic DC-to-DC converters. *IEEE Transactions on Power Electronics*, *23*(1), 387–401. doi:10.1109/TPEL.2007.911829

Zhang, F., Xiao, L., & Yan, Y. (2004, June). *Bi-directional forward - flyback DC-DC converter*. Paper presented at the Power Electronics Specialists Conference PESC 2004. Aachen, Germany.

ENDNOTE

1 Hyper Light Load is trademark of the Micrel, Inc.

Chapter 7
DC/AC Conversion

BASIC INDICATORS

The operational principle of converters from DC into AC electrical powers, so-called inverters, is based on the idea studied in Chapter 3. In this chapter, the autonomic or independent inverters are studied, whose operational frequency is determined in general by their control system (Bedford, 1964). These inverters can be separated in the following basic groups:

1. **Current-source inverters:** Their special feature is a presence of an inductance with a very high value connected in series with the DC source and thus the source become a current generator. As a result, the current through the devices and in the non-branched part of the output circuit have waveforms close to a square one. Voltage across the load is specified by its type and parameters.
2. **Voltage-source inverters:** Their name comes from the waveform of their output voltage – it has a specific form which depends only on the control algorithm of the power devices of the inverter circuit and on the type of the circuit at the side of the DC source. More often a capacitor with a high value is connected in parallel to the DC source and thereby the source is a voltage generator. Since a voltage with a specific waveform is applied to the inverter output, waveform of the inverter output current depends on type and parameters of the load.

DOI: 10.4018/978-1-61520-647-6.ch007

3. **Resonant inverters:** Their common feature is the use of resonance processes to switch the power devices. Dependent on the power device type and a topology of the power schema, switching at zero current or zero voltage may be realized decreasing the commutation losses and making the use of the resonant inverters in high frequency appropriate. The supplies of the resonant inverters are more often voltage-sources. The inverter load is a part of the resonance circuit, in which resonance processes develop. These processes are developed at the switching of the power devices and thus the DC source voltage appears to be applied to a series oscillating circuit causing the currents through the power devices and the load to have waveforms close sinusoidal.
4. **Z-source inverters:** Their special feature is a presence of inductances and capacitors connected in Z-circuit in the input circuit at the DC source side. A special control algorithm is used as a result of which the total source voltage applied to the inverter is higher than the voltage of the DC source.

Different power devices have implementation in inverter realizations. The relative part of the devices has been change during the time. Nowadays, thyristors are applied in the current-source inverters and resonant inverters in comparatively high powers. Power IGBTs are the basic devices applied in the voltage-source inverters and Z-source inverters. The IGBTs have also found a wider implementation in resonant inverters. Power MOSFETs are applied in high frequencies in the resonant inverters. The MOSFET implementation is more limited in the voltage-source inverters.

In general, a tendency of expanding the use of the transistors in DC/AC conversion regardless of the inverter applied is imposed. Therefore, the main attention in this chapter is paid to the transistor inverter circuits.

A fundamental indicator of a conversion is efficiency coefficient showing the ratio of the AC power in the load to the DC power consumed from the source:

$$\eta = \frac{P_L}{P_d} = \frac{\sum_{K=1}^{\infty} U_L^K . I_L^K . \cos\phi_K}{U_d . I_d} \tag{7.1}$$

where in U_L^K, I_L^K are effective values of the k-th harmonics of the load voltage and load current, respectively, and ϕ_K - is displacement angle between them. In this definition, it is assumed that in general both the load voltage and load current are of non-sinusoidal waveform. Applications and loads, which allow this assumption to be made, exist. In other cases, it is important to supply the load with a voltage whose waveform is as closed as possible to the sinusoidal one. Then only a multiplier of k=1 rest in the numerator of (7.1). This is a characteristic mostly of the voltage-source inverters. Therefore, it is usually necessary to separate only the fundamental harmonic from a non-sinusoidal voltage at the inverter output. In this case, the harmonic spectrum of the output inverter voltage is important. The harmonics following the fundamental are preferable to be of a higher possible frequency to simplify their filtering.

Additional important parameters of the inverters in particular applications may be indicators as volume, size, level of noise, electromagnetic radiations in the environment, etc.

SINGLE-PHASE AND THREE-PHASE CONVERTERS

Single-Phase Converters

Current-Source Inverter

Electrical schematic of a single-phase bridge current-source inverter implemented by thyristors, waveforms describing inverter operation and vector diagram for the first harmonics of inverter currents and voltages are presented in Figure 1. When thyristors *VS*1 and *VS*3 conduct after the commutation, the current flowing through them ($i_{VS1} = i_{VS3}$) is equal to the current flowing through the non-branched part of the output circuit *i*. Part of this current charges the capacitor i_C and the rest of it flows through the load i_L. At the beginning of the next half-period (in our case the negative half-period) for the current *i*, firing pulses are passed to turn on the thyristors *VS*2 and *VS*4. For a short time all four thyristors are on; the capacitor voltage is applied as a reverse voltage to *VS*1 (through *VS*2) and to *VS*3 (through *VS*4). Capacitor discharging current flows through the circuit including *VS*1 and *VS*2 in an opposite direction to the anode current of *VS*1, causing a turning off of the last thyristor. Analogical process is developed in the circuit including *VS*3 and *VS*4 leading to a turning off of *VS*3. Recharging of the capacitor begins through *VS*2 and *VS*4 with current $i_{VS2} = i_{VS4}$ and to a voltage whose polarity is opposite to this of the previous half-period. The current in the output circuit only changes its direction. It appears in Figure 1.b and Figure 1.c that the first harmonic of the output current leaves behind the output voltage at a leading angle, β which determines the circuit reverse recovery time of the thyristors t_{qc} ($\beta = \omega.t_{qc}$) if the commutation processes are neglected.

Therefore, part of the capacitor value has to compensate the inductive type of the load and the other part to secure reactive type of the output circuit and circuit recovery time t_{qc} for a pair of the diagonally connected thyristors that has been conducting till that moment. If this time is less than the required catalogue time for the chosen type of the thyristors - t_q, a breakdown regime happens. In this mode the four thyristors are on simultaneously and the inverter is a short - circuit for the supply source. The current through the thyristors in a steady state would be limited only by the parasitic resistance in the circuit and thus an irresistible break of the devices takes place. The velocity of the current increase is limited by the large value of the input inductance. To preserve the break in this breakdown mode, special methods and protection schemas are provided.

Using vector diagram for the first harmonics of the voltages and currents in Figure 1.c, basic characteristics connected with the inverter operation are derived:

$$P_d = P_L \tag{7.2}$$

$$U_d.I_d = U_L.I.\cos\beta \tag{7.3}$$

After transforming (7.3), it is found:

$$\frac{U_L.I}{U_d.I_d} = \frac{1}{\cos\beta} \tag{7.4}$$

Figure 1. A single-phase bridge parallel current-source inverter: a) electrical circuit; b) operational waveforms; c) vector diagram

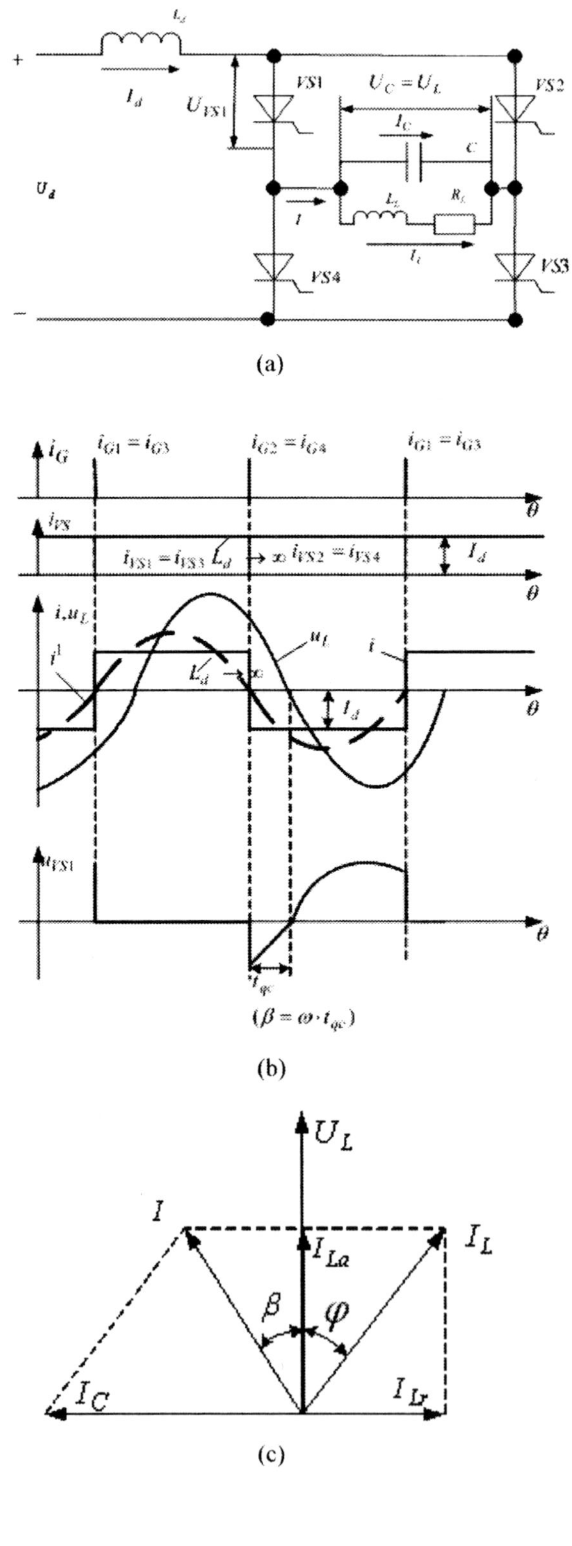

Figure 2. Block schemas illustrating a) independent excitation and b) load dependent excitation in current-source inverters

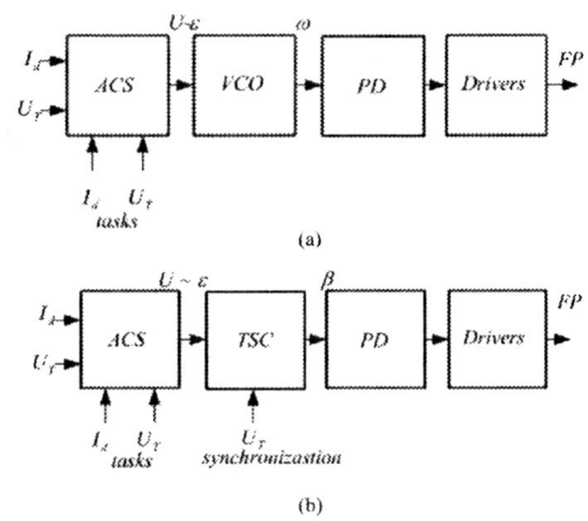

Since the current waveform in the inverter diagonal is a square one with a maximum value of I_d and I is the effective value of its first harmonic, then using Furrier series the following equation is valid:

$$\frac{I}{I_d} = \frac{2.\sqrt{2}}{\pi} = K_I \tag{7.5}$$

where in K_1 is an inverting coefficient. Its substitution in (7.3) leads to:

$$\frac{U_L.K_I}{U_d} = \frac{1}{\cos\beta} \tag{7.6}$$

Therefore, the output voltage can be regulated by changing either the supply voltage or the angle β.

Two methods for excitation exist in the thyristors current-source inverters – independent excitation and load dependent excitation or self-excitation. The methods are illustrated in Figure 2 by block schemas.

In both methods, it is possible to realize stabilization of some of variables – input current I_d or load voltage U_L. The type of the stabilizing variable depends on the particular purpose of the inverter. To be able to stabilize a variable, an automatic control system (ACS) is provided. In the independent excitation – Figure 2.a, the firing pulses are generated by a voltage controlled oscillator VCO. The control voltage is proportional to the error ε in ACS. Thus, the output parameter is inverter operational frequency ω, and a consequence from it is the angle β, the input current and the load voltage, which of course also depend on load parameters and power inverter circuit. In self-excitation – Figure 2.b, a synchronization to the load voltage (more rare to the current through the bridge diagonal) is required. The firing pulses are generated by so-called time set circuits, which assistance determines the angle β, which is an output parameter. The operational frequency ω, the input current and the load voltage that also depend on the load parameters and the power schema are consequences of the angle β. Pulse distributors to separate the series of the pulses in different channels for simultaneously turning on of the thyristors, as well as drivers used to form the required power of the firing pulses so the thyristors to be able to turn on, are needed in both methods. It is worthy to be mentioned, that the control systems with independent excitation have higher inertia which at a sharp change of the load may lead to break of the inverter operation and to turning on of the protection systems. The control systems with self-excitation are characterized with faster operation but with significantly more complex schema and technology. Special interest in them is the possibility to regulate and stabilize the load voltage within some range, without a need of regulation of the controlled rectifier to change the supply voltage U_d of the inverter, by regulation and stabilization of the angle β because of the equation (7.6).

Resonant Inverter

The electrical schematic of a resonant inverter is shown in Figure 3. The value of the input inductance L_C is lower than this in the current inverter. Here, the thyristors of the pairs of the diagonally connected devices $VS1, VS3$ or $VS2, VS4$ are simultaneously turned on to conduct. The control system secures the switching with a control frequency ω_C. The self-resonance frequency of the load circuit is $\omega_L \approx \frac{1}{\sqrt{L_L C_L}}$. Dependent on the ratio between the two frequencies, the inverter operates in resonance - $\omega_C = \omega_L$, inductive mode $\omega_C < \omega_L$ or capacitive mode - $\omega_C > \omega_L$. The load circuit is presented by different equivalent schematic in the different operational modes. Regardless the pair of the conducting thyristors, a resonance circuit of L_C, C_C and the equivalent load circuit is connected to the supply voltage. Let us mark the elements of this series formed circuit as *L*, *C* and R. The parameters of this circuit should satisfy the following equation for the condition of a oscillation mode to be valid:

$$R < 2 \cdot \sqrt{\frac{L}{C}} \qquad (7.7)$$

Figure 3. Electrical schematic of a bridge resonant inverter loaded with parallel resonance circuit

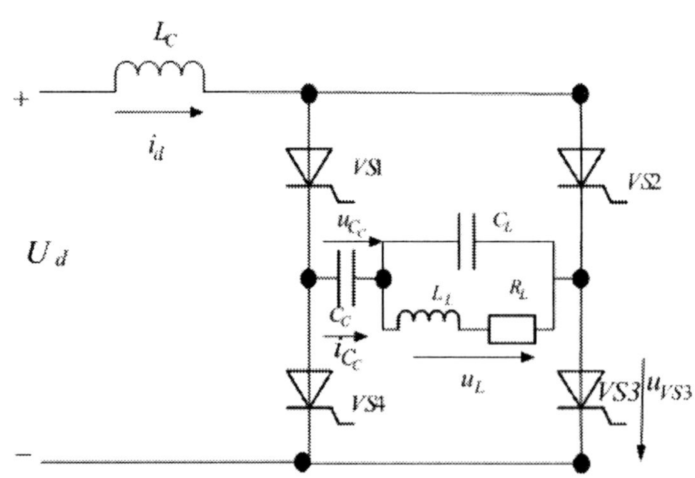

Figure 4 displays the waveforms illustrating the inverter operation in capacitive (Figure 4.a) and inductive (Figure 4.b) modes of the parallel load circuit.

Input current i_d with average value I_d is formed of the thyristor currents $i_{VS1} = i_{VS3}$ and $i_{VS2} = i_{VS4}$. The duration of each current pulse is $\frac{\pi}{\omega}$, $\omega_0 = \sqrt{\frac{1}{L.C}} - \delta^2$, after a pulse, a pause follows till the firing of the next thyristor pair. Therefore, the distance between the beginnings of the current pulses is $\frac{\pi}{\omega}$. The second waveform is the current through the bridge diagonal i_L, feeding the parallel oscillating circuit. Dotted line presents its first harmonic. An assumption for a sinusoidal waveform of the current pulses through the thyristors is usually made. Thus, the voltage across the series connected capacitor U_{C_C} changes according to a cosinusoidal law during the conduction intervals and it stays DC with its maximum value $U_{C_C \max}$ during the pauses. The waveform of the load voltage is illustrated at the fourth diagrams. In capacitive mode it lags in phase to the first harmonic of the load current (Figure 4.a) and in the inductive mode it goes ahead of the first harmonic (Chapter 4, Figurc 6).

The waveform of the thyristors in both modes is of interest. A special feature of series resonance circuit load is the change of the voltage across the power devices during the pauses $\theta_1 \div \theta_2$ and $\theta_3 \div \theta_4$. This happens because the load voltage changes at these moments. It appears in Figure 4 that in capacitive mode the transitory value of the load voltage at θ_3 is negative, while in inductive mode it may become positive. Therefore, in inductive mode, the voltage across the thyristhor, when a reverse voltage is ap-

Figure 4. Waveforms illustrating the inverter operation in two modes: a) capacitive mode; b) inductive mode

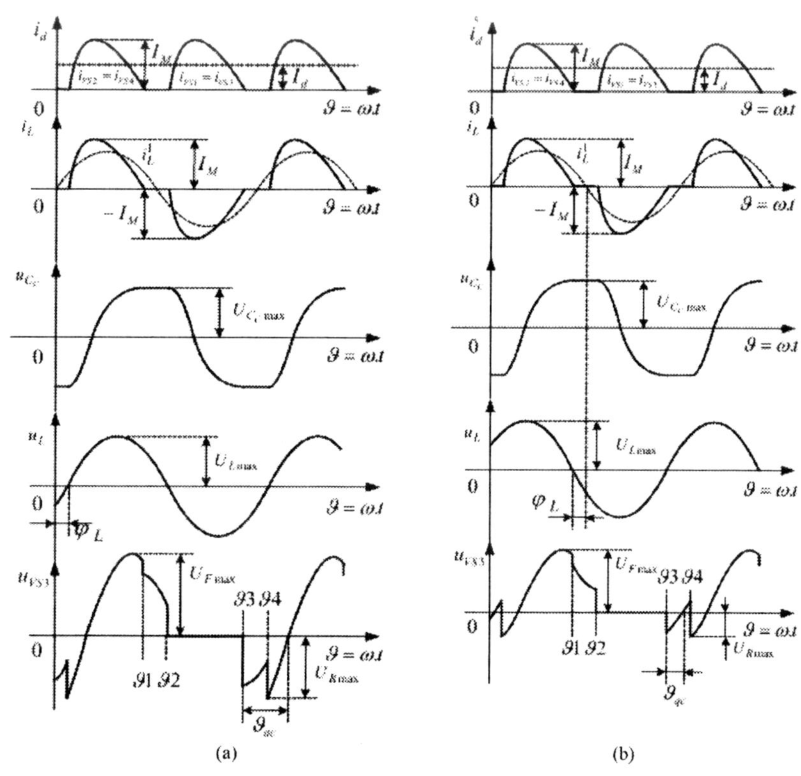

plied, is lower and also a possibility the voltage to become positive during the interval $\theta_3 \div \theta_4$ exists. This worse case is shown in Figure 4.b. Thus, the angle θ_{qc} and the reverse recovery time t_{qc} are reduced. In capacitive mode, the reverse recovery time increases but the thyristor voltages also increases, which is not appreciated because a possibility of break arises.

Voltage-Source Inverters

Different schemas of single-phase voltage-source inverters, as well as different algorithms for controlling the power devices and forming a waveform of the output voltage exist. They may be controlled and uncontrolled inverters dependent on the permission of the used algorithm to change the effective value of the load voltage. Several of the most widely spread voltage-source inverters are:

- Single-phase inverters with a square form of the output voltage;
- Single-phase inverters with sinusoidal pulse-width modulation of the output voltage;
- Single-phase inverters with selective elimination of harmonics in the waveform of the output voltage;
- Single-phase inverters with multi-leveled modulation of the output voltage – multilevel inverters;
- Resonant DC-link single-phase inverter.

The last mentioned inverters have not found a wide implementation and thereby they are not examined in this chapter.

Figure 5 displays a bridge electrical schematic of a single-phase voltage-source inverter, where Z_L is the impedance of the load. It may be of different type but it is more often of active-inductive type as it is additionally shown in Figure 5. This is because between points A and B, where the output voltage is obtained, a primary winding of a transformer or input of a smoothing filter to remove the higher harmonics of the output voltage are usually connected. Each pair of lower and upper transistors forms so-called phase-leg transistors and it is offered by the producer companies in a common body. Sometimes all four transistors are produced in a common body. The legs are marked as leg A and leg B. The transistor symbols include subscript of the leg to which it belongs and upperscript "+" or "-" indicating whether the transistor has a common point with the positive or the negative source lines. The gate signals to the transistors in a leg are inverted to one another because the two phase-leg transistors can not be on simultaneously. This would be a break-mode represented a short - circuit through the transistors for the supply source.

Figure 6 depicts the waveforms illustrating the schema operation in both cases – uncontrolled inverter Figure 6.a and controlled inverter with a waveform of the output voltage undependable on the load Figure 6.b. At the top of the load voltage, the power switches, which conduct during the corresponding intervals are marked. Now, the uncontrolled inverter will be studied. The transistors turn on two by two and diagonally connected ones. For example, when VT_A^+, VT_B^- are on, the voltage at point A regarding point B is positive and equal to the source voltage (the voltage drop across the transistors are neglected). The current flows from left to right through the load. At $\vartheta = \pi$ control pulses to turn off VT_A^+, VT_B^- and to turn on VT_B^+, VT_A^- are delivered. VT_A^+, VT_B^- turn off but because of the stored energy in the load inductance VT_B^+, VT_A^- do not immediately turn on. A certain time interval during

Figure 5. Bridge schema of a single-phase voltage-source inverter

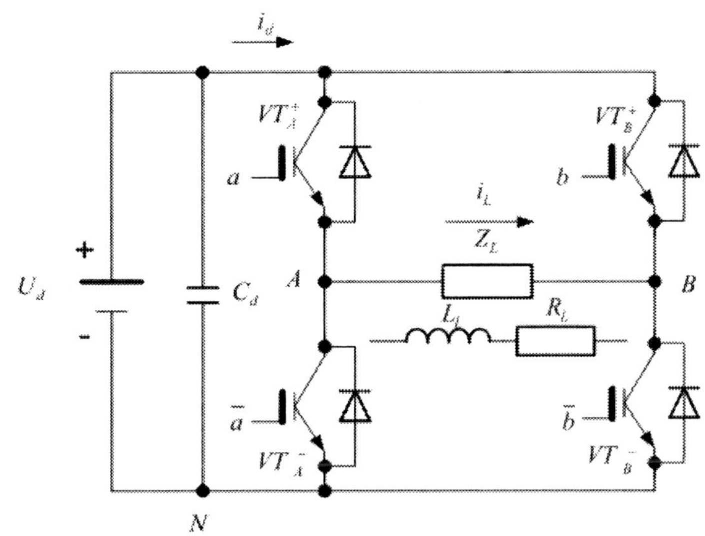

which the current maintain its direction is necessary. During it the diodes VD_B^+, VD_A^- conduct, the load voltage changes its sign and through the diodes the load energy is accumulated in capacitor C_d. Only when the load current decreases to 0, the transistors VT_B^+, VT_A^- begin to conduct (they have turning on pulses since $\vartheta = \pi$) and the load current begins to flow from right to left. Analogical processes are also repeated afterwards, i.e. before the next pair of the diagonally connected transistors star conducting, their antiparallel diodes conduct. The conducting intervals of the diodes depend only on the load parameters – time constant and value of the load current. Figure 6 displays the waveform of the current i_d, whose place in the inverter schema is marked in Figure 5. This current always flows from left to right when a pair of diagonally connected transistors conducts, and it is always flows from right to left when a diagonally connected diodes conduct. The average value of the current i_d is equal to the DC current consumed from the supply source U_d. Its value corresponds to the DC power from the source. If the load is active, then the waveforms of the current and the output voltage will be identical and the conducting intervals of the antiparallel diodes will miss. In practice, the ideal pure active load is hardly met. It appears from the schema operation that the output voltage value regulation can not be fulfilled in the chosen control algorithm for the power devices. The output voltage is equal either to $+U_d$ in the positive half-period or to $-U_d$ in the negative half-period.

The effective value of the output voltage for uncontrolled inverter can be derived from the graphic in Figure 6.a as:

$$U_L = \sqrt{\frac{1}{\pi}\int_0^{\pi} U_d^2 . d\vartheta} = U_d \tag{7.8}$$

The effective value of the first harmonic of its output voltage is:

$$U_L^1 = \frac{U_{LM}^1}{\sqrt{2}} = \frac{2.\sqrt{2}}{\pi} . U_d \tag{7.9}$$

Figure 6. Waveforms of a) uncontrolled inverter; b) controlled inverter

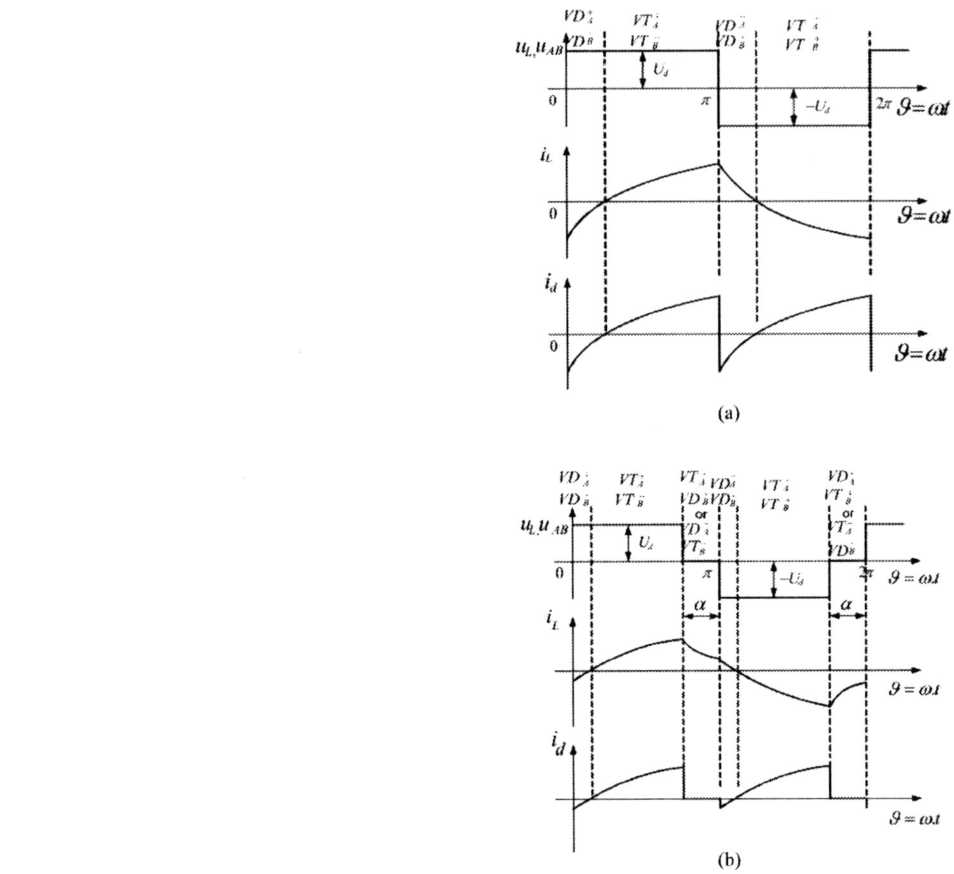

Single-phase uncontrolled inverter is applicable when changes of the output voltage caused by changes of the source voltage stay within a range, which is permitted for particular consumer. The voltage drops across the power devices, which increase with an increase of their current, as well as the load current, have been neglected in the already made study. Very often a range of the change of the source voltage is wide, for example accumulation batteries, fuel cells, photovoltaics, and causes high change in the output voltage values in a combination with a change of the load current. Thus, normal operation of the consumer will be disturbed and because of that a regulation and stabilization of the output voltage is required. The primary idea is to make a delay between a moment of turning off of a transistor pair and a moment of turning on the next pair, as it is shown in Figure 7. For example, if transistors VT_A^+, VT_B^- are turning off at $\vartheta = \vartheta 1$ then VT_B^+, VT_A^- are turning on at $\vartheta = \vartheta 2$. Thus a controlled voltage-source inverter with an output voltage waveform dependable on the load parameters is realized. As it appears, at every turning off of transistors, which have been conducting till the moment, antiparallel diodes of the other two transistors are turning on because of the stored energy in the load inductance. The output voltage alters its sing and stays in that state till the load current decreases to 0. These intervals depend only on the load parameters and in practice both harmonic spectrum and regulation are disturbed because the pauses in the waveform of the output voltage may not appear in particular loads.

Figure 7. Waveforms of the output variable dependent on the load

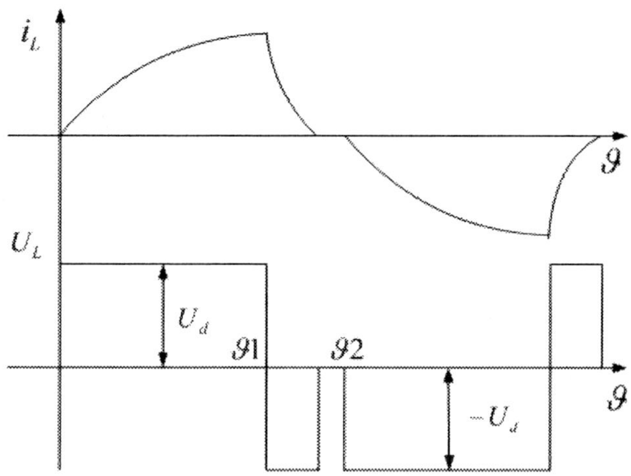

A controlled inverter with a waveform of the output voltage undependable on the load parameters is realized using a control algorithm for the power devices operating as it is illustrated in Figure 6.b. Duration of a pause, which is changeable to be able to implement the regulation, is marked as α. As it is seen form the figure, only one of the transistors of a diagonally conducting pair is turned off after their interval of conductivity. The load current flows through the continuing conducting transistor and the antiparallel diode of its neighbor in the other leg. The output voltage becomes equal to 0 and also a circuit for the load current is secured. At the passing of pulses to turn on the other pair diagonally connected transistors, the processes are analogical to those studied in the uncontrolled inverter.

The effective value of the output voltage may be derived from the waveform shown in Figure 6.b as:

$$U_L = \sqrt{\frac{1}{\pi}.\int_0^{\pi-\alpha} U_d^2.d\vartheta} = U_d.\sqrt{1-\frac{\alpha}{\pi}} \tag{7.10}$$

The effective value of the first harmonic of its output voltage is:

$$U_L^1 = \frac{U_{LM}^1}{\sqrt{2}} = \frac{2.\sqrt{2}}{\pi}.U_d.\cos\frac{\alpha}{2} \tag{7.11}$$

Two additional topologies of single-phase inverter with square form of the output voltage are shown in Figure 8. Figure 8.a displays a half-bridge electrical schematic, and Figure 8.b depicts the schema with a middle point of the output transformer or so-called push-pull schematic.

The half-bridge inverter schema contains only one transistor leg and the supply voltage is divided in half by the use of two capacitors. The purpose of the antiparallel diodes is the same as this in already studied bridge schematic. The transistors are turned on in sequence to form each of the half-period of the output voltage. In active-inductive load, when a control pulse to turn on a transistor is passed, its antiparallel diode conducts at first securing a return of the part of the reactive load energy on the corresponding

Figure 8. Schema topologies of single-phase voltage-source inverters: a) half-bridge and b) push-pull schematics

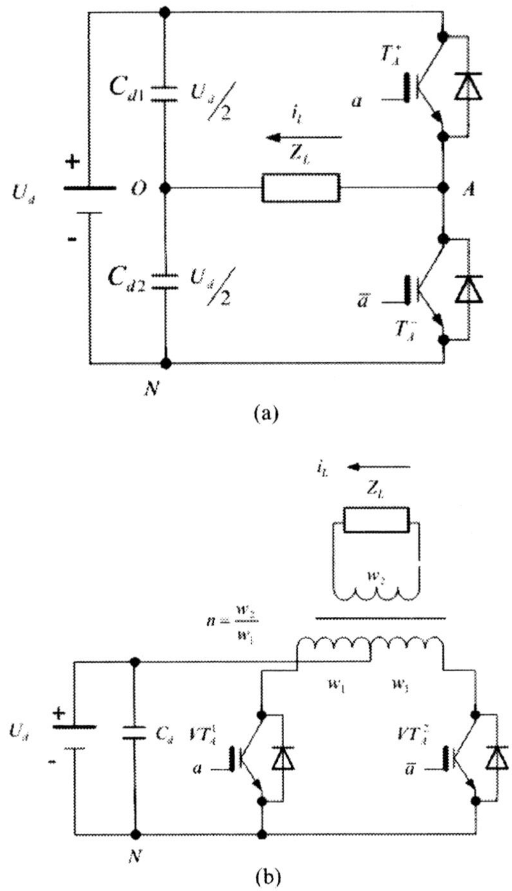

input capacitor. Only when the current decreases to 0, the transistor begins to conduct and the load current alters its direction. When an upper device VT_A^+ or VD_A^+ is on, the voltage of the upper capacitor C_{d1} is applied to the load with the positive terminal at the right end of the load. When a lower device VT_A^- or VD_A^- is on, the voltage of the lower capacitor C_{d2} is applied to the load with the negative terminal at the right end of the load. The effective value of the output voltage and its first harmonic are:

$$U_L = \sqrt{\frac{1}{\pi}\int_0^{\pi}\left(\frac{U_d}{2}\right)^2 .d\vartheta} = \frac{U_d}{2} \qquad (7.12)$$

$$U_L^1 = \frac{U_{LM}^1}{\sqrt{2}} = \frac{2.\sqrt{2}}{\pi}.\frac{U_d}{2}. \qquad (7.13)$$

In this schema a regulation of the form of the output voltage dependent on the load parameters can be realized. Also, a regulation undependable on the load parameters can be implemented if a bidirectional controlled device is connected in parallel to the load. The controlled device is turned on when a main transistor turns off and it maintains the same load current direction and zero voltage across it. In practice, the already studied bridge schematic is preferable.

The output transformer with a middle point is a special feature of the push-pull inverter schematic – Figure 8.b. The transistors turn on in sequence and thus the supply voltage is applied to each half of the primary winding. Therefore, the secondary voltage alters its sign each time and changes its value according to the corresponding transform ratio *n*. The effective value of the output voltage and its first harmonic are:

$$U_L = \sqrt{\frac{1}{\pi}\int_0^{\pi}\left(n.U_d\right)^2.d\vartheta} = n.U_d \tag{7.14}$$

$$U_L^1 = \frac{U_{LM}^1}{\sqrt{2}} = \frac{2.\sqrt{2}}{\pi}.n.U_d \tag{7.15}$$

The purpose of the antiparallel diodes is analogical to this in the already studied schematics. The diodes begin to conduct at the moments when control pulses for turning on the transistors are delivered. The transistors turn on only when the current through the antiparallel diodes decreases to 0. The above written about the regulation of the half-bridge schema is also valid here with the only difference that bidirectional controlled power switch has to be connected in parallel to the primary winding of the transformer. However, this schema is usually used to implement uncontrolled inverter. The maximum value of the voltage across each power device in the push-pull schema is equal to $2.U_d$. For example, when VT_A^1 turns on the sum of the supply voltage and the inductive voltage of the right half of the primary winding, also equal to the source voltage is applied to VT_A^2.

Low-number harmonics following the fundamental one – all odd harmonics 3,5,7… characterize the waveforms of the output voltage of the uncontrolled inverter and of the controlled inverter in фиг.7.6. This is disadvantage because the inverters require filters of a large volume and size.

A method of sinusoidal pulse-width modulation is used to eliminate the lower number harmonics and to improve the harmonic spectra (Mohan, 1995). This method is illustrated in Figure 9 for the half-bridge schema in Figure 8.a. Here, the modulation is bipolar because within a half-period the output voltage changes its polarity. Turning on and off of the power switches is based on a comparison in the control system of a control sinusoidal voltage u_C with frequency equal to the frequency of the first harmonic of the output voltage (with a period T_1) to a saw voltage u_{SAW} (with a period T_S), as it is shown in the figure. If the control voltage is higher than the saw one, the upper device turns on, otherwise the lower switch turns on.

When a switch is on, the load voltage is equal to the half of the source one. The regulation of the first harmonic value of the output voltage is obtained by changing the ratio of the amplitudes M_A, which can be between 0 and 1. For this purpose, the amplitude of the control voltage U_{C_M} is changed. The maximum value of the first harmonic of the output voltage is derived using Figure 9, as:

$$U_{LM}^1 = M_A.\frac{U_d}{2} \tag{7.16}$$

Figure 9. Waveforms explaining bipolar sinusoidal pulse-width modulation for the half-bridge schema in Figure 8.a

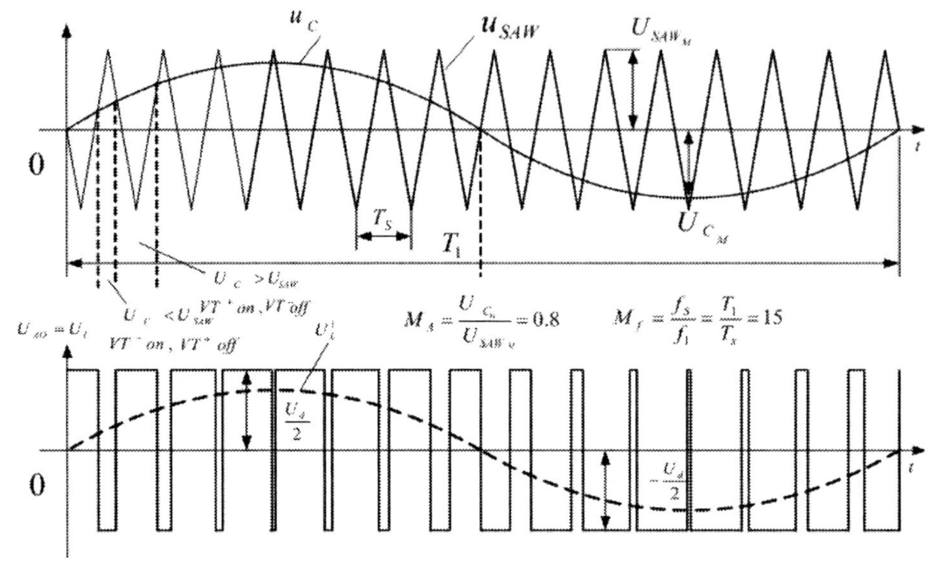

Therefore, the maximum value of the first harmonic of the output voltage depends on the value of the source voltage and the voltage amplitude ratio.

An important parameter determining the harmonic spectrum is the frequency ratio M_f, which must be an odd number in bipolar modulation. The harmonic spectrum depends on the coefficient M_f and it is presented in Figure 10 when $M_A = 0.8$. Lower number harmonics miss. It appears that with an increase of the switching frequency of the power devices (increase of M_f), the harmonics shift to the direction of higher frequencies, i.e. their filtering becomes easier.

Figure 10. Harmonic spectrum for bipolar sinusoidal pulse-width modulation

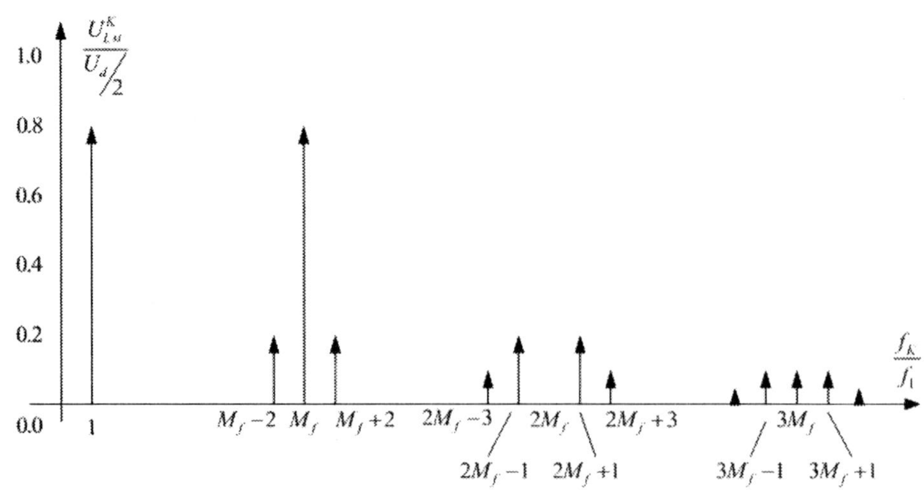

The harmonic values change with a change of M_A, i.e. at regulation of the output voltage. Therefore the shown ratios in Figure 10 are valid only when $M_A = 0.8$. The most influence has always the harmonic with frequency M_f. In this amplitude ratio, it is almost equal to the fundamental harmonic, but in a low values of M_A, it may appear higher in value than the fundamental harmonic. Bipolar sinusoidal pulse-width modulation can be also realized in the bridge schema in Figure 5. The operational algorithm is the following – when the control voltage is higher than the saw voltage VT_A^+ and VT_B^- turn on, otherwise VT_B^+ and VT_A^- turn on. When a pair of power devices is on the whole supply voltage is applied to the load. The maximum value of the output fundamental harmonic is:

$$U_{LM}^1 = M_A.U_d \qquad (7.17)$$

Further improvement of the harmonic spectrum can be achieved by using uni-polar modulation applied in the bridge voltage-source inverter circuit – Figure 11. Here, the coefficient M_f has to be an even number. The operational algorithm is: when the control voltage is higher than the saw voltage VT_A^+

Figure 11. Waveforms explaining unipolar pulse-width modulation for the bridge schema in Figure 5

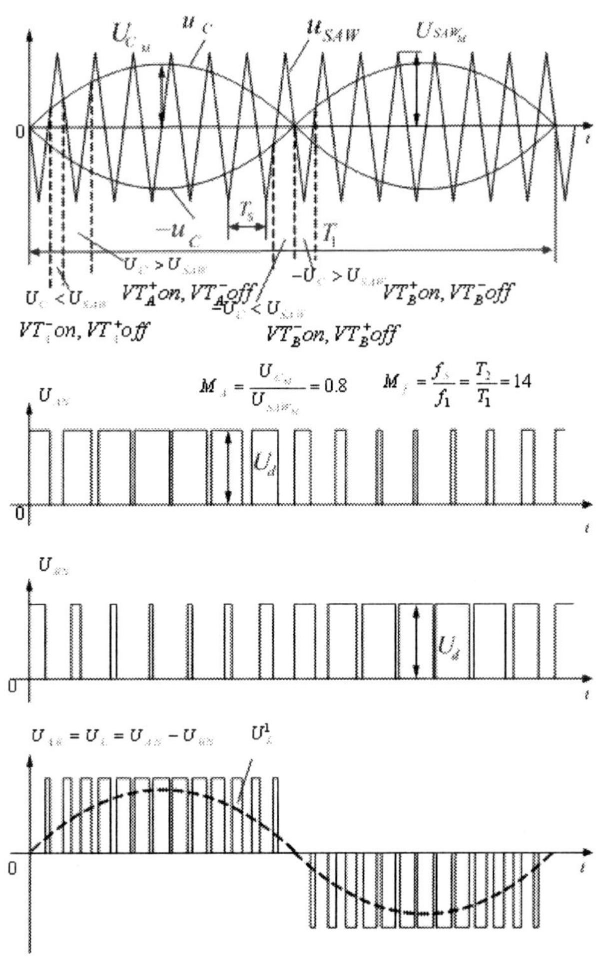

turns on, otherwise - VT_A^-. When the control voltage displaced at 180° is higher than the saw voltage VT_B^+ turns on, otherwise - VT_B^-. There are intervals during which the two upper or the two lower devices are simultaneously on. These are the intervals of the pauses in the output voltage when a power devise conducts with an antiparallel diode of the devise connected to the same pole of the source voltage dependent on the load current direction. The equation (7.17) for the maximum value of the first harmonic of the output voltage is also valid in this case.

Figure 12 displays the harmonic spectra for uni-polar sinusoidal PWM. It appears in Figure 12 that the harmonics are shifted to the higher frequencies and thus their filtering is simplified. Compared to Figure 10, a frequency bandwidth around M_f, $3.M_f$, etc, misses.

The method of a selective elimination of harmonics will be examined for the bridge schema in Figure 5 (Chiasson, 2004). The method is designed to eliminate bad harmonic spectrum of the square waveform of the output voltage and high commutation losses of the sinusoidal pulse-width modulation. The purpose of this method is to obtain comparatively good harmonic spectrum of the output voltage using comparatively low commutations and some of lowest odd harmonics typical for the square waveform of the output voltage to be selectively removed during this commutations – see Figure 13.

Figure 13 depicts the waveform of the load voltage. The commutation number and their angles are chosen in such a way that specific harmonics of the waveform of the output voltage to be eliminated. Three cuts into negative directions are made at the positive half-period – 1, 2, 3. Analogical, three cuts are made into positive directions at negative half-period - 4, 5, 6. The fifth and seventh harmonics of the output voltage waveform are eliminated using three cuts in a half-period. Using five cuts in a half-period – the fifth, seventh, eleventh and thirteenth harmonics are eliminated. This method may be applied also in three-phase inverters. As it will be clarified at the study of the three-phase inverters, the third harmonic and harmonics divisible to three are mutually compensated and do not exist in the output voltages of the three-phase inverter because of the properties of a symmetric three-phase system. A special ratio among the angels ϑ_1, ϑ_2, ϑ_3 in Figure 13 has to be observed for the so-studied harmonic spectrum at regulation of the output voltage to be kept the same. Figure 14 explains this ratio. The ratio of the maximum value of the first harmonic to its maximum possible value is put at X-axis in Figure 14.

Figure 12. Harmonic spectrum for uni-polar sinusoidal pulse-width modulation

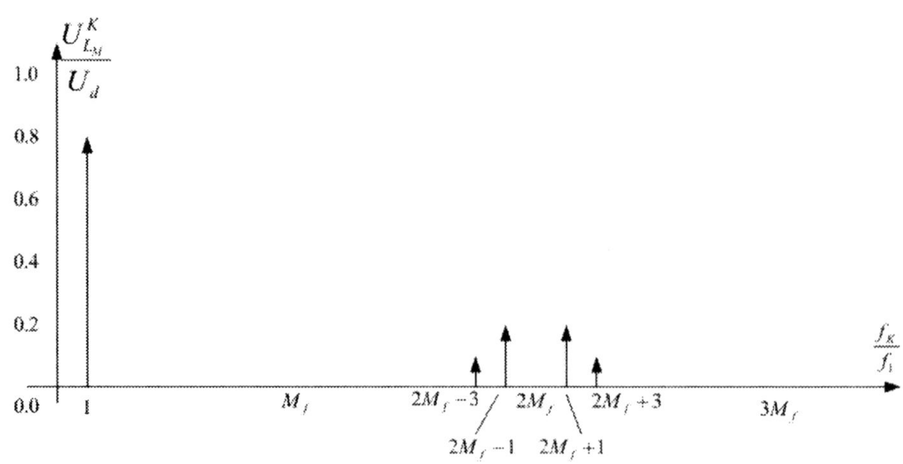

Figure 13. Waveform of the load voltage in selectively harmonic elimination

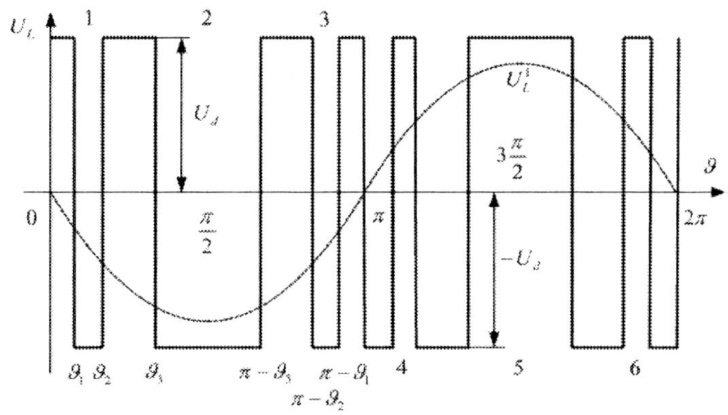

The maximum possible value of the first harmonic is:

$$(U_L^1)_M = 1{,}188.U_d \tag{7.18}$$

Disadvantage of the method of the selective elimination of harmonics is that different ratio among the angles (see Figure 14) is required in order to maintain the harmonic spectrum constant at every change of the output voltage. Contemporary control systems implemented on the basis of microcontroller or programming logic allow removing this disadvantage. Also, artificial neuron nets are used (Haque, 2005).

Another method to improve harmonic spectra in voltage-source inverters is modulation of the output voltage in several levels (Skvarenina, 2002). The electrical schematic of a single-phase inverter with modulation of the output voltage in several levels is shown in Figure 15. Their number is five - $(-U_d)$, $(-\frac{U_d}{2})$, 0, $\frac{U_d}{2}$, U_d. Figure 16 depicts the waveform of the output voltage across Z_L.

Disadvantage of the method of the modulation of the output voltage in several levels is the increased number of power devices and because of that it has not found a wide implementation in

Figure 14. Change of the output voltage dependent on angles $\vartheta_1, \vartheta_2, \vartheta_3$ shown is Figure 13

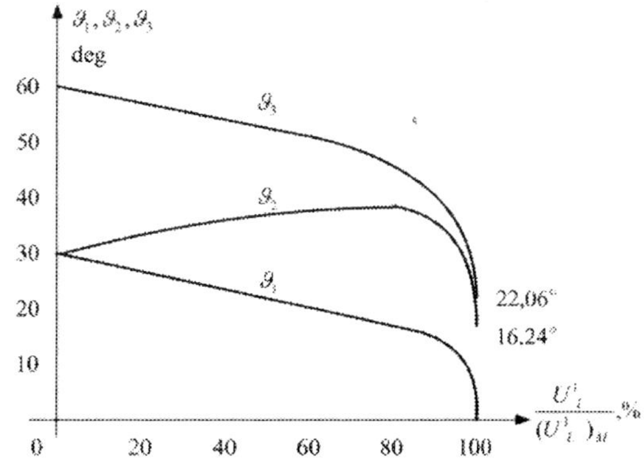

Figure 15. Bridge schema of a single-phase voltage-source inverter with five levels of modulation and limiting diodes

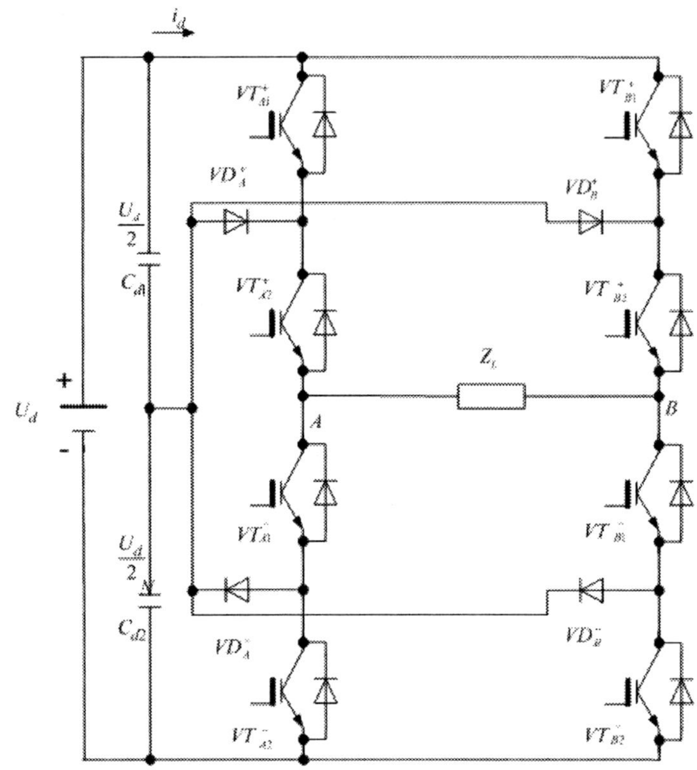

single-phase inverters. The implementation of the method is wider in three-phase inverter schemas and in higher powers.

Three-Phase Converters

Figure 17 displays an electrical schematic of a three-phase voltage-source inverter containing 6 transistors; their antiparallel diodes are also presented. The legs are leg R, leg S and leg T corresponding to the common point of the two transistors. The transistor symbols are as those in the single-phase inverters - symbols include subscript of the leg to which it belongs and upperscript "+" or "-" indicating whether the transistor has a common point with the positive or the negative line with the source. Some producers offer the six transistors in the same body. The control signals of the transistors are marked with low letters corresponding to the leg. According to the control algorithm the following inverters can be realized by the shown schematic:

- Three-phase voltage-source inverters with a square waveform of the output voltage;
- Three-phase voltage-source inverters with sinusoidal pulse-width modulation of the output voltage;
- Three-phase voltage-source inverters with a space-vector pulse-width modulation of the output voltage;

Figure 16. The output voltage of the inverter shown in Figure 12

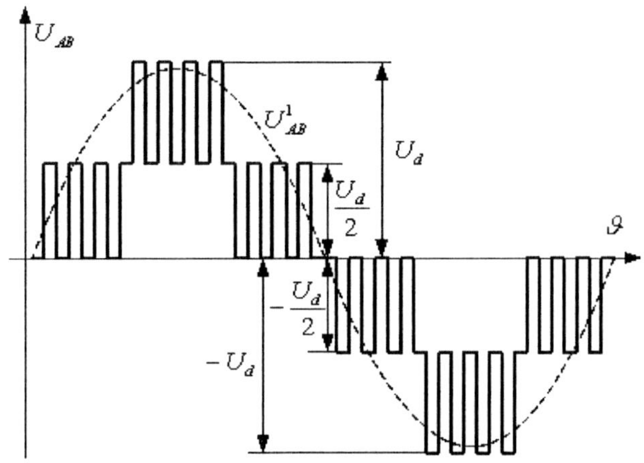

- Three-phase voltage-source inverters with hysteresis-current control;
- Using a change in the power schema or a combination of several schemas, three-phase voltage-source inverters with modulation of the output voltage in several levels is implemented – multi-level inverters.

Three-phase inverters with square waveform of the output voltage. Let r, s, t as binary variables are the control signals for the upper transistors. When a signal is 1 than the corresponding transistor is on,

Figure 17. Electrical schematic of a three-phase voltage-source inverter

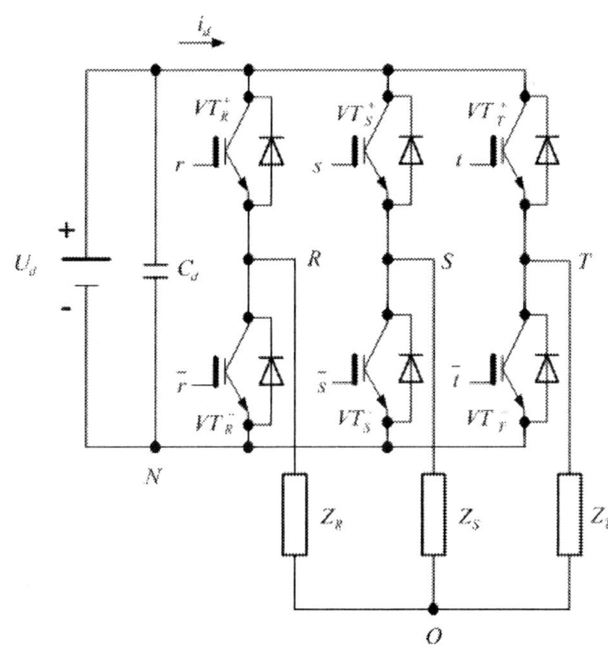

and when a signal is 0 – the transistor is off. Since it is impossible both phase-leg connected transistors to be simultaneously on, because this would be a break-mode represented a short-circuit through the transistors to the supply source, $\bar{r}, \bar{s}, \bar{t}$ are the signals for controlling the lower transistors. Of course, the logical levels of 1 and 0 correspond to the required voltage values to turn on and to maintain the off states of the transistors – they are usually +15V and -5V. Possible combinations of these three binary variables are 8, but two of them are not used in the case of a square waveform of the output voltage. These combinations are 111 – all upper transistors are on, and 000 – the three lower transistors are on.

Figure 18 depicts the six possible combinations of the control signals. Also, the first harmonics of the corresponding phase and phase-to-phase voltages forming a symmetrical three-phase system are shown.

It has to be mentioned that the waveforms of the phase voltages regarding point O can be monitored only if the load is with a taken out star center, which is not available in all loads.

Figure 19 shows the equivalent connection schemas of the three-phase loads to the source voltage corresponding to the six possible combinations of the control signals. Taking in mind that the load is symmetrical, the values of the phase voltages in each interval are easily found from the equivalent schematics. For example, in a combination 001 the voltage across Z_R is $\frac{2}{3}.U_d$, and the voltages across Z_S and

Figure 18. Voltage waveforms for the schema in Figure 17

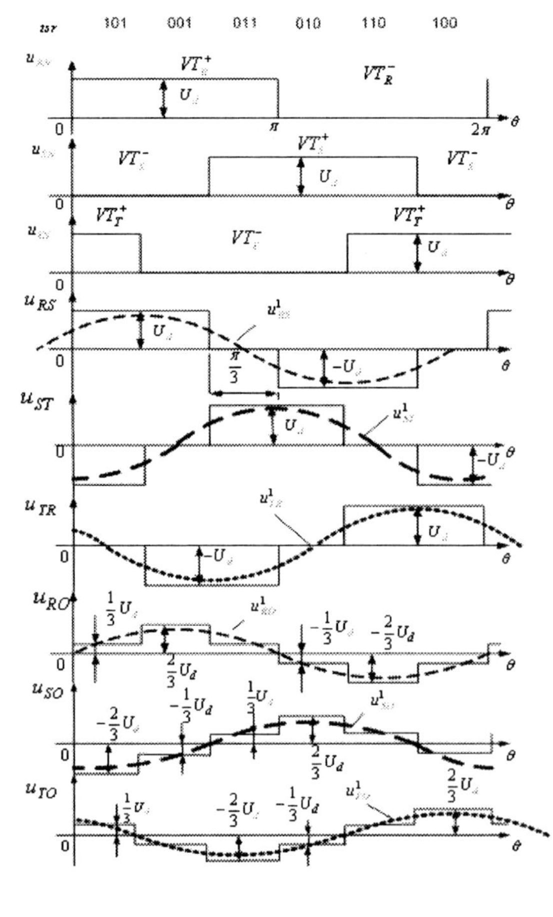

Z_T are $(-\frac{1}{3}\cdot U_d)$. If only a voltage across particular loads is studied consequently in all intervals, then the corresponding waveforms in Figure 18 can be obtained. The maximum value of the first harmonic of the phase-to-phase voltage can be derived as:

$$U^1_{ST_M} = \frac{4}{\pi}.U_d.\cos\frac{\pi}{6} = \frac{4}{\pi}.\frac{\sqrt{3}}{2}.U_d = 1,1.U_d \qquad (7.19)$$

The maximum value of the phase voltage is:

$$U^1_{RO_M} = \frac{U^1_{RS_M}}{\sqrt{3}} = \frac{4}{\pi}.\frac{\sqrt{3}}{2}.\frac{1}{\sqrt{3}}.U_d = \frac{2}{\pi}.U_d \approx 0,637.U_d \qquad (7.20)$$

Based on the three phase voltages a space voltage vector can be defined. It is made by multiplying each phase voltage by an operator giving its space orientation. The space voltage vector is:

$$\vec{V}_S(t) = V_{RO}(t).e^{jo} + \vec{V}_{SO}.e^{j\frac{2\pi}{3}} + V_{TO}.e^{j\frac{4\pi}{3}} \qquad (7.21)$$

Figure 19. Equivalent schematics of a three-phase load for the schema in Figure 17

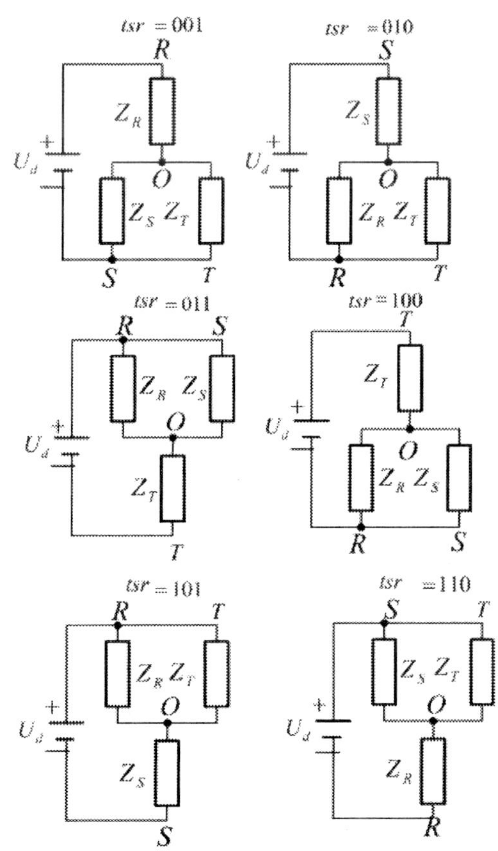

It is known that:

$$V_{RO} = V_{RN} + V_{NO}$$

$$V_{SO} = V_{SN} + V_{NO}$$

$$V_{TO} = V_{TN} + V_{NO} \tag{7.22}$$

The space voltage vector can be also defined on the basis of the three voltages in respect to the minus of the supply source point N after substituting (7.22) in (7.21) and taking in consideration that $e^{j0} + e^{j\frac{2\pi}{3}} + e^{j\frac{4\pi}{3}} = 0$:

$$\vec{V}_S(t) = V_{RN}(t).e^{j0} + V_{SN}e^{j\frac{2\pi}{3}} + V_{TN}.e^{j\frac{4\pi}{3}} \tag{7.23}$$

The values of the space voltage vector in the possible combinations of the control signals of the upper transistors are put in Table 1. As it has already been mentioned the first and the last combinations are not used in the inverters with square waveform of the output voltage. Thus, the space vector always takes one of the set six positions and its apexes form a regular hexagon shown in Figure 20. The module of the vector is always equal to the source voltage:

$$\left|\vec{V}_S(t)\right| = U_d \tag{7.24}$$

The maximum value of each phase voltage is equal to $\frac{2}{3}$ of the space vector module or it is $\frac{2}{3}U_d$. This coincides with the waveforms in Figure 18.

In symmetrical systems the harmonics divisible by 3 are mutually compensated, thus they are missing in Figure 21 that displays the harmonic spectrum of the phase-to-phase voltage of the three-phase inverter with square waveform of the output voltage. The harmonic numbers in the figure are determined in accordance with:

Figure 20. Positions of the space vector in the case of a square shape of the output voltage

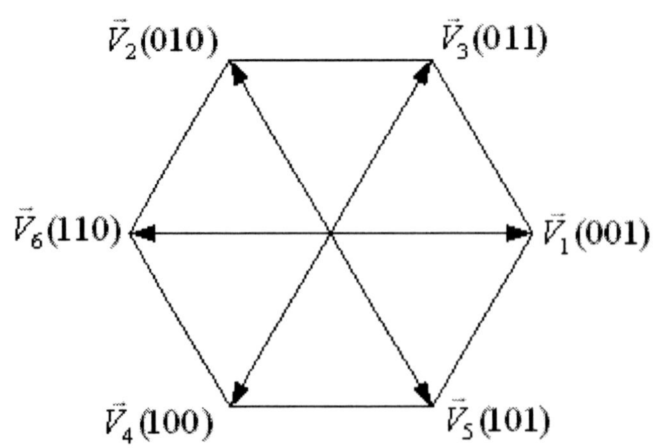

$$k = 6.n \pm 1, \; n = 1, 2, 3 \ldots\ldots \tag{7.25}$$

The maximum and effective values of the harmonics are:

$$U^K_{AB_M} = \frac{U^1_{AB_M}}{k} \qquad U^K_{AB} = \frac{U^1_{AB}}{k} \tag{7.26}$$

Figure 21. Harmonic spectrum in the case of a square shape of the output voltage of a three-phase voltage inverter

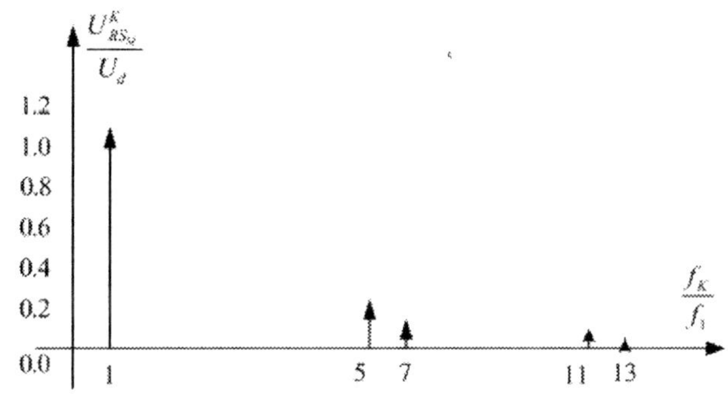

Table 1.

Control variable			Basic vector	Value
t	s	r		
0	0	0	$\vec{V}_0(000)$	0
0	0	1	$\vec{V}_1(001)$	$U_d.e^{j0}$
0	1	0	$\vec{V}_2(010)$	$U_d e^{j\frac{2\pi}{3}}$
0	1	1	$\vec{V}_3(011)$	$U_d e^{j\frac{\pi}{3}}$
1	0	0	$\vec{V}_4(100)$	$U_d e^{j\frac{4\pi}{3}}$
1	0	1	$\vec{V}_5(101)$	$U_d e^{j\frac{5\pi}{3}}$
1	1	0	$\vec{V}_6(110)$	$U_d e^{j\pi}$
1	1	1	$\vec{V}_7(111)$	0

The method of the sinusoidal pulse-width modulation illustrated in Figure 22 is applied to improve the harmonic spectra of the output voltage of the electrical schematic in Figure 17. The saw voltage with a high frequency is compared to three control voltages displaced at 120° to each other (one for a phase) with a frequency of the first harmonic of the output voltage.

When the control voltages U_{CA}, U_{CB}, U_{CC} are higher than the saw voltage, the upper transistors of the corresponding phase turn on and when they are lower – the lower transistors turn on. In these inverters, the first and the last combination of the control signals are used – 000 – pulses to turn on the three lower transistors and 111 – pulses to turn on the three upper transistors. Figure 22 also depicts the voltages in points A and B in accordance with the minus of the supply source point N, as well as the phase-to-phase voltage found as a difference between the two phase voltages. There are three types of conductivity of the power devices for the three-phase inverters with a sinusoidal pulse-width modulation and active-inductive load. The two of these types are as in the inverters with a square waveform of the output voltage. The first is when only transistors conduct – the current i_d flows from left to right and active energy is consumed from the supply source. The second is when antiparallel diodes conduct – the current i_d flows from right to left and reactive energy is passed back to the supply source. The third is when all phase voltages in accordance with point N are equal to U_d or 0. In this conductivity type, simultaneously conduct only devices connected to the positive pole of the source or only devices connected to the negative pole of the source. Thus the current i_d is 0, therefore reactive power of the load circulates in the inverter schema.

Figure 22. Waveforms of a sinusoidal pulse-width modulation in a three-phase voltage-source inverter

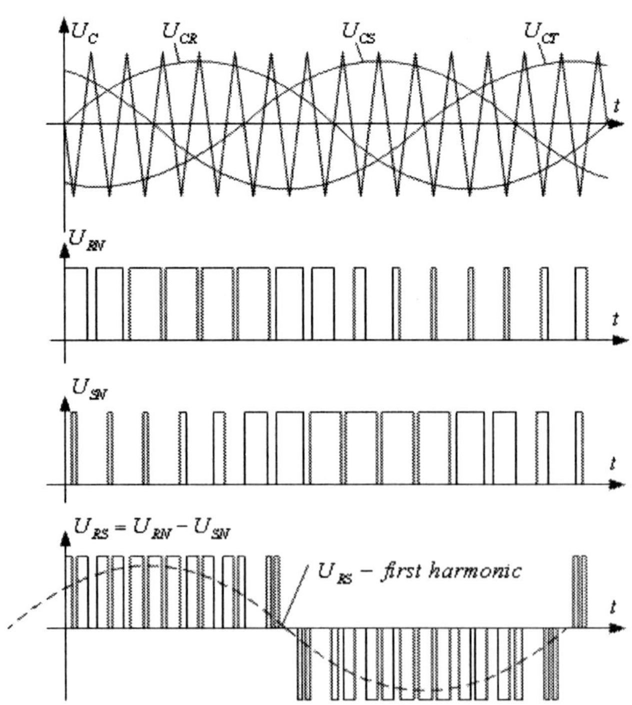

Figure 23. Positions of the space vector in sinusoidal pulse-width modulation of the output voltage

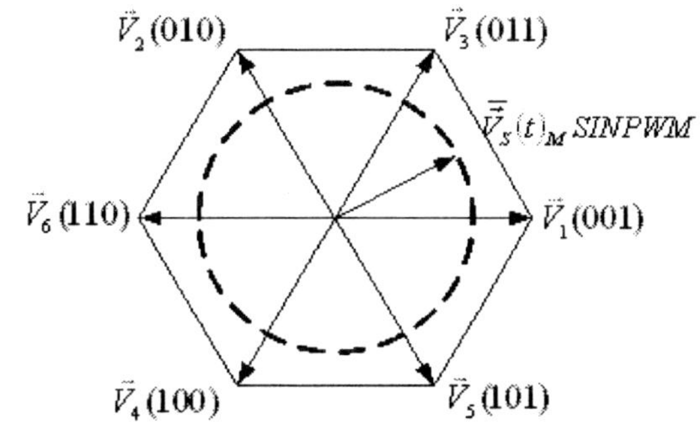

Similar to the single-phase inverter with sinusoidal pulse-width modulation, here, also the coefficient M_f has to be odd and moreover divisible to 3. Therefore, the harmonic spectrum of each voltage of point A and point B are identical to those of the single-phase inverter with bipolar pulse-width modulation – Figure 10.

The maximum value of the first harmonic of the phase voltages in accordance with point N is:

$$U^1_{RN_M} = M_A \cdot \frac{U_d}{2} \text{ and } U^1_{SN_M} = M_A \cdot \frac{U_d}{2} \tag{7.27}$$

The first harmonics are displaced at 120°. The effective value of the first harmonic of the phase-to-phase voltage is:

$$U^1_{RS} = \frac{\sqrt{3}}{\sqrt{2}} .M_A \cdot \frac{U_d}{2} = 0.612.M_A.U_d \tag{7.28}$$

Since the switching is with a high frequency, the space vector rotates in spite of setting in one of the fixed six positions.

Rotating, the apex of the space vector forms a circumference shown in Figure 23. In the figure the hexagonal shape of the output voltage is also shown. The value of the space vector for the six fixed position in square waveform of the output voltage of inverter is U_d, and in sinusoidal pulse-width modulation it is $\frac{3}{4}.U_d$.

Space-vector pulse-width modulation of the output voltage is realized using the basic schematic in Figure 5 applying an appropriate control. The voltage space vector is defined by its average value in a switching period for the power devices T_S:

$$\vec{\bar{V}}_S(t) = \bar{V}_{RN}(t).e^{j0} + \bar{V}_{SN}(t).e^{j\frac{2\pi}{3}} + \bar{V}_{TN}(t).e^{j\frac{4\pi}{3}} \tag{7.29}$$

The six vectors corresponding to the voltage space vector positions in square waveform of the output voltage form six sectors shown in Figure 20. Also, the combination of the control variables corresponding to zero vector $\vec{V}_0(000) = 0$ and $\vec{V}_7(111) = 0$ are used. See Figure 24.

Figure 24. Explanation of the principle of the space-vector pulse-width modulation of a three-phase voltage-source inverter

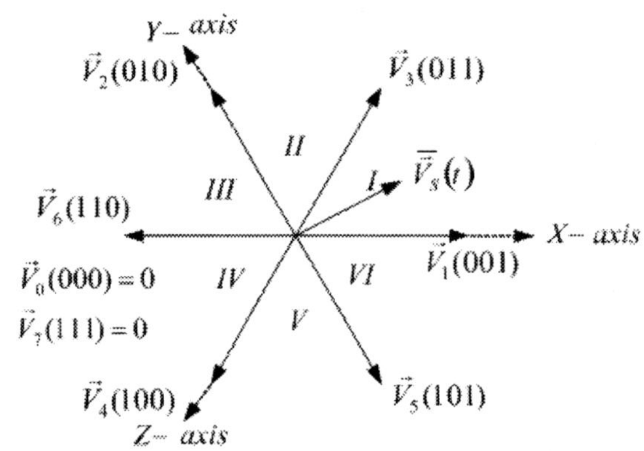

Let us mark as M_f the ratio of the period of the first harmonic of the output voltage to the switching period:

$$M_f = \frac{T}{T_S} \tag{7.30}$$

Then, the space vector will take positions over $\frac{2\pi}{M_f} rad$. For a sector the positions will be over $\frac{2\pi}{6.M_f} rad$. How to determine the module and phase of the space vector in each sector is explained in Figure 25 for sector *I*.

The combinations of the control variables corresponding to the basic vectors $\vec{V}_1$ and $\vec{V}_3$ bordering the sector are applied for times $x.T_S$ and $y.T_S$, respectively. The combinations of the control variables

Figure 25. Explanation of determining the module and phase of the space-vector

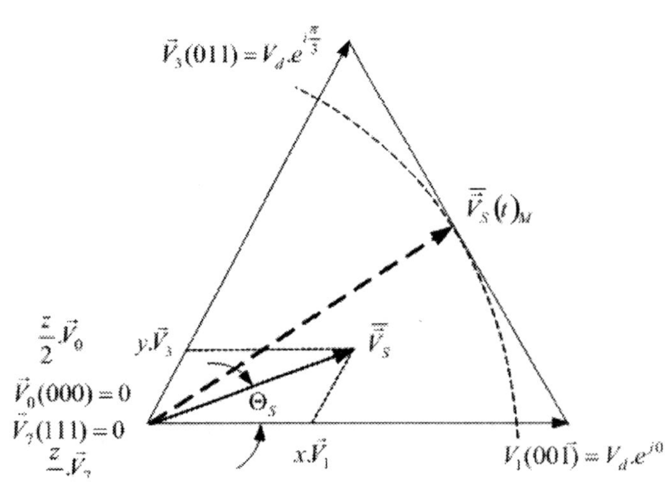

corresponding to $\vec{V}_0$ and $\vec{V}_7$ are applied for $\frac{z}{2}.T_S$ each. The following equation is valid:

$$x + y + z = 1 \tag{7.31}$$

The average value of the space vector for interval T_S is:

$$\vec{\bar{V}}_S = \frac{1}{T_S} \int_{nT_S}^{(n+1)T_S} \vec{\bar{V}}_S(t).dt = \frac{1}{T_S}\left(x.T_S.\vec{V}_1 + y.T_S.\vec{V}_3 + z.0\right) = x.\vec{V}_1 + y.\vec{V}_3 \tag{7.32}$$

After substituting the basic vectors with their values in Table 1, it is found:

$$\vec{\bar{V}}_S = V_d.\left(x.e^{j0} + y.e^{j\frac{\pi}{3}}\right) \tag{7.33}$$

It appears from the equations and in Figure 25 that changing $z = 1 - (x + y)$ will change the module of the vector and through changing the ratio $\frac{y}{x}$ - its phase angle Θ_S will change.

It is obvious that the space vector has a maximum value in $x + y = 1$. Figure 25 depicts its position in $x = y = \frac{1}{2}$. After substituting in (7.33) and transformation, the maximum value is found as:

$$\vec{\bar{V}}_S(t)_M = \frac{\sqrt{3}}{2}.U_d \tag{7.34}$$

The apex of this maximum in value space vector will trace a circumference which tangents in each sector are the lines connected the apexes of the basic vectors of the corresponding sector. So-formed circumference is shown in Figure 26.

This circumference is the inscribed one in the regular hexagon which apexes are the apexes of the basic vectors or the positions of the space vector in the square waveform of the output voltage.

Figure 26. Positions of the space vector in space-vector pulse-width modulation of the output voltage

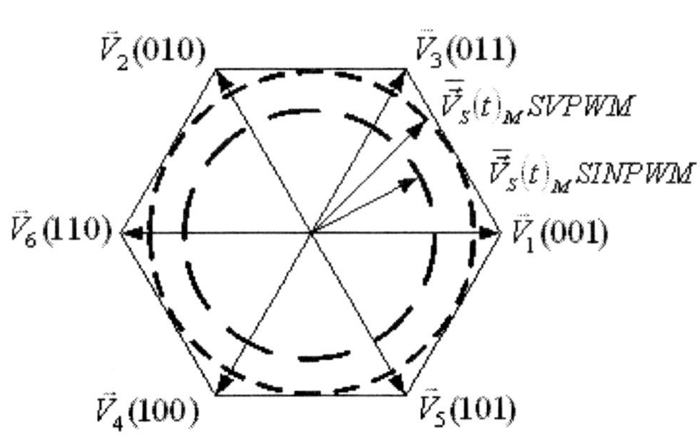

The maximum value of each phase voltage is $\frac{2}{3}$ of the value of the space vector or from (7.34) it is found:

$$\bar{V}_{ROM} = \frac{1}{\sqrt{3}}.U_d \tag{7.35}$$

Since the maximum value of the space vector in sinusoidal modulation is $\frac{3}{4}U_d$, using the space-vector pulse-width modulation secures 15% better use of the supply voltage.

In practice values of *x* and *y* are found corresponding to a sequence of values of Θ_S. The following equations are derived using the triangle in Figure 22:

$$x = \frac{\bar{\vec{V}}_S.2}{\sqrt{3}}.\cos\left(\Theta_S + \frac{\pi}{6}\right)$$
$$y = \frac{\bar{\vec{V}}_S.2}{\sqrt{3}}\sin\Theta_S \tag{7.36}$$

If a maximum value of the space vector is maintained and taking in consideration from (7.31) that it referees to the value of the supply voltage of $\frac{\sqrt{3}}{2}$, then from (7.36), it is found:

$$x = \cos\left(\Theta_S + \frac{\pi}{6}\right)$$
$$y = \sin\Theta_S \tag{7.37}$$

Shifting consecutively from sector into sector it is known which basic vectors corresponds to *x* and *y*. Calculation and forming of the control signals in real time is made using appropriate software based on digital signal processors. Therefore, it is very convenient that the component *x* to be identify as axis *r*, *s*, *t* in series.

In the three-phase inverters with a hysteresis-current control the transistor control is realized on the base idea explained in Figure 27 (Mohan, 1995). The purpose in this control is to obtain sinusoidal currents in the three-phases. The basic part of the control system for the transistors in phase R is shown. The control channels for the other two phases contain the same blocks. Three sine waves forming a symmetrical three-phase system with a frequency equal to the fundamental harmonic are passed to the corresponding channels as reference sinusoidal waveforms. The sine wave i_{Rref} is the reference one for the channel R. It is continuously compared to the transitory value of the current i_R gained through a current sensor CS_R. The comparison is made with a hysteresis and it is explained in Figure 28. If the current has to increase in accordance with the transitory value of the reference sinusoidal wave, the upper transistor turns on. At the reach of the upper border of the hysteresis, the lower transistor turns on because the current should already to be reduced in accordance with the reference sine wave. The control signal for the upper transistor is shown in Figure 28 – signal *r*. This operational principle is realized during the whole period also for the other two phases. The switching is with a high frequency.

Other methods to control and regulate the voltage inverters are known, for example, sliding mode control (Carpita, 1994) and predictive current control (Abu-Ruh, 2004).

Figure 27. Schema to realize hysteresis-current control

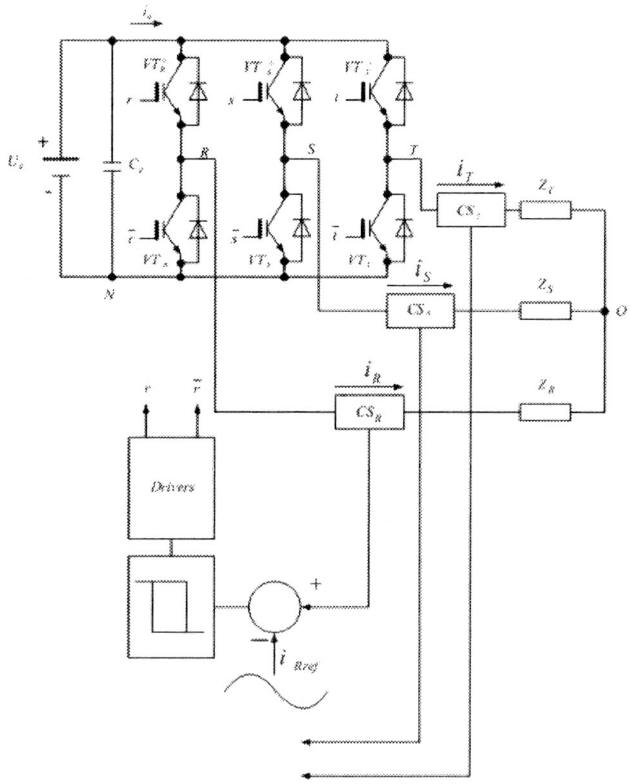

Regardless the implementation and the element base – analogue, digital, based on microcontrollers, special integrated circuits or digital signal processors, the secured functions from the control systems connected to the voltage-source inverter operations are the same.

Figure 29 displays a general structural schema of a control system describing the required basic function. The pulses for the gates of the power transistors in the inverter schema are delivered from a driver

Figure 28. Waveforms illustrating the hysteresis-current control

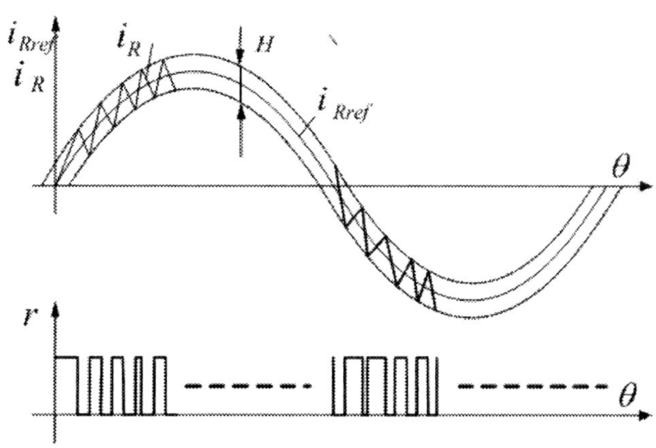

blocks securing the required levels of signals to turn on and off the device and also securing a galvanic isolation. The electrical schemas of these blocks are many and different. Moreover, in the "intelligent power modules" the drivers are embedded in the same body with the transistors. The basic signals connected with the operation of the power schema and the load supplied by the inverter are monitored using feedbacks – current and voltage transformers, Hall sensors, optoelectronic circuits, current and voltage transducers, etc. Of course, their number and type are defined by the particular implementation.

The signals received from the feedbacks are processed by the regulator system including appropriate regulators and regulators algorithms. The output signals of this system characterizing the voltage-source inverters such as regulation of the first harmonic of the output voltage and its value regulation are conventionally shown. The functions of the generator block are general. It generates pulse-width pulses, as well as a signal with the fundamental harmonic of the output voltage but with different value in the two half-periods. These two signals after logical operations are distributed to the transistor drivers. As it is in each power electronic converter, here, also special measurements are provided for break-down mode protections, as well as a suitable indication of the basis states and parameters. The last mentioned functions are obligatory but they are not separately shown in Figure 29. Part of the protection measurements are sometimes provided in the drivers.

The comparatively easiest implementation of the control system is that of the single-phase bridge inverter in Figure 5.

Figure 30 displays a block schema, which operation is the closest to the implementation way of the shown in Figure 8 – a comparison of a saw voltage with a high frequency to a sinusoidal waveform, whose frequency is equal to this of the first harmonic of the output voltage. The symbols in Figure 30 are in accordance with the symbols in Figure 9.

The sinusoidal voltage generator (SVG) may be implemented using analogue integrated circuits or combination of analogue and digital methods and it has two output signals. The first is the sinusoidal voltage, which is passed to the non-inverting input of the comparator C, and the second – a voltage with a saw waveform with a different sign in its two half-periods. The second signal can be obtained of the sinusoidal voltage using a comparator with a zero threshold. A phase locked loop (PLL) circuit is available. The pulses in its output are with frequency M_f times higher than this of the sinusoidal voltage because of the connected frequency devisor (FD). The saw voltage generator (SawVG) generates a voltage of a saw waveform and it passes the voltage to the inverting input of the comparator. The operation of SawVG is synchronized to the output pulses of PLL-circuit. At the output of the comparator, a sequence of pulses is produced, as it is shown in Figure 30. In applications where a regulation of the output voltage

Figure 29. General structural schema of a control system

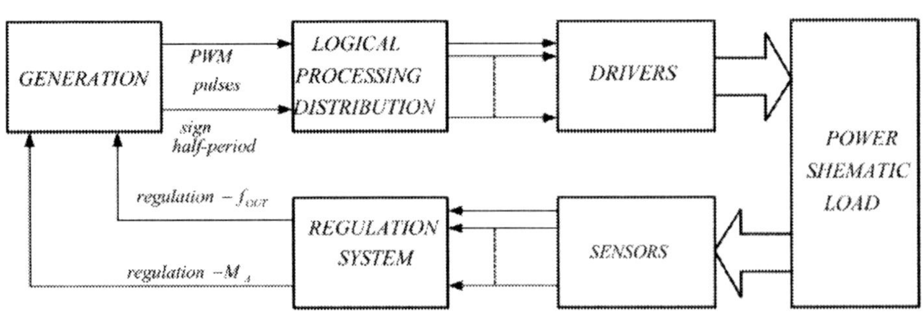

Figure 30. Structural schema to implement a sinusoidal pulse-width modulation

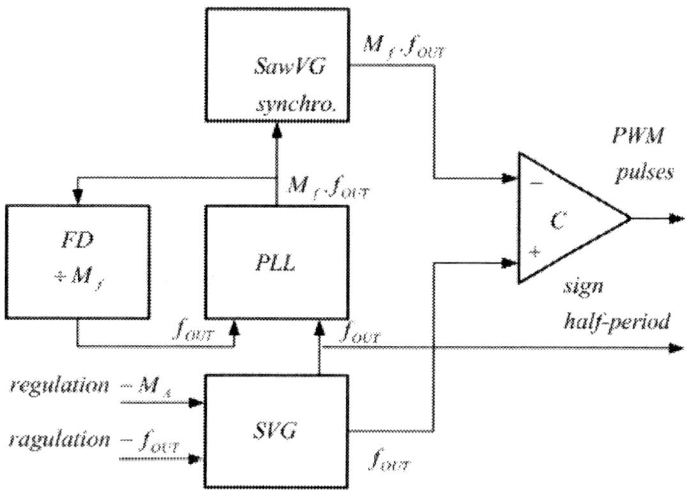

frequency is not required, the schema can be simplified and the PLL block can be removed. The necessity to regulate the frequency of SVG, as well as the input signal for this regulation, is also eliminated. When the ratio M_f is very high, i.e. the frequency of the saw voltage is many times higher than this of the sinusoidal voltage, the error of the lack of a precise synchronization does not significantly reflect on the waveform of the output voltage and SawVG can be without a synchronization input. All these simplifications can be made after an assessing of the requirements of the inverter load.

The shown principle is applicable also in three-phase inverters with pulse-width modulation, as it has been described in Figure 22. In this case, SVG has to generate three sinusoidal waveforms displaced to each other at $2\pi/3$ rad. The three sine waves have to be passed to three comparators at the other inputs of which a signal of a common SawVG should be delivered.

A possible conception at using microprocessor systems is to bring the function of generation of the pulse-width modulation signals in a special peripheral chip to use minimum resources of the control processor. The firm GEC Plessey has offered at the market the integrated circuit MA838 of a single-phase generator of pulse-width modulation pulses in accordance with a sinusoidal law (GEC Plessey Semiconductors, 1995). The IC is compatible with microprocessors and microcontrollers of Intel and Motorola. The structural schema of the IC is shown in Figure 31. It has two outputs with TTL-levels. They are used to control upper and lower transistor of an inverter leg via a driver. Regardless of inner register states, both outputs are in low level when a low level signal is passed to the input $\overline{TRIP}$. This input is used to protect the inverter of current overloading. At the same time at the output $SETTRIP$ a high level is set that indicates the prohibition of the pulses to the power transistors. This state is maintained till a circuit reset is performed by a low level at input $\overline{RST}$.

MA838 generates pulses with a required waveform for both transistors directly from inner memory ROM and requires influence of a control processor only to change the parameters of these signals. The change is made through a standard 8-bit multiplexed bus for address/data $SYSTEM\ BUS$ of $MOTEL$[1] type, compatible with the most widely spread microprocessors and microcontrollers. Using this bus, signal $\overline{CS}$ and signal for interface $MOTEL$[1], the IC is programmed as a peripheral chip for the micro-

Figure 31. Structural schema of a special integrated circuit MA838. Source: ©Gec Plessey Semiconductor 1995, Publication No DS3798 Issue No 3.1, July 1995

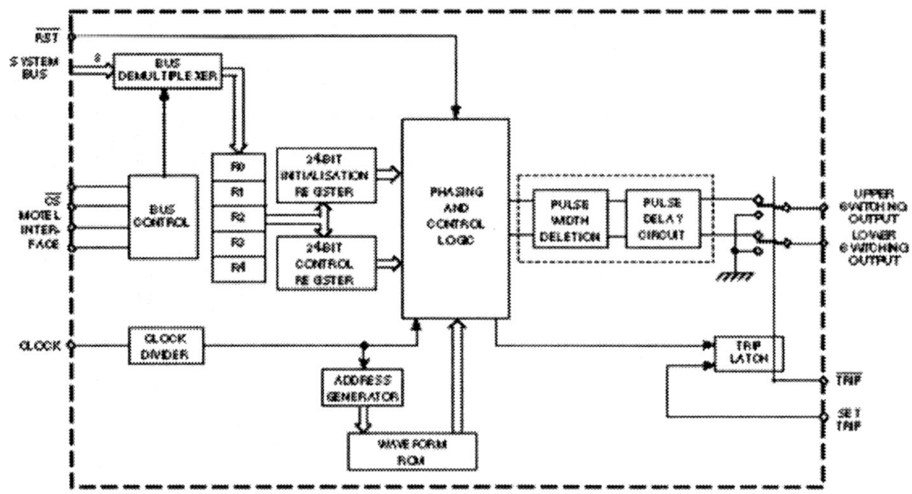

processor system. The programming is made using special inner registers. There is a possibility to set the output frequency and the modulation coefficient, minimum duration of the pulses and "dead" time, as well as to change the carrying frequency at modulation. The accuracy of setting the output frequency corresponds to 12-bit. The frequency can be changed to $4\,kHz$; the carrying frequency can be chosen to $24\,kHz$. The written data in the inner permanent memory corresponds to a bipolar sinusoidal pulse-width modulation as one shown in Figure 8.

The contemporary microprocessors combine most of the functions to generate pulse-width modulation pulses sequences through very well developed timer systems and with wide abilities of programming control. This has become possible by the simultaneous increase of their fast reaction to be good enough to obtain a required accuracy and discreet level. The series of the width modulated pulses may be preliminarily written in the series of memory cells, as it is proposed in (Maurice, 1992).

A special feature of the three-phase voltage-source inverters with a space-vector pulse-width modulation is a presence of six active vectors $\vec{V}_1 - \vec{V}_6$ (at which a DC voltage is applied to the load) and two zero vectors $\vec{V}_0, \vec{V}_7$ (at which a DC voltage is not applied to the load). Other feature both to single-phase and three-phase voltage-source inverters is a prohibition of simultaneously turning on of both phase-leg devices, because of the input circuit impedance this would lead to a short-circuit for the source. A simultaneously turning on of both phase-leg devices is allowed only in resonant DC-link inverters where a resonance process is developed caused by the elements connected in the input.

In the Z - source inverters one additional zero vector is added to the two zero vectors. The new zero vector corresponds to a simultaneously turning on of both phase-leg devices and no voltage is applied to the load. This is made without a danger of breaking the power devices because of the impedance of the inner circuit. Additionally, in this case as a result a DC supply voltage to the inverter is gained and also the voltage is higher than the output voltage of the DC source. I.e. if the DC source voltage is low, the medial boost high-frequency DC/DC converter or a connection of boost low-frequency transformer in the output of the inverter is not necessary. The principle is applicable also in the single-phase inverters.

An electrical schematic of a three-phase Z-source inverter is shown in Figure 32 (Huang, 2006). It contains passive elements in its input - L_1, L_2, C_1, C_2, which elements form so-called Z-source. The connection of diode *VD* is obligatory in using DC voltage, which current should not change its direction, for example fuel cell. Since the voltage in the input of the Z-source may become higher than the DC source voltage U_{DC} in particular intervals, the diode is connected.

If a switching period T_S is divided into two parts T_a and T_b, where in T_a corresponds to the time during which both phase-leg devices are on, and taking in consideration (7.28), the following equation is found:

$$\begin{aligned} &T_S = T_a + T_b \\ &a = \frac{T_a}{T_S}, b = \frac{T_b}{T_S} \\ &x + y + z = b \\ &a + b = 1 \end{aligned} \qquad (7.38)$$

Equivalent schemas illustrating the operation of the inverter shown in Figure 32 are presented in Figure 33. The first equivalent schema Figure 33.a corresponds to the third zero vector when both phase-leg devices are on, and the second one – Figure 33.b – to the rest eight vectors.

The values of the two inductances are equal as well are the values of the two capacitors. Therefore, they can be made in a common core also because of the inductance voltage polarities.

Using Figure 33.a the following equations can be written:

$$\begin{aligned} &U_{C1} = u_{L1} = L_1 \frac{di_d}{dt} \\ &U_{C2} = u_{l2} = L_2 \frac{di_d}{dt} \\ &U_{C1} = U_{C2} = U_C; u_{L1} = u_{L2} = u_L \end{aligned} \qquad (7.39)$$

Figure 32. Electrical schematic of a three-phase Z-source inverter

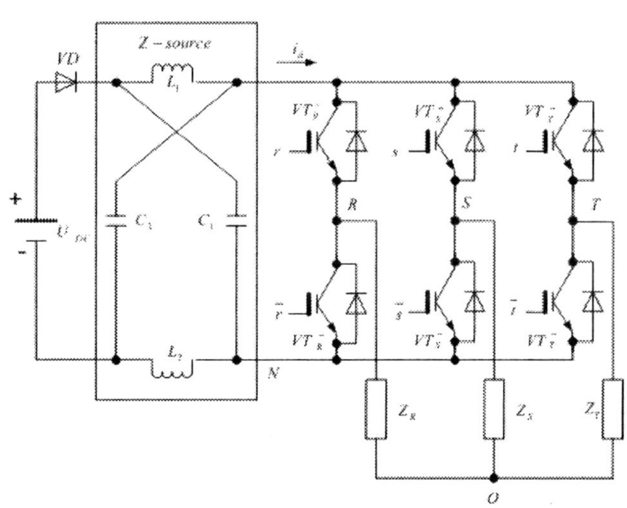

Figure 33. Equivalent schemas illustrating the operation of the inverter in Figure 29

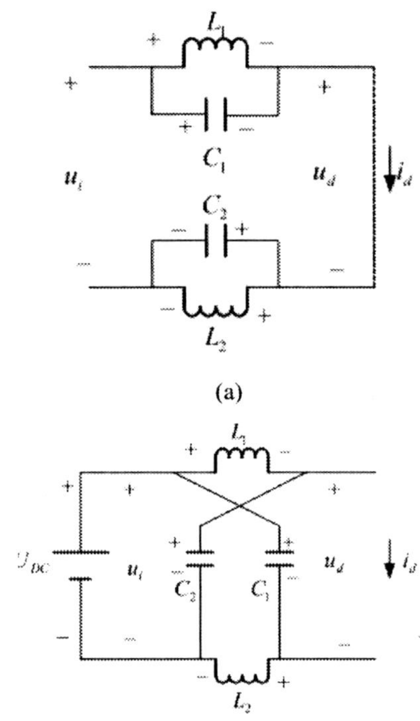

(a)

(b)

Since the DC capacitor voltages are applied to the inductances, their currents increase in accordance with a linear law.

Using Figure 33.a for T_a is found:

$$u_i = 2.U_C$$
$$u_d = 0 \qquad (7.40)$$

For the rest intervals included in T_b, it is found from Figure 33.b:

$$u_L = U_{DC} - U_C$$
$$u_d = U_C - u_L = 2.U_C - U_{DC} \qquad (7.41)$$

Across the inductances a DC voltage is applied again and the inductance currents decrease in accordance with a linear law.

Since the voltage across each inductance during a switching period is zero, it is found:

$$U_L = 0 = \int_0^{T_S} u_L.dt = \frac{T_a.U_C + T_b.\left(U_{DC} - U_C\right)}{T_S}$$

$$U_C = \frac{T_b}{T_b - T_a}.U_{DC} = \frac{1 - \frac{T_a}{T_S}}{1 - 2\frac{T_a}{T_S}}.U_{DC} \qquad (7.42)$$

Using similar approach, the average value of the input inverter voltage is obtained as:

$$U_d = \int_0^{T_S} u_d.dt = \frac{T_a.0 + T_b.\left(2U_C - U_{DC}\right)}{T_S} \qquad (7.43)$$

After substituting U_C from (7.42) in (7.43) and transformations, the following equation is derived:

$$U_d = \frac{T_b}{T_b - T_a}.U_{DC} = U_C \qquad (7.44)$$

Therefore, the voltage of one of the capacitors can be monitored as a feedback for the value of the input inverter voltage. Using (7.41) and (7.44), the highest value of this voltage during interval T_b is found as:

$$U_{dM} = 2.U_C - U_{DC} = \frac{T_S}{T_b - T_a}.U_{DC} = K.U_{DC} = \frac{1}{1 - 2\frac{T_a}{T_S}}.U_{DC} \qquad (7.45)$$

As it appears form (7.45), $K \geq 1$ and it may be called an increasing coefficient, i.e. during T_b at the input of the inverter an increased voltage dependent on the time T_a will be applied. Therefore, a medial high-frequency transformation of the DC source voltage is not required. During T_b a modulation to regulate the output inverter voltage is made, and, for example, if the modulation is a sinusoidal one, the maximum phase voltage is:

$$U^1_{ANM} = M_A.U_{dM} = M_A.K.U_{DC} \qquad (7.46)$$

It appears from (7.46) that the regulation of the output inverter voltage can be obtained by the two coefficients K and M_A. But one have to take in mind that the two coefficients are mutually connected by the equation (7.38) $a + b = 1$.

METHODS TO IMPROVE POWER EFFICIENCY IN DC/AC CONVERSION

Use of zero voltage switching and zero current switching is also possible in inverters to reduce power losses across the power switches and to increase the efficiency coefficient (Heumann, 1994). As an example, transistor resonant inverter with active load shown in Figure 34 may be examined.

Figure 34. Electrical schematic of a bridge transistor resonant inverter

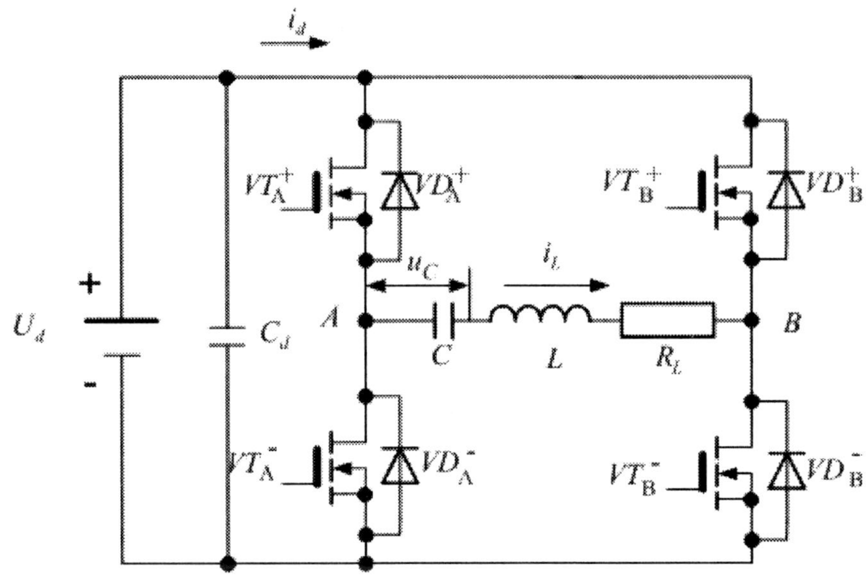

Of course, other load variants are possible dependent on the inverters purpose – more often a parallel oscillating circuit. It is also possible the use of output transformer, whose primary inductance is a part of the inductance *L*. Other possibility is to connect a high-frequency rectifier at the transformer side thus DC/DC converters with resonance switching are obtained. The load may be connected in different ways to the resonance circuit. These converters are not a subject of examination in this chart, therefore, it is assumed at the description of operational features of the inverter that elements connected between points A and B are only the ones shown in the Figure 34. Figure 35 depicts waveforms of current through the bridge diagonal i_L and capacitor voltage u_C in several typical operational modes dependent on the ratio of switching frequency of the transistors ω_S to resonance frequency of the series circuit ω_0. Devices that conduct during different intervals are marked. It has to be mentioned that the first three modes are realizable also in thyristor inverters with antiparallel diodes. After a decrease of a current through diagonally connected transistors, their antiparallel connected diodes are turned on. Discontinuous-conduction current mode with turning off at zero current both the transistors and diodes is achieved if the next transistor pair is turned on with a delay in accordance with the decrease of diode current to 0. If the switching frequency is half of the resonance one $\omega_S = 0.5\omega_0$ - second diagrams, the switching also happens at zero current, but the load current is continuous. In the third mode when $0.5\omega_0 < \omega_S < \omega_0$, zero current switching is only for the transistors, because they are turned on before the decreasing of the diode current to zero. The forth mode is resonance mode and it is achieved only in the transistor inverters because in them a reverse recovery time is not necessary, as it is in the thyristor inverters. The next pair of transistors is turned on as soon as the current through the pair, which has been conducting till that moment, decreases to zero. And in ideal case, the antiparallel diodes are not necessary. Although, since delays and inaccuracies of the transistor switching are always possible to appear, the diodes are required. Furthermore, the resonance mode is maintained with a specified accuracy and there is time to achieve the mode in transient processes. The last operational mode is very interesting – switching above the resonance frequency. Pulses to turn off the conducting transistor pair and to turn on the next one are passed before the transistor current decreases to zero. At first, the antiparallel connected diodes to the transistor pair, which has been received the on signals, begin to conduct. The transistors begin to conduct only when the diode current decreases to zero. For example, after the turning off of VT_A^+, VT_B^- and the passing of turning on signals to VT_B^+, VT_A^-, at first, diodes VD_B^-, VD_A^+ begin to conduct and after the diode current decreases to 0 the transistors VT_B^+, VT_A^- begin to conduct. It is clear that a mode of turning off at zero current is realized for the diodes and also a mode of turning on at zero voltage is realized for the transistors because the transistor antiparallel connected diode has been conducting at the start of the transistor conductive interval. In this mode it is possible to use as clamp elements for the devices only capacitors in parallel to the devices because at the turning on of the transistors the capacitors will be discharged and it is not necessary to limit their discharged current. Thus, dissipating of active power in other elements of a clamp group is avoided.

It is clarified from the study made and from the waveforms in Figure 35 that load power of the inverter can be regulated using the described operational modes. For this purpose, the control pulses frequency of the power devices in respect to the resonance frequency has to be changed using the control system. This is so-called pulse frequency modulation (PFM) or frequency regulation.

Figure 36 describes another way to change the load power - pulse density modulation (PDM). The voltage of the output of the inverter bridge – between points A and B u_{AB}, as well as load current i_L are shown in the figure - u_{AB}. A special feature of this modulation is the appearance of an interval during which the voltage across the resonance circuit becomes equal to 0. This interval is achieved by simultaneously delivering of on signals to both lower or both upper transistors in the schematic. Thus, conditions for free fading oscillations of the load current flowing through a transistor and an antiparallel diode of the other transistor are created in one of the half-periods, conditions for a change of the transistor and diode of the other half-period are also secured. During the rest of the time a resonance mode is maintained when the diagonally connected transistors are turning on. To be able to be reached and to be maintained the resonance mode, the control frequency is required to be changed in case of a load change and also during transient modes. Therefore, the control system should have the possibility for frequency-pulse modulation. It is necessary to mention the difference between this mode and the mode of discontinuous conduction current through the load examined in the first diagrams shown in Figure 35. The voltage between the point A and B is also zero in the previously examined mode (see Figure 35) but current through the load does not flow. At a scrutiny look upon the operation of the transistor inverter in the PDM mode, it can be seen that since the antiparallel diode of one of the next to conduct diagonally connected transistors has been conducting, the transistor turns on at zero voltage after the interval of zero voltage in the bridge diagonal. This transistor is the one that has been used during the voltage pause. There are two possibilities to regulate load power: the first is to maintain the interval of a normal operation of the resonant inverter constant (for example, fixed period number) and to change the duration of the interval during which the voltage of the bridge diagonal is zero; the second is to change the interval of a normal operation of the resonant inverter constant (for example, changeable period number) and to maintain the duration of the interval during which the voltage of the bridge diagonal is zero constant.

Use of resonance modes in voltage-source inverters is also possible. More detailed description can be found for example in (Skwarenina, 2002; Bose, 2007).

Figure 35. Waveforms illustrating the operation of the bridge transistor resonant inverter shown in Figure 34

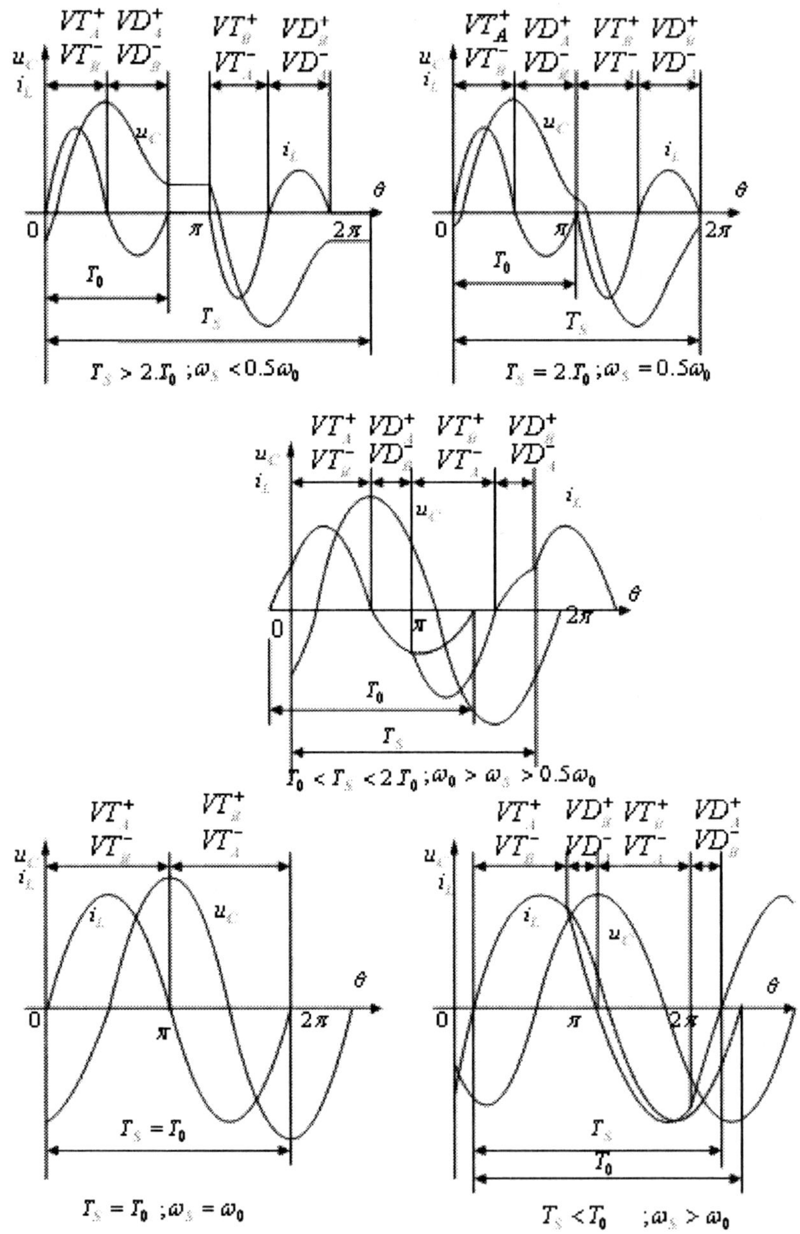

As it has been clarified by the voltage inverter operational principle and by the output voltage waveform dependent on the modulation, the waveform is characterized by different harmonic spectrum. This presumes some special features at measurement of the output active power corresponding to (7.1) and the efficiency coefficient connected to this power, especially in the three-phase inverters. The measurement of input and output powers is convenient to be made using a special digital measurement device (Marrinan, 2007).

Figure 36. Waveforms illustrating pulse density modulation

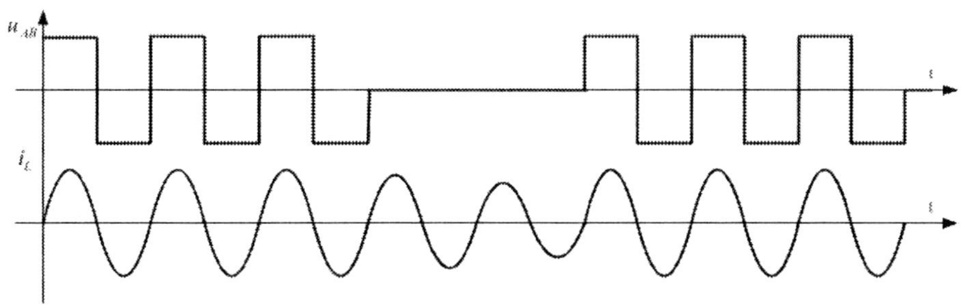

REFERENCES

Abu-Ruh, H., Guzifiski, J., Krzeminski, Z., & Tolijat, H. A. (2004). Predictive current control of voltage-source inverters. *IEEE Transactions on Industrial Electronics, 51*(3), 585–593. doi:10.1109/TIE.2004.825364

Bedford, B. D., & Hoft, R. G. (1964). *Principles of inverter circuits.* New York: J. Willey & Sons.

Bose, B. K. (2007). Need a switch? *IEEE Industrial Electronics Magazine, 1*(4), 30–39. doi:10.1109/MIE.2007.909539

Carpita, M. (1994). Sliding mode controlled inverter with switching optimization techniques. *European Power Electronics and Drives Journal., 4*(3), 30–35.

Chiasson, J. N., Tolbert, L. M., McKenzie, K. J., & Du, Z. (2004). A complete solution to the harmonic elimination problem. *IEEE Transactions on Power Electronics, 19*(2), 491–499. doi:10.1109/TPEL.2003.823207

Heumann, K., & Ying, J. (1994). Small loss series resonant inverter. *European power Electronics and Drives Journal, 4*(1), 21-27.

Huang, Y., Shen, M., Peng, F. Z., & Wang, J. (2006). Z – Source inverter for residential photovoltaic systems. *IEEE Transactions on Power Electronics, 21*(6), 1776–1782. doi:10.1109/TPEL.2006.882913

Marrinan, T. (2007, November). Digital measurement of inverter efficiency. *Bodo's power . System,* 32–33.

Maurice, B., Bourgeois, J. M., & Saby, B. (1992). Versatile and cost effective induction motor drive with digital three phase generation. In *Designers' Guide to Power Products* (pp. 309-322). Italy: SGS – Thomson.

Mohan, N., Undeland, T. M., & Robbins, W. P. (1995). *Power Electronics Converters, Applications and Design.* New York: John Wiley & Sons.

Plessey Semiconductors, G. E. C. (1995, July). *MA 838. Single pulse width modulation waveform generator.* Publication No DS3798.UK. Haque, T.M., & Taheri, A. Using neural network for execution of programmed pulse width modulation (PPWM) method. *IEEE Transaction on Engineering . Computing and Technology*, *6*, 58–61.

Skvarenina, T. (2002). *The Power Electronics Handbook*. New York: John Wiley & Sons.

ENDNOTE

[1] MOTEL is trademark of Motorola Corp. and Intel Corp.

Section 3
Applications of Electronic Energy Converters

Chapter 8
Conversion of Electrical Energy in the Processes of Its Generation and Transmission

CONVERSION IN THE PROCESS OF ELECTRICAL GENERATION

Electrical energy can be obtained by burning coal (thermal power plant), by using nuclear fuel (nuclear power plant) or by using the power of water (hydro power plant). In these cases, the energy obtained by the sources put a shaft of an electrical generator in motion. The generator generates electrical energy – see Figure 1. In the installation, excitation system for the generator is used. The system turns on an uncontrolled rectifier, thyristor-controlled rectifier or AC thyristor regulator dependent on the generator type.

The obtained energy is transmitted using a transmission system towards consumers. The transmission yet is made mainly in high-voltage AC energy form (HVAC). In different parts of the transmission network the voltage value may be different. There are so-called high-voltage (420 kV, 220 kV, 110 kV, etc) and medium voltage (20 kV, 6.6 kV, etc.) systems. General consumers consume electrical energy from so-called low-voltage systems (230V, 50Hz or 110V, 60Hz). During the transmission the type of energy does not change, only the value of the voltage changes using transformers.

The processes of obtaining electrical energy using renewable energy sources have their features studied in the next Chapter 9.

DOI: 10.4018/978-1-61520-647-6.ch008

Figure 1. Generation of electrical energy

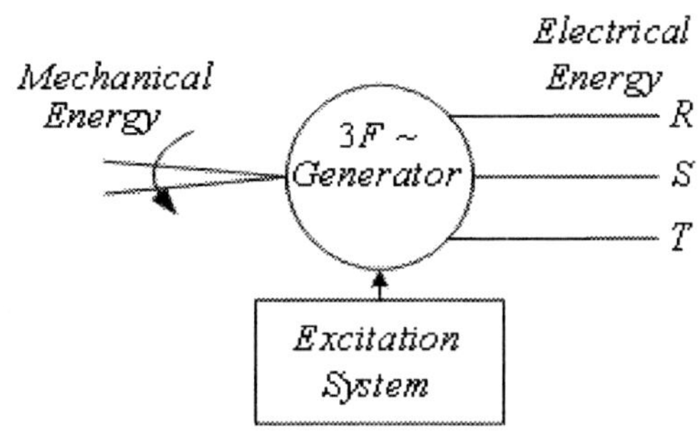

The processes, occurring in the transmission lines for AC electrical energy, are connected with the type and the mode of operation of the consumers. Also, these processes often cause different in types disturbances of the quality of electrical energy. Furthermore, some limitations have to be monitored. The limitations are connected with both energy quality and capabilities of the corresponding transmission system, and they are:

- Steady-state power transfer limit
- Contingency limit
- Voltage stability limit
- Dynamic voltage limit
- Transient stability limit
- Power system oscillation damping limit
- Inadvertent loop flow limit
- Thermal limit
- Short-circuit current limit

The benefits of control in the process of electrical energy transmission can be marked as:

- Increase loading and more effective use of transmission corridors
- Added power flow control
- Improved power system stability
- Increased system security
- Increased system reliability
- Added flexibility in sitting new generation facilities
- Elimination or deferral of the need for new transmission lines

The achievements in the field of Power Electronics allow the use of power electronic converters in the processes of transmission of electrical energy with a purpose to observe the mentioned limits (Asprund, 2008). Incorporating such converters gives certain flexibility of energy transmission; therefore, the systems using such converters are called Flexible AC Transmission Systems (FACTS).

Thus, FACTS enhance controllability and increase power transfer capability. Thoroughly, FACTS are systems containing semiconductor converters, information and control technologies (software), and interconnecting conventional equipment that creates intelligence into the network by providing enhanced-power system performance, optimization, and control (Hingorani, 2000). The advantages of FACTS to a construction of new transmission lines are less infrastructure investment, environment impact, and implementation time.

Name FACTS is also applicable in the thyristor use in the power electronic converters designed for this purpose. A new class of systems – Advanced FACTS, is formed with implementation of fully controlled transistors. Figure 2 displays a comparison among the different systems on the converters incorporated into the systems (Tolbert, 2005).

For each problem a conventional solution, e.g. shunt reactors or shunt capacitor, is also provided, as well as dynamic applications of FACTS in addressing problems in transient stability, dampening, post contingency voltage control and voltage stability. FACTS devices are required when there is a need to respond to dynamic network conditions. The conventional solutions are normally less expensive than FACTS devices, but limited in their dynamic behavior. Moreover, usually one FACTS device can sole several problems, which would otherwise need to be solved by several different conventional solutions. In conclusion, FACTS devices are more flexible than most conventional solutions.

Figure 2. Comparison among different FACTS

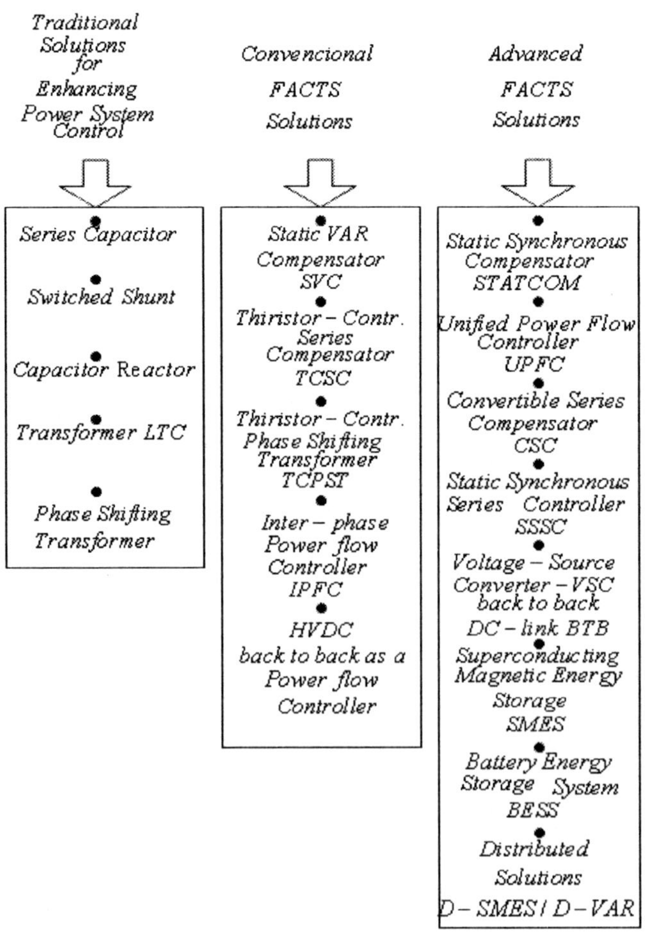

Figure 3. Mechanically switched capacitor and reactor

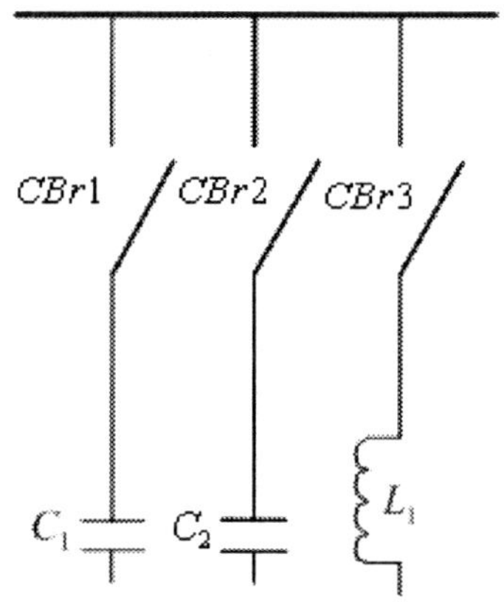

In the following presentation, the operational principle and features of the different power electronic converters used in FACTS are studied.

STATIC VAR COMPENSATORS (SVC)

SVC is a shunt-connected static var generator or absorber whose output is adjusted to exchange capacitive or inductive current so as to maintain or control specific parameters of the electrical power system.

SVC are mainly used in AC/AC conversion to improve power efficiency. The detailed operation of them has been already described in the final chart of Chapter 5. The already studied SVC can be complimented with a mechanically switched capacitors and reactors. The way they are connected to the network is shown in Figure 3.

Using an appropriate control of SVC, the following results can be obtained:

- Damping of system swings
- Non-linear gain – to increase response during large disturbances
- TCR winding DC current elimination
- Negative sequence balancing
- TCR overcurrent protection
- Secondary overvotlage protection

STATIC SYNCHRONOUS COMPENSATOR (STATCOM)

Using Figure 4 the operational principle of static synchronous compensator can be illustrated. The STATCOM consists of a coupling transformer, voltage source converter (VSC) with a control system. DC source can be either capacitor or accumulator battery. More often capacitor is used. The control of the converter is made in such a way that its output current I_{OUT} either leading or lagging with angle $\frac{\pi}{2}$ to

Figure 4. Schema used to illustrate the operational principle of STATCOM

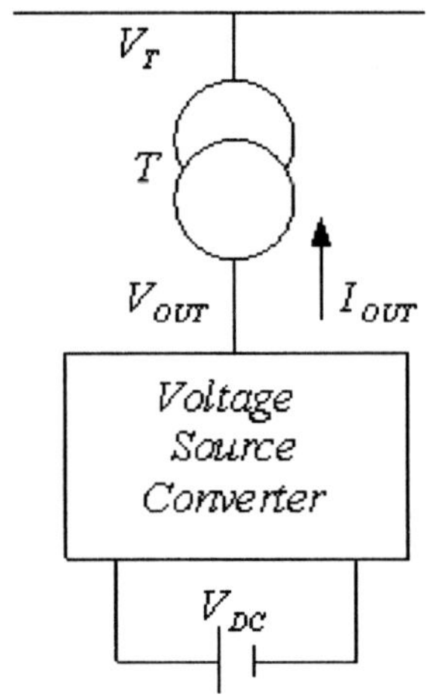

the voltage V_T. When the current leaves behind the voltage, the STATCOM appears as a reactive energy source, while in the second case, it is a consumer of reactive energy. Here, the losses of active power in the power devices of VSC are neglected. The displacement of the output voltage V_{OUT} to V_T determines the operation mode. The voltage V_T may be changed using the control system of VSC. Also, using the control system of VSC, the value of reactive energy may be changed by the value of the voltage V_{OUT} in a particular operational mode (consumer or source of reactive energy).I.e. the value of reactive energy may be changed within the range of 100% capacitive energy to 100% inductive energy using the voltage, which depends on the control system of VSC.

Taking in consideration the description of operational principle of the Static Var compensator and STATCOM, the main difference is seen, namely, the output voltage of the STATCOM is undependable of the value of the system voltage V_T, while the output voltage of the SVC is determined by the system voltage (Chen, 2008).

Another important implementation of the STATCOM in three-phase symmetrical system uses the capability of the STATCOM to produce negative sequence output voltage in addition to positive sequence output voltage by controlling independently the magnitude and phase angle of each phase output voltage. Thus, the negative sequence component included in AC current can be reduced by adjusting the magnitude and the phase angle of the negative sequence component of the STATCOM output voltage, even when negative sequence component is included in the utility system voltage.

Important feature of the STATCOM is the probability to maintain the value of its output current constant in both operational modes – capacitive and inductive modes regarding the system voltage V_T, even in significant decrease of the value of the system voltage. This is done by a change of the output voltage V_{OUT} using the control system of VSC.

Briefly, the STATCOM has the following features:

- Maintain its reactive current at low voltage since it has an essentially constant current characteristic while a thyristor SVC has constant impedance
- Reduce real estate use to about 40% of a thyristor SVCs requirement

Figure 5. Schematic illustrating the operational principle of a TCSC

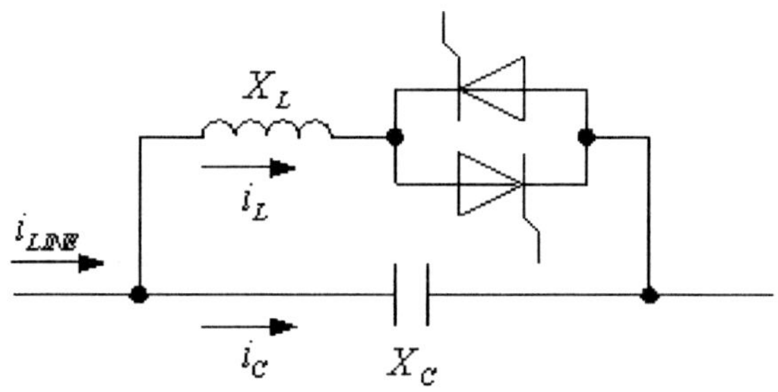

- Store energy, if batteries replace capacitors
- Be applied as an active filter because each step can be switched in response to a harmonic

A novel fuzzy-PI-based direct output-voltage control strategy for STATCOM used in utility distribution system is proposed (Luo,2009).

THYRISTOR CONTROLLED SERIES COMPENSATOR (TCSC)

The operational principle of a TCSC is based on the idea to change the line impedance above or below its natural impedance value using electronic way.

A schematic of a TCSC is shown in Figure 5. As it is seen, the TCSC consist of a capacitor with a constant value and thyristor controlled inductance which value depends on the firing angle α of the thyristors. The inductance value is:

$$X_L(\alpha) = X_L\left(\frac{\pi}{2.\pi - 2.\alpha - \sin(2.\alpha)}\right) \tag{8.1}$$

Therefore, the value of equivalent impedance is:

$$X_{TCSC} = \frac{X_L(\alpha).X_C}{X_L(\alpha) + X_C} \tag{8.2}$$

Since the value of $X_L(\alpha)$ is changeable from X_L to ∞, the value of X_{TCSC} may be changed from $\frac{X_L.X_C}{X_L + X_C}$ to X_C using the change of the firing angle α from the control system of the thyristors.

The change of the impedance may be used to decrease oscillations during or after occurring of disturbances.

This approach is used to suppress sub-synchronic resonance or electromechanical oscillation within the range of 0.5-2 Hz.

STATIC SYNCHRONOUS SERIES CONTROLLER (SSSC)

The operational principle of a SSSC is illustrated in Figure 6. It consists of a converter, capacitor and transformer. The voltage is always maintained displaced with angle $\frac{\pi}{2}$ to current *I*- the voltage either leaves behind or lags to the current. Thus, a series compensation is made. The value of the voltage V_q can be also regulated. Active power transmitted from Bus 1 towards Bus 2 is regulated using V_q. This is a result from the SSSC capability to injects voltage in the line regardless of the *I*. Namely, the reactive drop across the line is changed.

UNIFIED POWER FLOW CONTROLLER (UPFC)

A combined connection of a STATCOM and SSSC in the DC side and a presence of a capacitor allow bidirectional transferring of energy between the outputs of the STATCOM and SSSC.

Figure 7 depicts a schematic illustrating the operation of UPFC (Pinarrelli, 2001; Orizondo, 2006).

UPFC includes two voltage source inverters (VSI) connected to the line through transformers – the first inverter via shunt transformer, and, the second inverter via series transformer.

Converter 2 operates in so-called "Automatic Power Flow Control Mode". The inverter injects a symmetrical three-phase voltage system V_{C2} with probability to vary voltage magnitude and voltage phase angle in series with the line to regulate reactive power and active power flows in it. Therefore, active power P and reactive power Q are required as reference values to inverter control system.

Converter 1 operates in so-called "Automatic Voltage Control Mode". Current I_{C1} consists of two components – active one, which is in phase to or displaced with angle π to the line voltage, and reactive component, which leads behind or lags with angle $\frac{\pi}{2}$ to the line voltage. If the Converter 2 is required to supply active power, then the voltage across the capacitor will be decreasing. The control system of the Converter 1 maintains the value of the capacitor voltage stable, i.e. in this case the active component

Figure 6. Schematic illustrating the operational principle of a SSSC

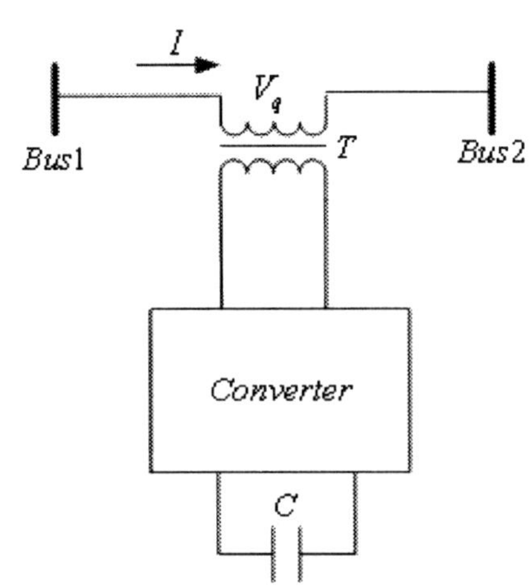

Figure 7. Schematic illustrating the operation of UPFC

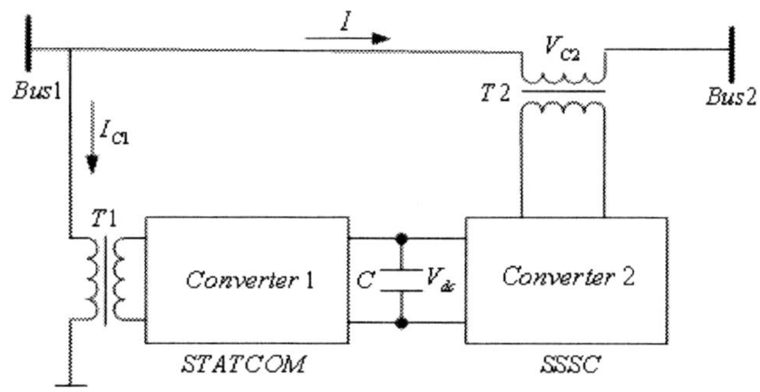

of the current I_{C1} will be in phase to the line voltage and active power will be transferred toward the capacitor C. If the Converter 2 has to take active power, then the capacitor voltage will be increasing. In this case the active component of the current I_{C1} will be displaced with angle π to the line voltage and active power will be transferred from the capacitor toward the line. The reactive component of the current I_{C1} is regulated undependably and in conformity with the reference value – inductive or capacitive values, to supply reactive power to or to absorb reactive power from the transmission line with a maintenance of the voltage value.

From the above-stated it is seen that a typical operational mode for Converter2 is as a SSSC, and for Converter1 – as a STATCOM. As an additional task, the control system of Converter1 maintains also the capacitor voltage V_{dc} unchangeable corresponding to the active power flow.

There are additional losses of active power in the system consisting of two converters, operating according the describe method. These losses are mainly from losses in the power converter schematics and the transformers.

INTERLINE POWER FLOW CONTROLLER (IPFC)

The idea to create an interline power flow controller has arisen from the necessity to control the flow of energies in multilinear transmission systems – more often one major line and several branches. For example, such a case is a compensation of several lines in a substation. Here, a possible solution is to use separated unified power flow controller UPFC in each line. Disadvantage of this solution is that an exchange of energy among the separate lines may happens only using the common line. Thus, controlling the flow of powers in one line may lead to a disturbance of the energy quality in other lines, because SSSC in each line may inject voltage which leads behind or lags to the current with angle $\frac{\pi}{2}$.

The idea of IPFC is to make possible the exchange of energy among the separate lines using a common DC source which supplies the separate SSSC included in each line – Figure 8.

The DC voltage can be obtained through a STATCOM incorporated in the Nth line – more often the major line. The power of Converter N is determined by maximum power of the rest converters. Thus, the following problems may be decided:

- Equalize both real and reactive power flow between the lines
- Reduce the burden of overloaded lines by real power transfer
- Compensate against resistive line voltage drop and the corresponding reactive power demand
- Increase the effectiveness of the overall compensating system for dynamic disturbances

Therefore, the IPFC can potentially provide a highly effective schema for power transmission management at a multi-line substation.

The development in the element base of Power Electronics – creation of power IGBT, as well as implementation of DSP in control systems, has made possible to reach a new level in solutions connected with the problems of the electrical power quality (Akagi, 2006). A new class of power electronic converters has been invented – active power filters, which serve to realize so-called Unified Power Quality Conditioner (UPQC).

The active power filters are in general shunt and series active power filters (Sigh, 1999; El-Habrouk, 2000).

The operational principal of a shunt active power filters is illustrated in Figure 9. The principle is valid for both single-phase and three-phase filters. The supply network is presented as a voltage network source U_S and predominant inductive character - L_S. The consumed current from the network is I_S. If an active power filter misses this current coincide to the load current I_L. In general, the load is

Figure 8. Schematic illustrating the operational principle of an IPFC

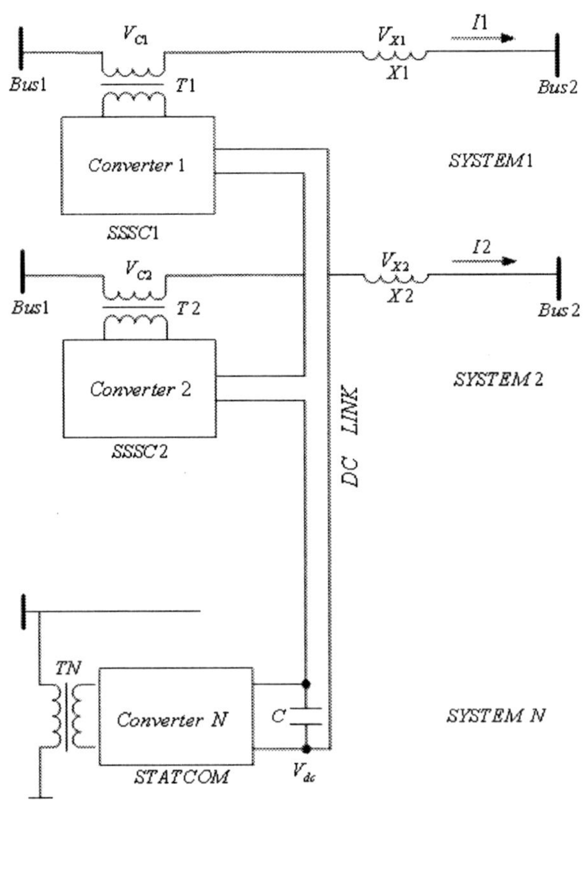

Figure 9. Schematic illustrating the operational principle of a shunt active power filter

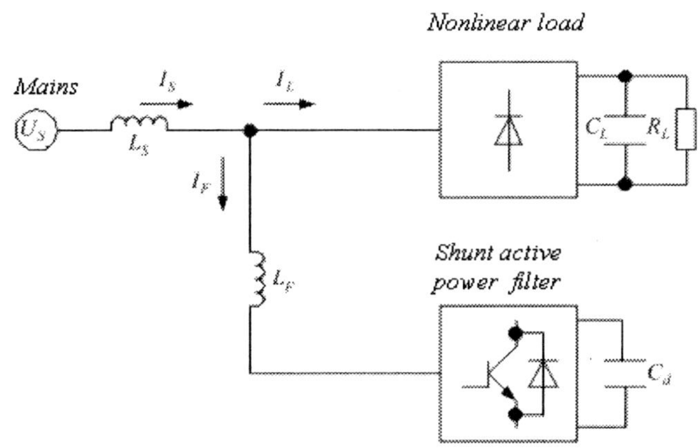

non-linear one. It is shown in Figure 9 as a most frequently met load – uncontrolled rectifier with active-capacitive type of load. In general, the non-linear load loads the network with non-sinusoidal current (a cause to decrease the distortion factor v and to load the network with distortion power D), whose first harmonic is displaced to the network voltage (a cause to decrease the displacement factor $\cos\varphi$ and to load the network with reactive power Q). The active power filter is shown as a most frequently used schematic – bidirectional converter supplying by a voltage source. The inductance L_F is required for the filter proper operation.

The basic idea is as follows: 1. Powers consumed by the load are continuously monitored using different algorithms in a control system. 2. Through mathematical transformations, a reference current waveform for the filter I_F is found. The reference waveform is calculated in such a way that if it is generated by the filter, it will compliment the load current to a sine wave current for the network. 3. Using an appropriate control for power switches, the reference current waveform found is monitored. This is made in real time which ensures minimum dependence of the changes of load type and its value. Moreover, two probabilities exist: 1. The filter corrects only the current waveform and it do not make the current in phase to the source network. 2. The filter corrects the current waveform to a sinusoidal one in phase to the network voltage. The second case is more often used. Here, the power factor approaches 1. Thus, the consumed reactive and distortion powers are minimized. The system active power filter- load has a behavior as an active resistance in respect to the supply network. Naturally, the consumed active power by the filter is increased on the expanse of the losses in the filter, which are no more than 5% in practice.

Let's study the case when the filter corrects only the current waveform to a sinusoid, i.e. it compensates only the distortion power. The displacement factor $\cos\varphi$ is assumed to be approximately 1. That is approximately valid for the load shown in Figure 9, for example. The effective value of the load current includes fundamental and higher harmonics:

$$I_L = \sqrt{I_{L1}^2 + \sum_{n=2}^{\infty} I_{Ln}^2} = \sqrt{\frac{\sum_{n=2}^{\infty} I_{Ln}^2}{K_H^2} + \sum_{n=2}^{\infty} I_{Ln}^2} = \sqrt{\sum_{n=2}^{\infty} I_{Ln}^2 \left(\frac{1+K_H^2}{K_H^2}\right)}$$

$$I_L = \sqrt{\sum_{n=2}^{\infty} I_{Ln}^2} \cdot \sqrt{\frac{1+K_H^2}{K_H^2}} \tag{8.3}$$

where in K_H is harmonic coefficient of the load current, and it is:

$$K_H = \frac{\sqrt{\sum_{n=2}^{\infty} I_{Ln}^2}}{I_{L1}} \tag{8.4}$$

The connection between K_H and the distortion factor is:

$$\upsilon = \frac{I_{L1}}{I_L} \tag{8.5}$$

$$\upsilon = \frac{1}{\sqrt{1+K_H^2}} \tag{8.6}$$

If the active power in the filter is neglected then its current contains only higher harmonics:

$$I_F = \sqrt{\sum_{n=2}^{\infty} I_{Ln}^2} \tag{8.7}$$

Substituting (8.7) in (8.4) and taking into consideration (8.6) leads to:

$$I_L = I_F \cdot \sqrt{\frac{1}{1-\upsilon^2}} \tag{8.8}$$

Total power consumed after the filter has been connected will be:

$$S = U_S \cdot I_L = U_S \cdot I_F \cdot \sqrt{\frac{1+K_H^2}{K_H^2}} = U_S \cdot I_F \cdot \sqrt{\frac{1}{1-\upsilon^2}} \tag{8.9}$$

In relative units, the relationship among the load power, the harmonic coefficient of the load and the required value of the current of single-phase and three-phase active power filters is respectively:

$$\frac{S}{\sqrt{3}.U_{LL}} = I_F \cdot \sqrt{\frac{1+K_H^2}{K_H^2}} \tag{8.10}$$

$$\frac{S}{U_S} = I_F \cdot \sqrt{\frac{1+K_H^2}{K_H^2}} \tag{8.11}$$

where in U_{LL} is effective value of the phase-to-phase voltage of the three-phase supply network, U_S - effective value of the voltage of the single-phase network, and, I_F - effective value of the filter current per phase.

The relationships (8.10) and (8.11) give the possibility to determine the effective value of the current of single-phase and three-phase shunt active power filters, under the above made assumptions, when the value voltages, total load power and the harmonic coefficient of the load are known. The values of S and K_H are usually determined after preliminarily made measurements, to chose an appropriate filter.

Bidirectional transistor converters, supplied by a voltage source, used to implement shunt active power filters are found the widest spreading. Figure 10 shows principle schematics of bridge and half-bridge single-phase topologies.

Their operational principle tracing a reference curve is studied in Chapter 4. Further down, how the reference curve is formed in active power filters is reported.

Figure 11 depicts a connection of a three-phase shunt active power filter to a three-phase three-wire supply voltage system, while Figure 12 depicts a connection to three-phase four-wire system.

From the above presented it is clear that due to the complicity of the tasks which have to be solved in real time, the control of the shunt active power filters may be realize on the basis of digital signal processors or programming logic. Two major tasks lay for implementation of the control of the shunt active power filters. The first is that for a proper operation of the bidirectional converters, voltage across the capacitor C_d - U_d, has to be stabilized to a value higher than the maximum value of the network voltage. This is usually implemented by incorporating a proportional-integral regulator influencing the value of the reference curve. The second task is the generation of the reference curve. Different methods for the generation exist in single-phase and three-phase active power filters. The choice of the method is also connected with the required number and type of the sensors used for feedbacks. One of them, as it has already been mentioned, is for voltage of the capacitor C_d. The second feedback is for the supply network voltage – its time curve and its value in single-phase networks, and, time development and value of two or three of the voltages in a three-phase system. It appears in Figure 9 that feedbacks for three currents may also be used, but which of the currents depends on the method used. It is clear, that the load current (or two of the phase currents in a three-phase system) has to be monitored. Methods with a predictive control exist which are limited their feedbacks only to the load current. However, a

Figure 10. Principle schematics of single-phase bidirectional converters used to implement shunt active power filters

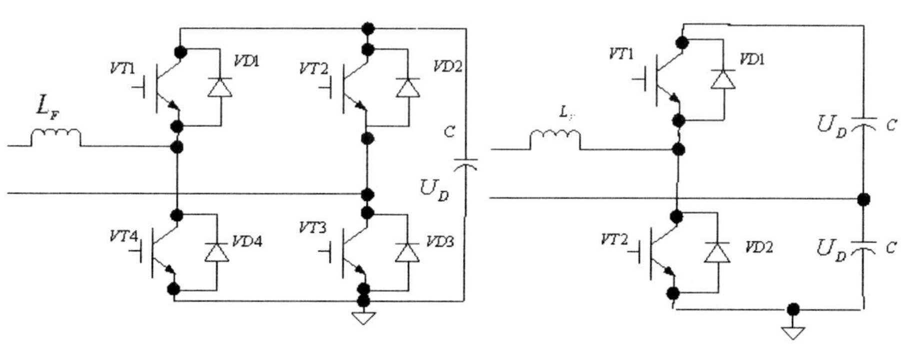

Figure 11. Connection of a three-phase shunt active power filter to a three-phase three-wire supply voltage system

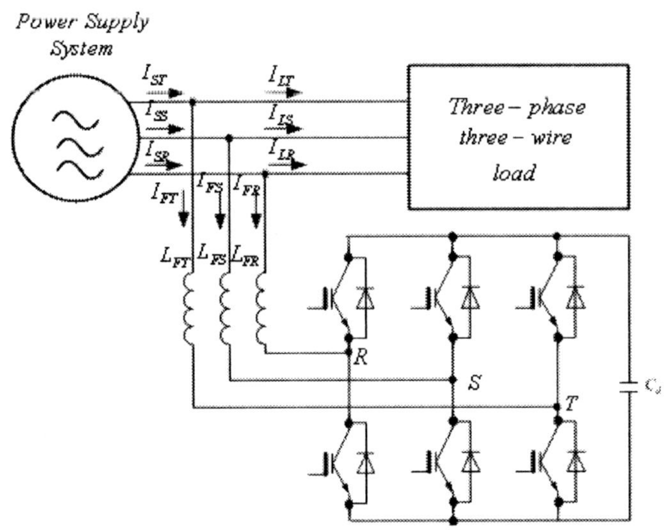

sensor for the filter current to be monitored in real time is used. Other probable combination is to use feedbacks of load current and total source current, which has to be of sinusoidal waveform and in phase to the source voltage. From the method using to control, more known are the following: pulse-width modulation, hysteresis-current control with constant or varying frequency, sliding mode control, control using phases – logic or neuron networks, genetic algorithms, etc. (Welsh, 2000; Antchev, 2007). One of the most spread methods is the method using the fast Fourier transformation in real time. In it, higher harmonics are extracted from the load current, and on their basis, the reference curve for the filter is formed. The reference waveform is traced using an appropriate way through filter devices control. An interesting method, in which besides the load current, the source current is monitored, is the so-called method of "equivalent sinusoidal" (Antchev, 2007). The method is applicable in both single-phase and three-phase filters. The Fourier transformation is avoided in this method and its nature is as follows. Reference sinusoidal curve for the source network of the system load – shunt active power filter has to be generated. This curve has to correspond only to active power consumed by the load when the filter does not switch on. This sinusoid (or three sinusoids in a three-phase system) has to be traced by power device switching. To implement the method, active power P_L consumed by the load is calculated using discreet value transformation of the waveforms of the source voltage and load current. Also, the effective value of the source voltage U_S is calculated from its discreet curve. In normal operation of the active power filter, the power factor of the system load-filter in respect to the source network becomes up to 1, furthermore, if the losses in the filter are neglected, total power after the switching on of the filter S_{L-APF} will be only active power, equal to the active power P_L:

$$S_{L-APF} = U_S \cdot I_S = U_S \cdot \frac{I_{SM}}{\sqrt{2}} = P_L \qquad (8.12)$$

Figure 12. Connection of a three-phase shunt active power filter to a three-phase four-wire supply voltage system

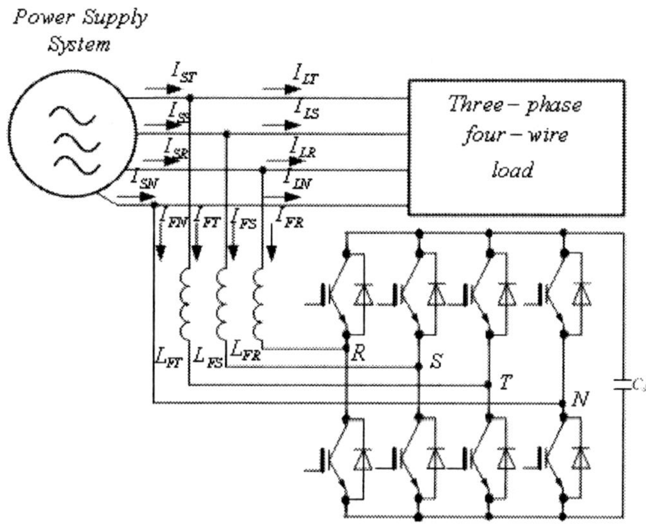

Using the equation (8.12), the maximum value of the "equivalent sinusoid" for the source current is calculated in real time for each period of the source voltage. Afterwards, different probabilities to trace this sinusoid are possible – hardware or software tracing.

On the basis of this method, also, a balance of three-phase asymmetrical loads may be made. First, active three-phase power P_{L3F} is calculated, afterwards, the maximum value of the "equivalent sinusoid": is calculated as the power is divided equally in the three phases:

$$\frac{P_{L-3F}}{3} = U_S \cdot \frac{I_{SM}}{\sqrt{2}} \tag{8.13}$$

A symmetrical system of supplying voltages is assumed in (8.13). From it, the three sinusoids for the three phase currents, which have to be consumed from the system load- three-phase shunt active power filter for the power factor to be closed to 1, may be obtained.

Different studies exist in the field of powers in the three-phase non-linear systems as well as of the possibilities to compensate non-active power (Sasdelli, 1994).

The theory of *"Akagi-Nabae"*, known also as "theory of transient values" or "p-q theory" (Akagi, 1993), is more frequently used in control of three-phase shunt active power filters. In it, a three-phase three-wire supply system – load is studied. Three-phase voltages and currents form space co-ordinates r – s - t are transferred in 2D coordinates $\alpha - \beta$ using the following equations:

$$\begin{bmatrix} e_\alpha \\ e_\beta \end{bmatrix} = \sqrt{\frac{2}{3}} \begin{bmatrix} 1 & -\frac{1}{2} & -\frac{1}{2} \\ 0 & \frac{\sqrt{3}}{2} & -\frac{\sqrt{3}}{2} \end{bmatrix} \cdot \begin{bmatrix} e_r \\ e_s \\ e_t \end{bmatrix} \tag{8.14}$$

$$\begin{bmatrix} i_\alpha \\ i_\beta \end{bmatrix} = \sqrt{\frac{2}{3}} \begin{bmatrix} 1 & -\frac{1}{2} & -\frac{1}{2} \\ 0 & \frac{\sqrt{3}}{2} & -\frac{\sqrt{3}}{2} \end{bmatrix} \cdot \begin{bmatrix} i_r \\ i_s \\ i_t \end{bmatrix} \tag{8.15}$$

In this theory, the terms "real transient power p" and "imaginary transient power q" are used. In accordance with the theory, the power p is defined into both co-ordinate systems, as:

$$p = e_r.i_r + e_s.i_s + e_t.i_t = e_\alpha.i_\alpha + e_\beta.i_\beta \tag{8.16}$$

The power q is defined only in 2D system:

$$q = e_\alpha.i_\beta - e_\beta.i_\alpha \tag{8.17}$$

The equations (8.16) и (8.17) may be presented as:

$$\begin{bmatrix} p \\ q \end{bmatrix} = \begin{bmatrix} e_\alpha & e_\beta \\ -e_\beta & e_\alpha \end{bmatrix} \cdot \begin{bmatrix} i_\alpha \\ i_\beta \end{bmatrix} \tag{8.18}$$

Since the power p in $\alpha - \beta$ co-ordinate systems is presented as a product of voltages and currents of the same subsymbols, it has also here unit of watt [W]. Besides the power q is a product of voltage and current, it has a different nature from those of p, because the voltages and currents are not of the same subsymbols. The unit of q is "imaginary watt" [IW]. Multiplication with the contrary matrix of the voltage in the left of the both sides of (8.18) leads to:

$$\begin{bmatrix} i_\alpha \\ i_\beta \end{bmatrix} = \begin{bmatrix} e_\alpha & e_\beta \\ -e_\beta & e_\alpha \end{bmatrix}^{-1} \cdot \begin{bmatrix} p \\ q \end{bmatrix} \tag{8.19}$$

The last equation may be presented as:

$$\begin{bmatrix} i_\alpha \\ i_\beta \end{bmatrix} = \begin{bmatrix} e_\alpha & e_\beta \\ -e_\beta & e_\alpha \end{bmatrix}^{-1} \cdot \begin{bmatrix} p \\ 0 \end{bmatrix} + \begin{bmatrix} e_\alpha & e_\beta \\ -e_\beta & e_\alpha \end{bmatrix}^{-1} \cdot \begin{bmatrix} 0 \\ q \end{bmatrix} = \begin{bmatrix} i_{\alpha p} \\ i_{\beta p} \end{bmatrix} + \begin{bmatrix} i_{\alpha q} \\ i_{\beta q} \end{bmatrix} \tag{8.20}$$

Taking in consideration (8.16) and (8.20), it is found:

$$\begin{bmatrix} p_\alpha \\ p_\beta \end{bmatrix} = \begin{bmatrix} e_\alpha.i_\alpha \\ e_\beta.i_\beta \end{bmatrix} = \begin{bmatrix} e_\alpha.i_{\alpha p} \\ e_\beta.i_{\beta p} \end{bmatrix} + \begin{bmatrix} e_\alpha.i_{\alpha q} \\ e_\beta.i_{\beta q} \end{bmatrix} \tag{8.21}$$

Therefore:

$$p = p_\alpha + p_\beta = e_\alpha.i_{\alpha p} + e_\beta.i_{\beta p} + e_\alpha.i_{\alpha q} + e_\beta.i_{\beta q} \tag{8.22}$$

After substituting (8.20) in (8.21), it is derived:

$$p = \frac{e_\alpha^2}{e_\alpha^2 + e_\beta^2} \cdot p + \frac{e_\beta^2}{e_\alpha^3 + e_\beta^2} \cdot p + \frac{-e_\alpha \cdot e_\beta}{e_\alpha^2 + e_\beta^2} \cdot q + \frac{e_\alpha \cdot e_\beta}{e_\alpha^2 + e_\beta^2} \cdot q \tag{8.23}$$

The sum of the last two members is zero. Therefore:

$$\begin{aligned} p &= e_\alpha . i_{\alpha p} + e_\beta . i_{\beta p} \\ 0 &= e_\alpha . i_{\alpha q} + e_\beta . i_{\beta q} \end{aligned} \tag{8.24}$$

The following conclusions are made using the last relationships:

- The first equation gives active power expressed in $\alpha - \beta$ co-ordinates in the three-phase.
- The second equation shows that the sum of the two written components is zero, i.e. they are mutually compensated and corresponds to the reactive power expressed in $\alpha - \beta$ co-ordinates in the three-phase system.

The purpose of the Akagi - Nabae's theory is that it shows the components $i_{\alpha q}, i_{\beta q}$ of the currents i_α, i_β, which the shunt active power filter has to eliminate.

The basic block-schema showing the realization of the method is given in Figure 13. The synchronization to the source voltage system is made using one of the voltages – in this case u_a, via PLL schema. The signals sinϑ и cosϑ are digitally generated and are required for forward and reverse Park's transformations. The three phase load currents - i_{Lr}, i_{Ls}, i_{Lt}, are monitored, and through them the components i_p, i_q are determined. At first, using (8.15) i_α, i_β are calculated; afterwards, forward Park's transformation is applied to them. Thus, the output currents of the block of the forward p-q transformation are:

$$\begin{bmatrix} i_p \\ i_q \end{bmatrix} = \sqrt{\frac{2}{3}} \cdot \begin{bmatrix} \cos\vartheta & \sin\vartheta \\ -\sin\vartheta & \cos\vartheta \end{bmatrix} \cdot \begin{bmatrix} 1 & -\frac{1}{2} & -\frac{1}{2} \\ 0 & \frac{\sqrt{3}}{2} & -\frac{\sqrt{3}}{2} \end{bmatrix} \cdot \begin{bmatrix} i_{Lr} \\ i_{Ls} \\ i_{Lt} \end{bmatrix} \tag{8.25}$$

The different harmonics of the phase currents, interacting to the phase voltages in the 2D co-ordinate system, will give different components of powers *p,q* (see equation (8.18)); furthermore, taking in consideration (8.19) and (8.25), these interactions will lead also to different components of the currents i_p, i_q.

The first harmonics of the phase currents, which are in phase to the phase voltages, interacting to the voltages will give uni-polar curves (similar to those after a full-wave rectification) with a DC component. Therefore, the DC components of the currents in (8.25) correspond to active power of the system.

The higher harmonics of the phase-currents, interacting to the phase voltages, will give AC components of the currents in (8.25). These are the components which have to be compensated by the active power filter. Because of that, the two currents are filtering using high-frequency filters (usually with a cut frequency of 10 Hz). Only AC components $i_{p\sim}, i_{q\sim}$ are obtained at the filter outputs. The outline of

Figure 13. Block-schema showing the realization of the control of a three-phase shunt active power filter in conformity with the Akagi - Nabae's theory

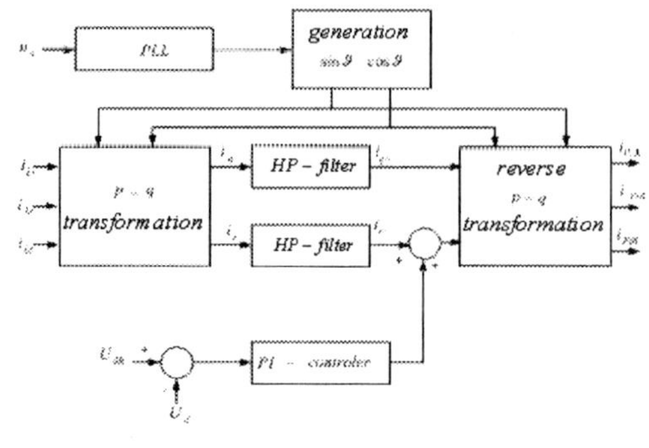

the stabilization of the voltage U_d across the capacitor C_d is additionally made. Moreover, the output signal of the PI-regulator finally determined the component $i_{p\sim}$ for the next block. In it, the reverse p-q transformation using reverse Park's transformation with a purpose to find the reference curves of the filter currents $i_{FrR}, i_{FsR}, i_{FtR}$ is made regarding the relationship:

$$\begin{bmatrix} i_{FrR} \\ i_{FsR} \\ i_{FtR} \end{bmatrix} = \sqrt{\frac{2}{3}} \cdot \begin{bmatrix} 1 & 0 \\ -\frac{1}{2} & \frac{\sqrt{3}}{2} \\ -\frac{1}{2} & -\frac{\sqrt{3}}{2} \end{bmatrix} \cdot \begin{bmatrix} \cos\vartheta & -\sin\vartheta \\ \sin\vartheta & \cos\vartheta \end{bmatrix} \cdot \begin{bmatrix} i_{p\sim} \\ i_{q\sim} \end{bmatrix} \tag{8.26}$$

Control signals for the filter power switches are generated from so-gained reference waveforms for the filter currents. The generation is made more frequently using a comparison of a saw signal with a usual frequency of 20 kHz.

The use of shunt active power filters in combination to a conventional means is probable. Figure 14 shows so-called hybrid filter – a combination of a shunt active power filter and passive filter (filters).

In this case, installation of a shunt active filter to already existed passive filter/s is possible. Moreover, a special attention has to be paid to the passive and active filter frequency ranges.

A use of series active power filters is probable to eliminate the harmonics of load current (Sigh, 1999; El-Habrouk, 2000). However, their more frequent and typical use is to improve the quality of the source voltage for critical consumers – elimination of voltage harmonics, disturbances, etc. Its function is possible to be mutual to those of an uninterruptible power source system. The operational principle of a series active power filter is illustrated by the use of the schematic shown in Figure 15. The connection to the source network is made via transformer Tr. In each time moment, voltage U_F is sumd to the source voltage U_S, in a way that the source voltage is complimented to a pure sinusoidal waveform. This voltage U_L is a source voltage to the load. Principle schematics to realize three-phase series active power filters are depicted in Figure 16 and Figure 17. Both schematics shown are supplied by a voltage source which is more frequently used; besides, a use of current source is also possible. The schematic shown in Figure 16 consists of three single-phase schematics, while this in Figure 17 is with a reduced

Figure 14. Connection of a hybrid filter as a non-linear load

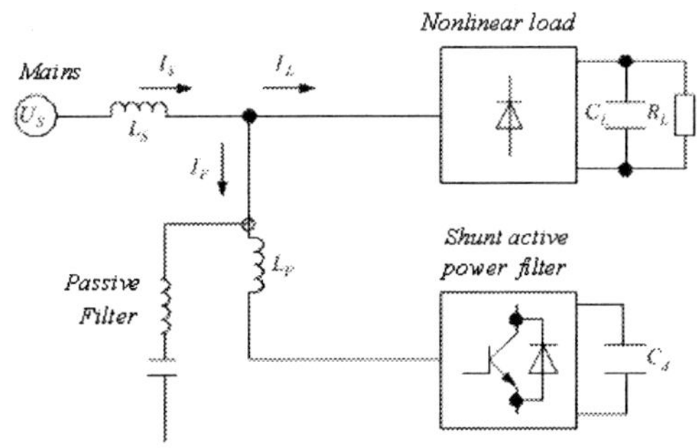

number of power devices. The last topology is based on a three-phase bidirectional transistor converter. In Figure 17, LC-filter is shown for each phase. The filter decreases the high-frequency components, caused by the switching of filter power devices, of the voltages between each two of the points r, s, t. The voltage of each capacitor is transformed towards the side of the source network and it is summed to the network voltage.

Unified power quality conditioner – UPQC is formed as a combined use a shunt and series active power filter (see Figure 18). The shunt filter improves the power factor of the load in respect to the supply network and eliminates the harmonics from the consumed source current. The series filter improves the quality of the voltage supplying the load and eliminates voltage disturbances. A model-based H_∞ control for PQC is described (Kwan, 2009).

The connection point of the active power filters is a task connected with a complex technical and economical analysis. It has to be solved thoroughly by a distributor and consumer of electrical power in conformity with the requirements of the standards.

Figure 15. Schematic illustrating the operation of a series active power filter

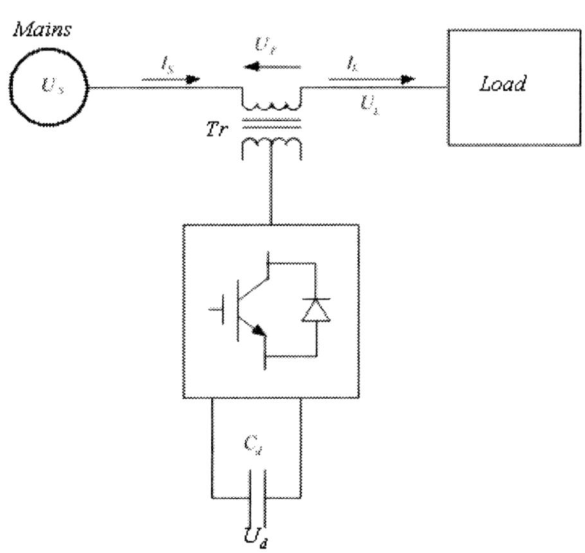

Figure 16. Principle schematic of a three-phase series active power filter implemented using three single-phase filters

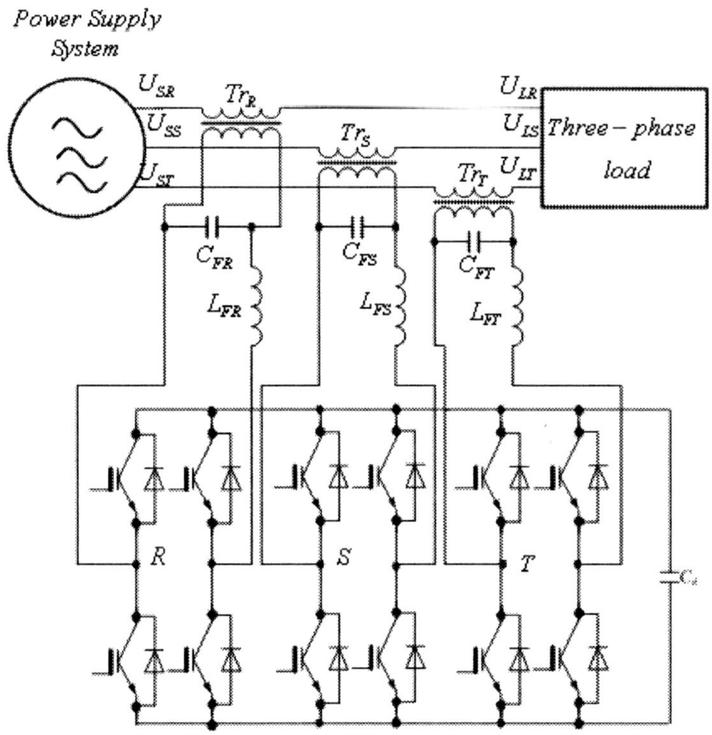

Figure 17. Principle schematic of a three-phase series active power filter implemented on the basis of a three-phase bidirectional transistor converter

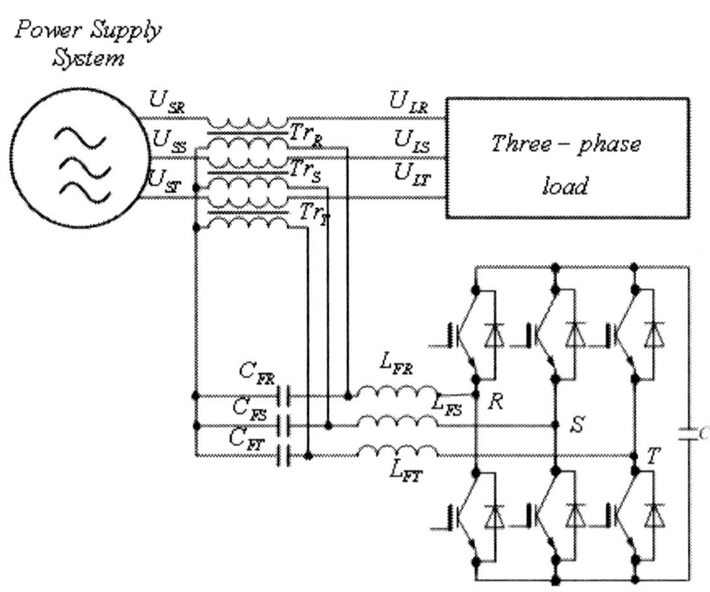

Figure 18. Structural schematic of unified power quality conditioner

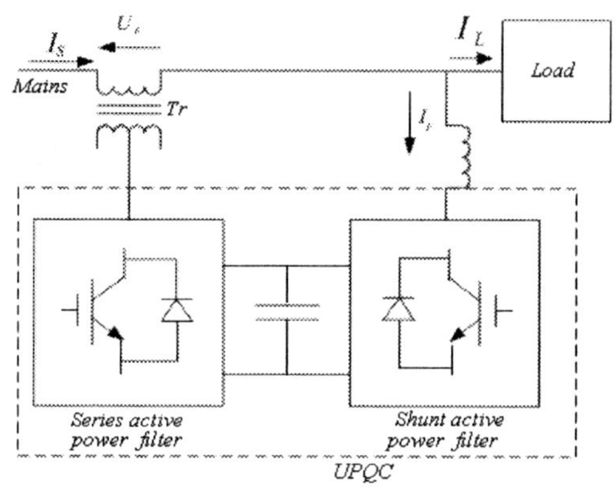

HIGH VOLTAGE DC TRANSMISSION

The HVDC technology is used to transmit electricity over long distances by overhead transmission lines or submarine cables (Funaki, 2006). It is also used to interconnect separate power systems, where traditional alternating current connections cannot be used.

In HVDC transmission, two types of power electronic converters of electrical energy are used–rectifiers and inverters (Bekink, 1994).

The rectifiers are usually implemented according the power schematic with a m =12 (see Chapter 4).

Further down, the three basic structures, which have found implementation in practice, are studied.

Monopolar HVDC is shown in Figure 19. It consists of the following:

- One high-voltage line for DC current transmission
- Return path is via earth ground or a low-voltage conductor
- Rating of up to 1500 MW

Figure 20 depicts bipolar HVDC. It consists of the following:

Figure 19. Monopolar HVDC

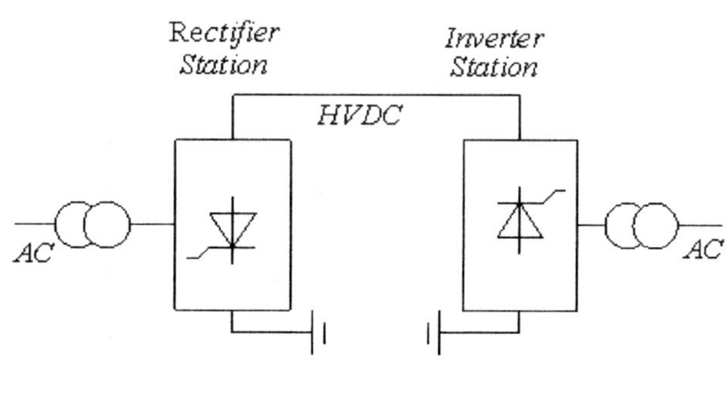

- Two DC lines with ± DC voltage level for transmission
- Rating up to 3000 MW

Figure 21 shows an HVDC back-to-back station. It is normally used to create an asynchronous interconnection between two AC networks, which could have the same or different frequencies. In these installations, both the rectifier and inverter are located in the same station. These can be used to interconnect two different systems that have the same frequency but are not synchronized or can be used to interconnect systems of different frequencies (50 Hz, 60 Hz).

Upgrading the nation's electrical transmission system infrastructure using FACTS ans HVDC can provide the following benefits:

- **Increasing capacity:** Increasing the real power capacity may reach up to 40%
- **Enhancing reliability:** FACTS can mitigate the effects of faults and make electricity supply more secure by reducing number line trips

Figure 20. Bipolar HVDC

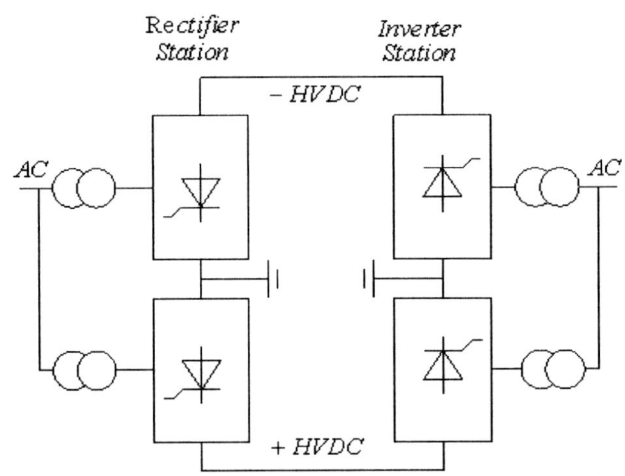

Figure 21. Back-to-back HVDC

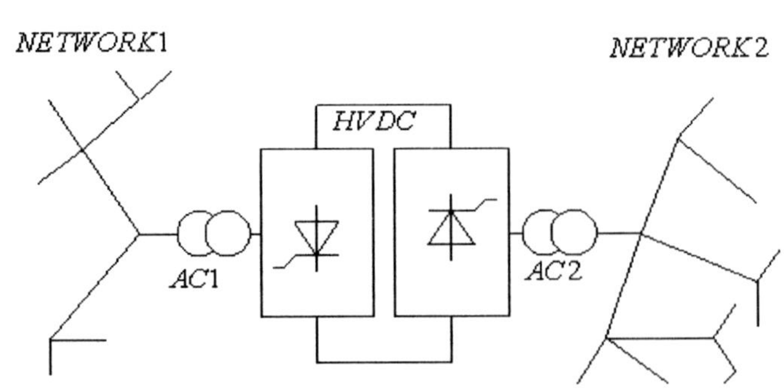

- **Improving controllability:** These converters can control the flow of power and the regulation voltages in the power network
- **Preserving the environment:** FACTS devices contain no hazardous materials and produce no waste or pollutant
- **Saving time and money:** These devices help distribute electrical energy more economically through better utilization of existing installations, thereby reducing the need for additional transmission lines

The development of the methods to convert energy from the (see Chapter 9) and their use in the distribution electrical energy generation directs the researches to improve the energy quality at the stage of its conversion from the renewable source (Lo, 2008; Molinas, 2008).

The creation of the Distribution Automated Systems allows the integration of the means of continuous power quality measurements in these systems (Munoz, 2008).

REFERENCES

Akagi, H. (2006). *Modern active filters and traditional passive filters.* Poland. *Bulletin of Polish Academy of Science*, *54*(3), 255–269.

Akagi, H., Kanazawa, Y., & Nabae, A. (1993). Generalized theory of instantaneous reactive power in three-phase circuits. In [Tokyo.Japan.]. *Proceedings of International Power Electronics Conference*, *93*, 1375–1386.

Antchev, M. H., Petkova, M. P., & Gurgulitzov, V. T. (2007). *Sliding mode control of a series active power filter.* Paper presented at the 5th IEEE Conference EUROCON 2007.

Antchev, M. H., Petkova, M. P., & Kostov, A. T. (2007). Hysteresis current control of single–phase shunt active power filter using frequency limitation. In *Proceedings of 9th IASTED PES Conference* (pp.233-238). Clearwater.USA.

Asprund, G. (2008, June). *Electric transmission systems in change.* Paper presented at the PESC 2008. Rhodes. Greece.

Bekink, J. S., & Colijn, A. J. J. (1994). High voltage DC power supplies. *European Power Electronics and Drives Journal*, *4*(2), 11–13.

Chen, B. S., & Hsu, Y. Y. (2008). A minimal harmonic controller for a STATCOM. *IEEE Transactions on Industrial Electronics*, *55*(2), 655–664. doi:10.1109/TIE.2007.896266

El-Habrouk, M., Darwish, M. K., & Mehta, P. (2000). Active power filters: A review. *IEEE Proceedings B. Electric Power Applications*, *147*(5), 403–413. doi:10.1049/ip-epa:20000522

Funaki, T. (2006). High power electronics in Japan – backgrounds and applications. *IEEE Power Electronics Society Newsletter*, *18*(4), 13–15.

Hingorani, L. G., & Gyugyi, L. (2000). *Understanding FACTS: Concepts and technology of flexible AC transmission systems.* New York: IEEE Press.

Kwan, K., Chu, Y., & So, P. (2009). Model-based H_∞ control of a unified power quality conditioner. *IEEE Transactions on Industrial Electronics*, *56*(7), 2493–2504. doi:10.1109/TIE.2009.2020705

Lo, Y. K., Lee, T. P., & Wu, K. H. (2008). Grid-connected photovoltaic system with power factor correction. *IEEE Transactions on Industrial Electronics*, *55*(5), 2224–2227. doi:10.1109/TIE.2008.921204

Luo, A., Tang, G., Shuai, Z., Tang, J., Xu, X., & Chen, D. (2009). Fuzzy-PI-based direct-output-voltage control strategy for the STATCOM used in utility distribution systems. *IEEE Transactions on Industrial Electronics*, *56*(7), 2401–2411. doi:10.1109/TIE.2009.2021172

Molinas, M., Suul, J. A., & Undeland, T. (2008). Low voltage ride through of wind farms with cage generators: STATCOM versus SVC. *IEEE Transactions on Power Electronics*, *23*(3), 104–1117. doi:10.1109/TPEL.2008.921169

Munoz, A. M., & De La Rosa, J. G. (2008). Integrating power quality to automated meter reading. *IEEE Industrial Electronics Magazine*, *2*(2), 10–18. doi:10.1109/MIE.2008.923520

Orizondo, R., & Alves, R. (2006).*UPFC simulation and control using the ATP/EMTP and Matlab/ Simulink programs.* Paper presented at the IEEE PES Transmission and Distribution Conference and Exposition Latin America. Venezuela.

Pinarrelli, A., Menitti, D., Martinis, U., & Andreotti, A. (2001, June). *Modeling of unified power flow controller into power systems using P-Spice*. Paper presented at the International Conference on Power Systems Transients (IPST) 2001. Rio de Janeiro. Brazil

Sasdelli, R., & Montanari, G. C. (1994). Compensable power for electrical systems in nonsinusoidal conditions. *IEEE Transactions on Instrumentation and Measurement*, *43*(4), 592–598. doi:10.1109/19.310173

Sigh, B., Al-Haddad, K., & Chandra, A. (1999). A review of active filters for power quality improvement. *IEEE Transactions on Industrial Electronics*, *46*(5), 960–971. doi:10.1109/41.793345

Tolbert, L. M., King, T. J., Ozpineci, B., Campbell, J. B., Muralidharan, G., Rizy, D. T., et al. (2005, December). *Power electronics for distributed energy systems and transmission and distribution applications.* USA: OAK Ridge National Laboratory. Retrieved from http://www.ornl.gov.

Welsh, M., Mehta, P., & Darwish, M. K. (2000). Genetic algorithm and extended analysis optimization techniques for switched capacitor active filters. *IEEE Proceedings B. Electric Power Applications*, *147*(1), 21–26. doi:10.1049/ip-epa:20000006

Chapter 9
Conversion of Electrical Power from Renewable Energy Sources

OVERVIEW

Towards the end of the previous century, the humanity understood very clearly two facts – first, the World supplies of fossil fuels (coal, oil, gas, uranium) are limited, and, second, industrial development and classical generation of electrical energy seriously endanger the environment.

Renewable energy sources (sun energy, wind energy, bio fuels, etc.) are based on the use of natural fluxes of energy (Masters,2004). That is why they are considered to be inexhaustible.

In specific cases of implementations, for example in lighting, a direct generation of electrical energy using photovoltaics is outlined as a long-term one.

The name photovoltaic comes from the combination of two Latin words – "photos" means light and "volt" means a measurement unit for electromotive force. The unit volt derives its name after the Italian scientist Alessandro Volta (1745-1827) who discovered battery (primary electrochemical source of electrical energy). The photovoltaic is used to convert sun energy into electricity. The official discovery of the photovoltaic is assumed to be made by French scientist Alexadre-Edmond Becquerel (1820-1891) in 1839. He discovered the effect upon silver plates. First data about photovoltaic effect in solid materials dated from 1877. Adams and Dew described variations, which they had monitored of electrical capabili-

DOI: 10.4018/978-1-61520-647-6.ch009

ties of selenium when it was left on sunlight, i.e. they discovered the photovoltaic effect in non-metals. In 1883, Charles Fritz, an electrician from New York, designed sun cell from selenium closed in quite high number of aspects to silicon cell used nowadays. The selenium cell consisted of a thin plaster of selenium, covered with very thin semi-transparent level of golden filaments and it was protected from a piece of glass. This cell was rather inefficient one.

The first solar cells had efficiency less than 1%. Nevertheless, the selenium cells began to be used in photography. The main reason for inefficiency of these devices became known later in first decades of XX century when physicists as Max Planck (1858-1947) developed new features in fundamental property of materials.

The break in high-efficient solar cell use started in 1950 in Bell Laboratories in New Jersey with an assistance of several scientists: Daryl Chapin, Calvin Fuller and Gerald Pearson. They made a research into influence of the light upon semiconductors.

In 1953 the team Chapin-Fuller-Pearson, relying on previously made discoveries in Bell laboratories about photovoltaic effect in silicon, created enriched silicon parts. These parts were more effective then the previously used one for obtaining electricity from sun light. The team succeeded to increase the efficiency of their cells up to 6%. Bell laboratories started to demonstrate the practical usage of solar cells for supplying provincial telephone stations in the midst 50th.

In 1958 however, solar cells started to be used for a supply of little radio-transmitters in the fourth American satellite Vanguard I. After these first successful demonstrations, photovoltaic usage as energy source turned to be universal for space ships. Fast headway in development of the photovoltaic cells and reduction of their price and their size contribute to their entering in cosmic and electronic industry during the last several decades. Networks of connected photovoltaic stations operate in the time being in USA, Germany, Italy, Switzerland, Spain, Japan, Mexico, etc. Photovoltaic supply networks are used in industry and business buildings.

The efficiency of the best silicon cell reached in laboratory environment is 24% while the efficiency of the best photovoltaic modules, which operate and are offered nowadays is about 16%.

"Spectrolab" under financing of the government was demonstrated solar cells with an optical concentrator of record-breaking efficiency of 40.7% (King, 2007). The technology is based on common structures of sun elements, thus, it allows to be caught a wider part of sun spectrum using several levels in each element. Each element catches a segment of sun light thus allowing more efficient conversion of energy.

There are announcements about efficiency coefficient reached 42.8% at solar energy conversion from the research time with University of Delaware (Jacquot, 2007).

Structural schemas of electrical supply systems, using renewable sources, are determined by the consumer's requirements. The common thing in the more cases is the requirement to have accumulating energy block in their structure. This requirement is imposed by the fact that the time during which energy may be generated not always coincide to the time when energy is consumed (sun shines during the day, but also electrical energy is necessary during the night). A presence of the accumulating block required, however, makes possible the achievement of uninterruptible supply. Therefore, the structural schemas used with renewable sources are compatible in general with the uninterruptible power supply systems.

Different structural (block) schematics for energy conversion using renewable sources are examined further down. There are several indications, which are significant in their study. The first is whether they have possibility to accumulate energy and what type is accumulated energy. The second is whether they have connection to a distribution supply network or not (Blaabjerg, 2004; Carrasco, 2006). The third indicator is the end electrical energy type – AC or DC power dependent on the consumer's type.

CONVERSION OF SOLAR ENERGY

At the period between 2002 and 2003, the total installed power of photovoltaic systems in the countries members of the International Energy Agency (IEA) increases with 36% and reaches 1.809 MW (Teodorescu, 2005). A change of an architectural structure of home buildings is noticed. The change is considered to tendencies of embedding photovoltaic systems in the buildings. Simultaneously, centralized photovoltaic energy networks are also implemented – for example, in Germany and USA with power up to 5MW. In contrast to eventually generated energy using wind, the solar energy is more precisely foreseen. It is assumed to be made with accuracy below 2% for a 5 year period. Three types of photovoltaic modules are mainly offered at the Market. They consist of:

- Mono-crystal Si elements
- Poly-crystal Si elements
- Amorphous Si elements

Spectrum sensibility of the different materials is different. Figure 1 displays the spectrum characteristics of photovoltaics made from amorphous- and mono- silicon materials.

A photovoltaic module is a combination of photovoltaic elements (photovoltaics), which are connected in groups in parallel or series to be able to obtain higher voltages and higher currents.

Figure 2 shows the relationship between the open-circuit voltage Voc, i.e. no current is consumed from the element, and the current of short-circuit - Isc for the photovoltaic. The open-circuit voltage dependent on the material is within the range of 0.7-0.9V. The short-circuit current is within the range of 50-100 mA/sm^2.

The power, which may be consumed from the photovoltaic, depends on the loading and it has maximum. Figure 3 illustrates a graphic for a change of power Psg, consumed from the photovoltaic and it also present a photovoltaic VA characteristic. The point corresponding to the maximum power (maximum power point) is specific one.

Figure 4 displays characteristics of power P, which is consumed from the photovoltaic module at different intensity of solar radiance Q.

Figure 1. The spectrum characteristics of photovoltaics made from amorphous- and mono- silicon materials

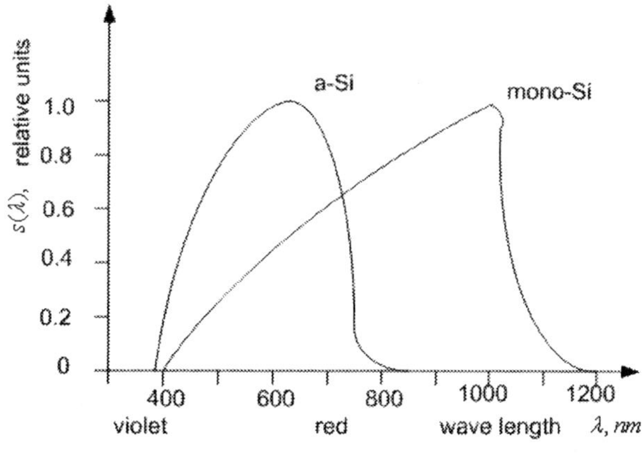

Figure 2. The relationship between the open-circuit voltage and the short-circuit current of a photovotlaic

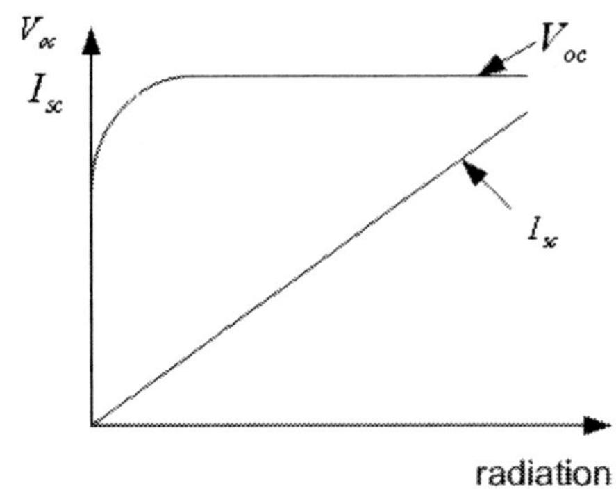

The area D corresponds to a maximum power. Figure 5 depicts volt-ampere characteristics of a module at different intensities of solar radiance (1000W/m^2, 500 W/m^2, 100 W/m^2). A change of temperature leads to a change of volt-ampere characteristic – dotted line in Figure 5. It has to be mentioned that based on the reasons of the change at the radiance and temperature changes, different algorithms for maximum power point tracing are provided in the control of the power electronic converters applicable in photovoltaic installations.

Figure 3. A graphic for a change of power Psg and photovoltaic VA characteristic

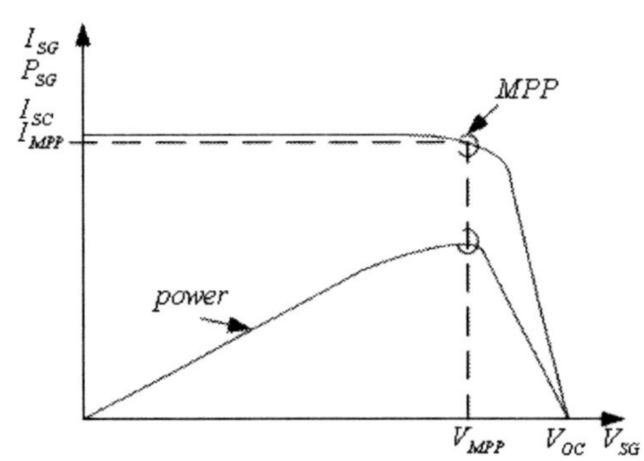

Figure 4. Characteristics of power P, which is consumed from the photovoltaic module at different intensity of solar radiance Q

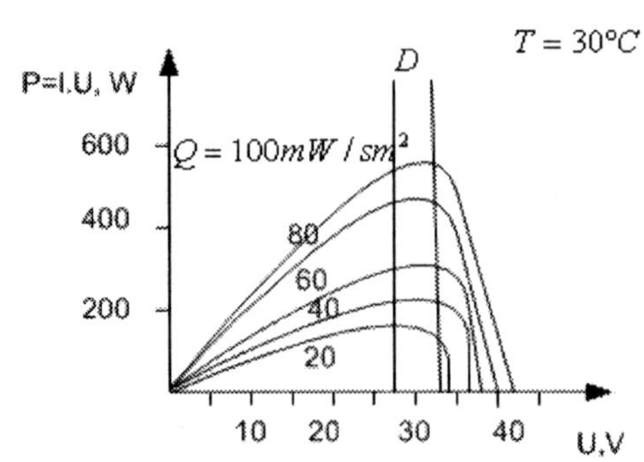

Figure 5. Volt-ampere characteristics of a module at different intensities of solar radiance

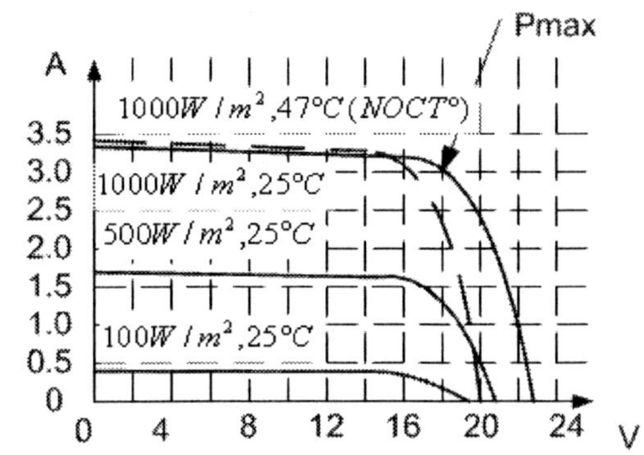

At first, the systems using photovoltaic modules for electrical supply for DC consumers will be examined. Figure 6 displays the structural schema of the system without an element for energy storage. Using DC/DC converter with appropriate characteristics, voltage of a desire value for the consumer is produced. Since there is no element for energy storage, naturally, the generation of desired voltage value is probable to happen at available conditions in regards to the sun shining.

To store energy is demanded in many cases of usage energy from renewable energy sources. The choice of an appropriate storage element is connected with different reasons in respect to power, delivery time, etc. (Holm, 2002).

Figure 7 shows the structural schema of the system with an element for energy storage. The first DC/DC converter has appropriate characteristics to charge the energy storage element, because of that it

Figure 6. Structural schema of the system without an element for energy storage

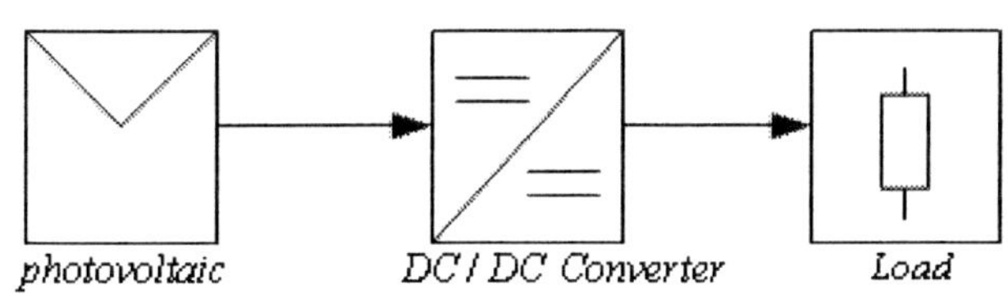

Figure 7. Structural schema of the system with an element for energy storage

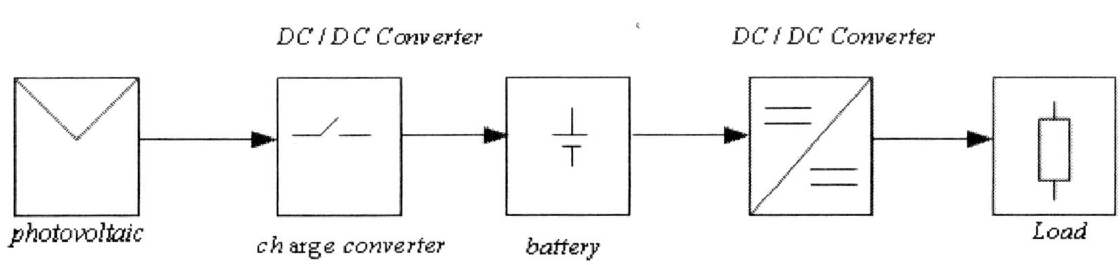

Figure 8. Structural schema of a system meant to supply an AC consumer

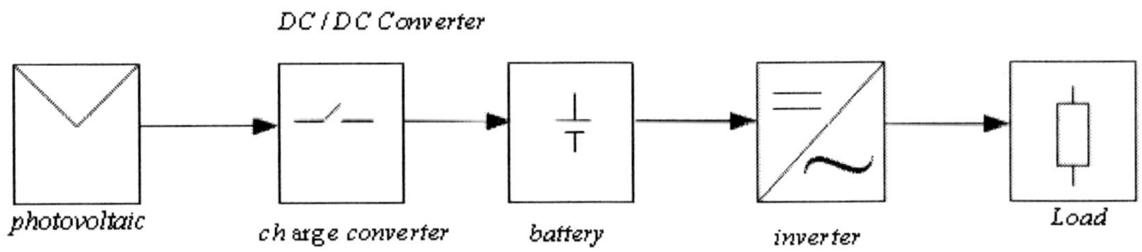

is sometimes called charge controller. The second DC/DC converter supplies voltage of a desire value for the consumer. This converter usually also stabilizes the voltage applied to the load within a definite range.

Figure 8 displays the structural schema of a system with an energy storage element which is meant to supply an AC consumer. This schema differs from this shown in Figure 7 in the second converter. Here it is a DC/AC converter – inverter.

The role of the Power Electronics is significant in distributed generation of electrical energy (Blaabjerg, 2004).

The requirements to the power electronic converters (inverters) finally connected to the distribution network are studied in connection with the usage of the RES and the development of the idea of the distributed electrical energy generation (Enslin, 2003).

Systems to convert solar energy and designed only to transfer the generated energy on the distribution supply network exist. Their general structure is shown in Figure 9. The inverter types used in them, as well as inverter control algorithms, may be different. Further down, a special classification of the inverters most often applied for the purposes of conversion at usage of solar energy is given.

Systems for solar energy usage, having possibilities to supply household AC appliances and to transfer energy on the distribution supply network, for example, are very attractive. These systems are designed using the structural schema shown in Figure 10.

Of course, only the main blocks are shown in the figure. One has to have in mind that such a system requires definite features regarding the inverter control. As it is seen, there is a possibility to monitor both the consumed energy for household from the distribution network and the delivered energy on the network. The payment may be done on these accounts. In the different countries, there are defined corresponding prizes for buying electrical energy delivered on the distribution network from types of

Figure 9. General structure of a system meant only to convert solar energy and transfer it on the distribution supply network

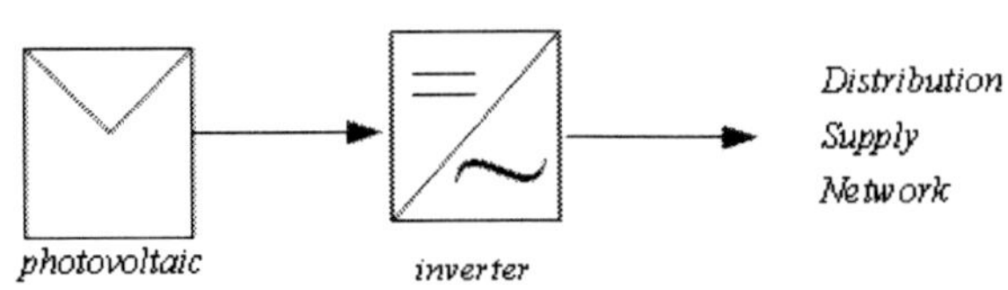

Figure 10. Systems having possibilities to supply household appliances or to transfer energy on the distribution supply network

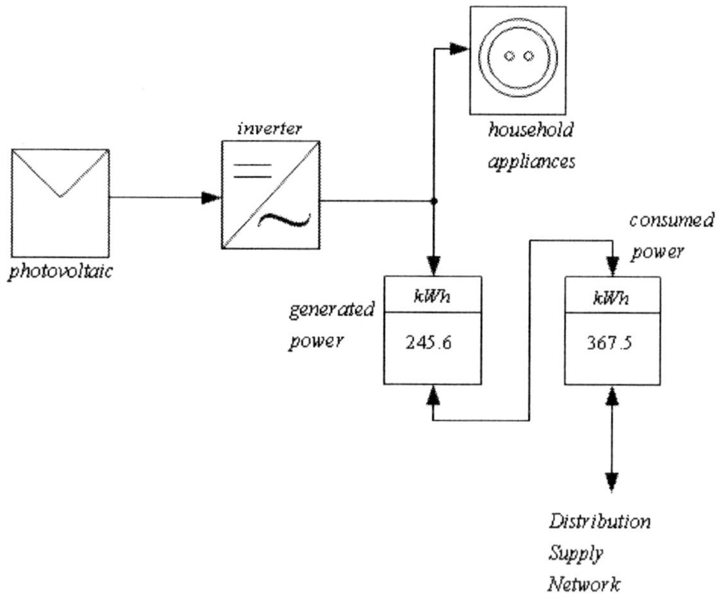

system such as shown in Figure 9 and Figure 10. However, the initial implementation of the systems requires significant investments.

Figure 11 illustrates special features of implementation of the inverter shown in Figure 9. Three versions are shown – 1. Central inverter, 2. String inverter and 3. Module integrated inverter. Special features of the three versions are the following:

The first version is applied in photovoltaic energy installations – more often in three-phase systems with power higher than 10 kW. The photovoltaic modules are connected in particular configurations – in series with or in parallel to the DC side of the inverter. Disadvantage is the dependence of the reliability of the entire installation on the single inverter. Therefore, very high requirements are imposed on the reliability and energy indicators of the inverter. In significant power (several MW), the inverter is designed using several inverters operating in parallel, each of them having tentative single power of 100 – 250 kW.

The second version is used in small house installation with a single power 1.5 up to 5 kW. Groups of photovoltaic modules are connected to each inverter. This allows monitoring the maximum power point of each of these groups. This version is a very flexible system permitting the separated groups to have a different space orientation – south, north, west, and east. The orientation depends on the possibilities of the architecture and household. The parallel operation of several inverters, characterizing the structure, increases the reliability.

In the third version, each photovoltaic module has its own inverter (Ho, 2005). The inverter is optimized to specific characteristics of its module – maximum power, temperature, etc. The operational conditions of the inverter embedded in the module, however, are very burdensome. The necessity to design and implement it with a certain reserve regarding these conditions does not permit optimizing of the inverters energy indicators. These kinds of systems require complicated wiring in the AC side.

Figure 11. Three versions for implementing the inverter shown in Figure 9

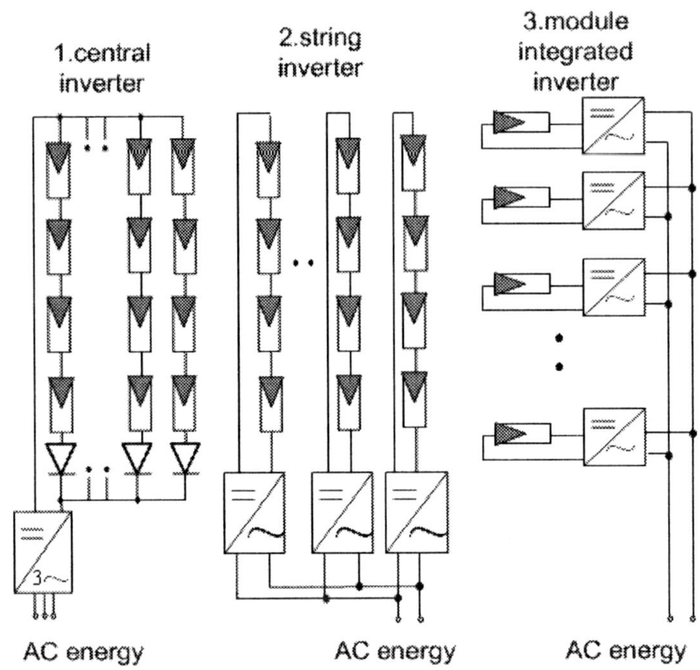

The installation process is very complicated and that is the reason why this version is usually orientated towards façade photovoltaic systems with power from 50 up to 400 W.

The structure chosen as well as the elements connected with the structure determine the reliability of the whole system. Researches into these topics are published in (Petrone, 2008).

As it is known, the inverter is converter of DC into AC electrical energies. A special classification of power electronic converters used in the solar energy systems is shown in Figure 12 (Teodorescu, 2005). Here, the name inverter is used in a wider mean, spread recently especially among people without a special knowledge in the field. The first mark of the classification is whether there is an interstitial DC conversion with a separate DC/DC converter or not. The second sign is whether there is a galvanic isolation or not. As it is seen in Figure 12, at the version without the interstitial DC conversion, the variants are two – with or without isolation, while at the version with the interstitial DC conversion, the variants are there – without isolation, with isolation in high-frequency side (in DC/DC converter) and with isolation in low-frequency side (50 Hz).

The following versions with an interstitial DC conversion are quite widely spread – first, with galvanic isolation – a push-pull DC/DC converter, performing also galvanic isolation in the high-frequency side, followed by a single-phase bridge inverter; second, without galvanic isolation – a boost DC/DC converter without galvanic isolation, followed by a single-phase bridge inverter.

In the inverters embedded in the photovoltaic modules, the variant of a flyback converter and single-phase bridge inverter is often used. Universal inverter for connection to the grid is described (Prasad, 2008).

Recently, a quicken activity in researches in the field of solar energy usage regarding both the schematic solutions of conversion and the systems and algorithms for indication and control is observed (Aglietti, 2009).

Figure 12. Special classification of inverters used in solar energy systems

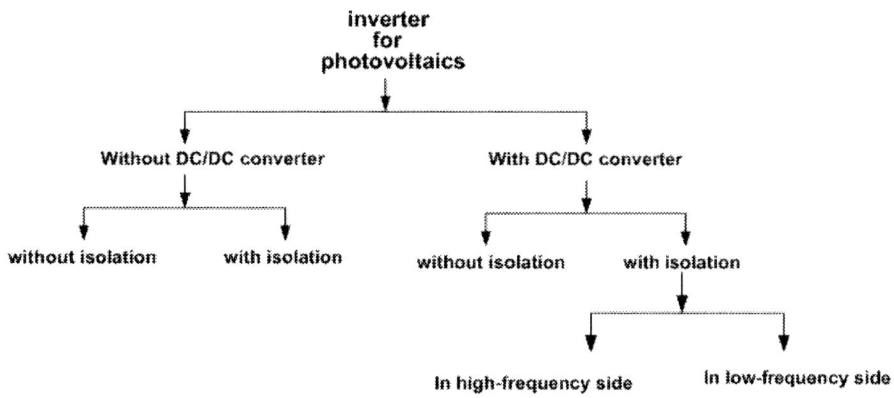

CONVERSION OF WIND ENERGY

The first wind turbine to generate electricity was designed in 1800. Almost a century later, about 1900, some rural properties used small wind turbines to be distributed with electricity. These turbines contained small generators, for example, an old auto generator.

Towards 1994 the total installed power in wind turbines is about 3500 MW. According to data from Global Wind Energy Council, the total power has increased from 7976 MW in 2004, and nowadays it is 47317 MW (Cheng, 2006). Countries as Germany, Spanish, USA, Denmark, the Netherlands, India, etc, have the major contribution to the increase. The wind energy is the fastest developing source of energy. At present, the biggest wind turbine offered at the Market has diameter of its propeller 104 meters and it generates 3.6MW electrical energy. Results from a research of General Electric into a turbine with a diameter 140 meters and capacity to7 MW are soon expected.

The development in wind energy usage is in two directions: the first is in small wind installations for an individual use and the second is in creation of centralized installations of turbines – wind generation farms with a high power (Gipe,2006). Sea areas closed to the main land are used for the centralized installations. Such an installation exists, for example, in Massachusetts, USA. It is situated on 25 square miles. It contains 130 turbines with a probable electrical power to be delivered of 420 MW.

Figure 13 depicts a typical relationship in tentative units between power of a wind turbine and the wind speed.

Figure 14 displays a common structural schema of wind generator of a comparatively high power meant to deliver energy on the distribution network.

Also, so-called autonomic wind generators exist. They supply objects not connected to the distribution network. These generators have a simpler structure.

On the point of view of the operational range of wind speed, at which the wind generators may operate, there are wind generators that operate with a constant speed of the turbine and wind generators that operate with a variable speed of the turbine.

The wind generators that operate with a variable speed of the turbine are more complicated than the other ones, but they can generate energy at low speed of the wind (typically under 7-9 m/s). This leads to

Figure 13. Typical relationship in tentative units between power of a wind turbine and the wind speed

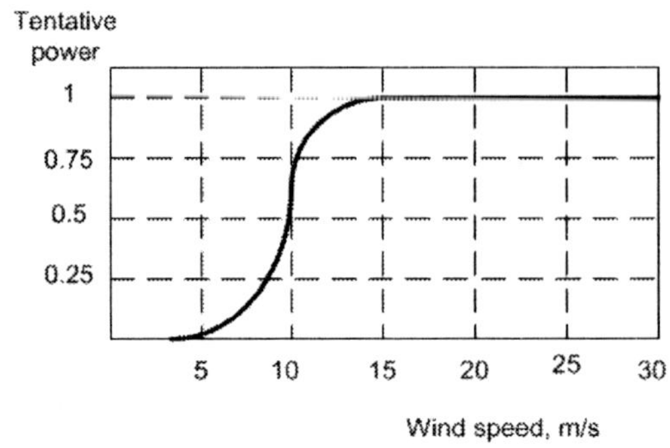

augmentation of generated power. The augmentation in a year period may be 12-18% and it is strongly dependent on the spot of the wind generation installation.

Generally, a structural schema of wind generator contains mechanical and electrical parts. The first part includes a turbine (rotor) and power transmission (gearbox). The second part includes an electrical generator, power electronic converters only in part of the implementations and transformer to connect to the distribution network (Stiebler,2008).

Wind generators that operate without a change of the speed of their turbine have found a wide implementation because of their simplicity. These generators are made using an asynchronous generator with a short-circuit rotor and is directly connected to the electrical network using a transformer. The generator delivers generated energy on this network. Owing to the features of the asynchronous generators used, the speed of the rotating of generators, and from that also the speed of the wind turbines, is practically unchangeable. The speed is determined by the frequency of the network to which the generators are connected. These generators can not operate at low speeds of wind. At such conditions, they turn off not to change their mode from generator to motor modes – i.e. the propeller of the turbine would be turned not by the wind but by the generator, which would consume energy from the electrical network.

Simple asynchronous generators with a short-circuit rotor may be used also independently – as autonomic ones. In such a use, they are not connected to the electrical network, but a block for self-excitation is added. This block consists of capacitor batteries. The implementation of these generators is limited and it is in low powers (Simoes, 2006).

The wind generators without a change of the speed of their turbine are not able to use the entire potential of the wind. If a generator operates with a change of the speed of its turbine, the generator is

Figure 14. Common structural schema of wind generator of a comparatively high power

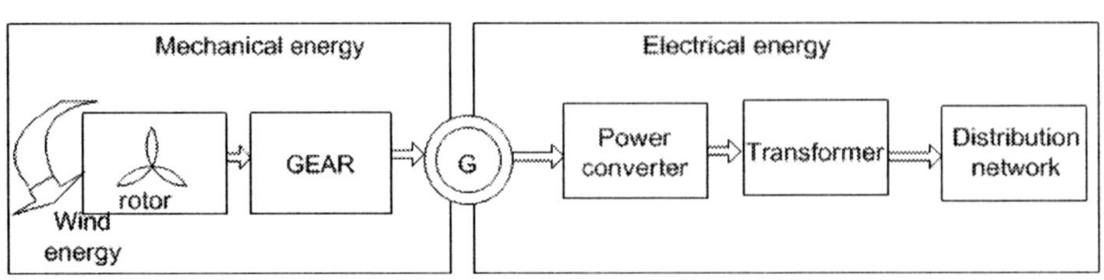

able to use the entire potential of the wind and the total annual production increases with 15%. These wind generators are more complicated ones but contemporary solutions are directed exactly to the wind generators with a variable speed. Two generators find place in widely used turbines with a speed change: double fed induction generator and synchronous generator (Cheng, 2006).

Figure 15 displays general blocks of the system with asynchronous generator. This system is used in medial powers and is known as Double – Fed Induction Generator (DFIG) system. Here, the stator windings are directly connected to the distribution electrical network via a transformer. Power electronic block is connected between the network and the rotor windings. This block consists of two three-phase bridge schematics with a common capacitor battery. They forms so-called back to back converters. Both converters are bidirectional ones. This block makes possible the operation of the generator with a variable speed of its turbine (Petterson, 2005).

If n_S is synchronous mechanical speed at the synchronous frequency f_S, the resulting output power comes from the induced voltage proportional to the relative speed difference between the electrical synchronous rotation and the mechanical rotation within a speed slip factor range, given by:

$$S = \frac{n_S - n_R}{n_S}, \tag{9.1}$$

where S is the slip and n_R is the rotor speed.

If the turbine, consequently the generator, operates with low speed (positive slip), the converters give towards the rotor electrical energy with a voltage having a frequency which complements the negative slip to the zero slip which is determined by the frequency of the distribution network. In this case, the direction of energy transmission using the converters is from the network to the rotor. The right converter in the figure operates as a rectifier, and the left converter – as an inverter.

If the turbine, consequently the generator, operates with low speed (negative slip), the converters consume energy from the rotor with a voltage having frequency which also compliments the negative slip to the zero slip which is determined by the frequency of the distribution network. The direction of energy transmission is from the rotor (and from the stator) to the distribution network. The left converter in the figure operates as a rectifier, and the right converter – as an inverter.

It has to be mentioned that in this configuration the power of the power converters is lower than the total power of the wind generator. Their power depends on the chosen range of the change of the speed

Figure 15. General blocks of the system with asynchronous generator

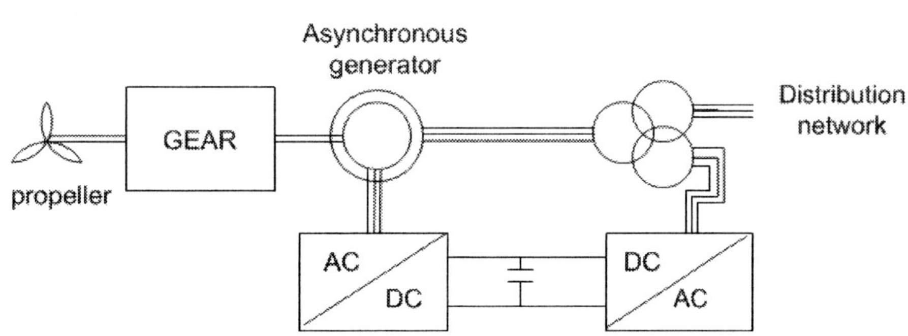

of the turbine. For a change of the speed with ±30% around the zero slip the power of the converters is also about ±30% of the power of the wind generator. The important is problem for reactive power in DFIG–system (Kayikci,2007).

There is a version of the system, in which the rotor windings are connected to an uncontrolled three-phase rectifier. The rectifier supplies a DC/DC converter. Load of the converter is a power resistance, in which the rotor winding energy is dissipated. Using an appropriate converter control leads to a change of the resistance of the rotor windings. Compared to the first version, this one has decreased energy efficiency.

The next three systems use synchronous electrical generators. In the three versions, power electronic converters between the generator and the network have to be designed for the total power of the wind generator.

The first version is with synchronous generator with a power transmission (gearbox) – Figure 16. The generator voltage is rectified and then it is inverted. The system contains also an additional converter with a lower power to control the excitation flux.

Figure 17 displays a version with a multipolar synchronous generator with *a* power transmission (gearbox) and without a control of excitation flux.

In the last version – Figure 18, a multipolar synchronous generator with *a* power transmission (gear-box) is also used but it is with an excitation using constant magnets.

Figure 16. Wind generator system with synchronous generator with a transmission housing

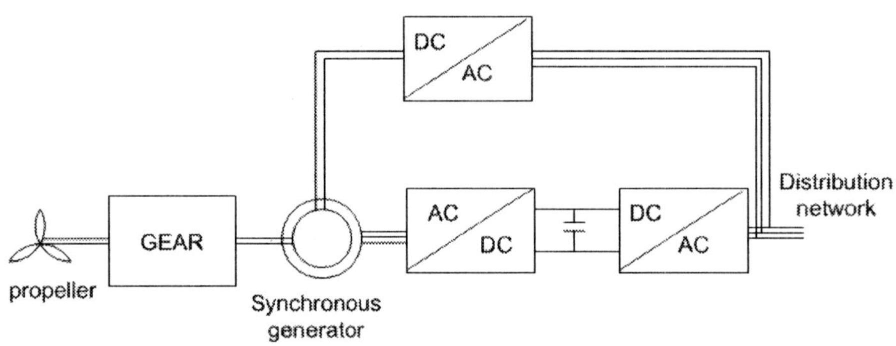

Figure 17. Version with a multipolar synchronous generator

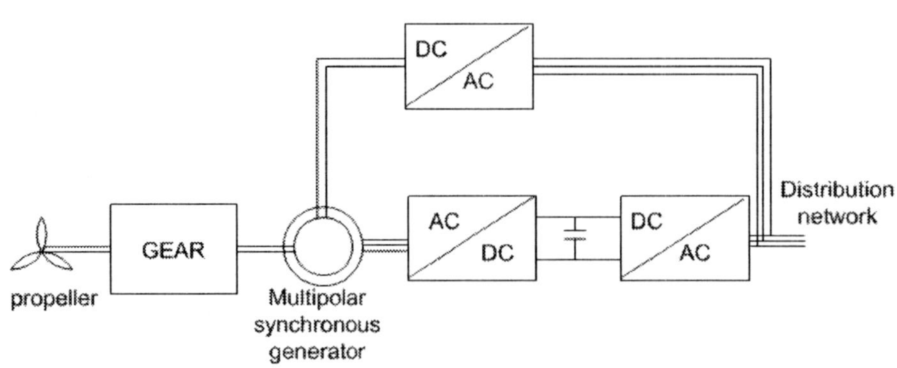

Figure 18. A multipolar synchronous generator with a transmission housing (gear) and with an excitation using constant magnets

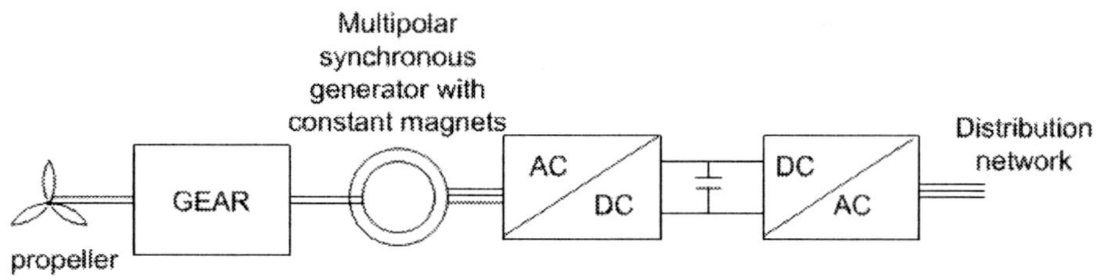

A special feature of all systems with synchronous generators is the probability for a quick control of active and reactive power because of the presence of the inverter connected to the network. Power management in wind/photovoltaic/fuel cell system is investigated (Caisheng, 2008).

The researches in the field of the systems using wind energy on one hand are connected with an improvement of the characteristics of electrical machinery used and also to creation of new machines particularly for this purpose. On the other hand, the researches are connected with an improvement of the characteristics of the power electronic converters based on novel power devices and appropriate algorithms for their control. Researches are made in turbine farms situated closed to the main land for an integration of electrical power produced from wind and electrical power produced from water. Other tendency is the usage of wind energy for a hydrogen generation.

A version using photovoltaic generator and inverter to transfer electrical energy on the distribution network is shown in Figure 19.

Figure 19. Version using photovoltaic generator and inverter to transfer electrical energy on the distribution network

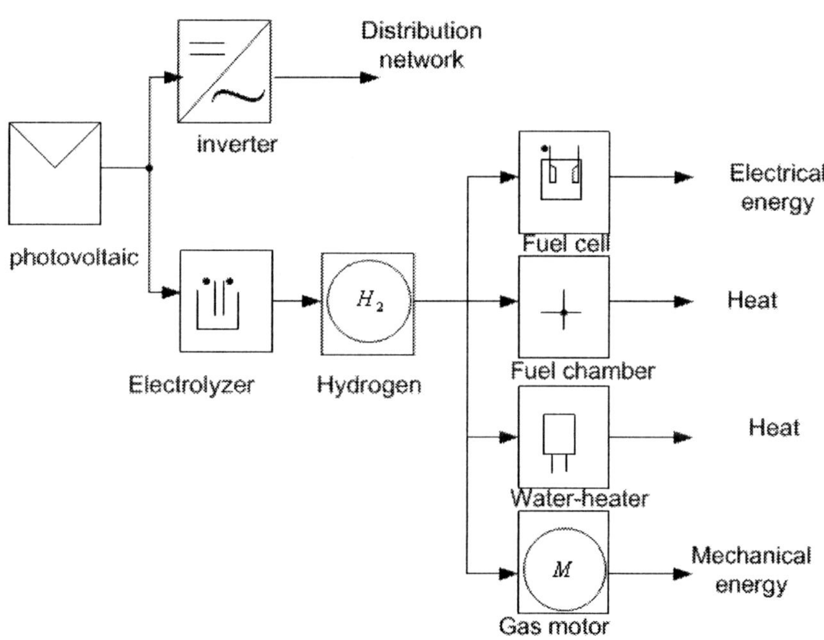

Figure 20. Distributed structure of an electrical energy supplying system

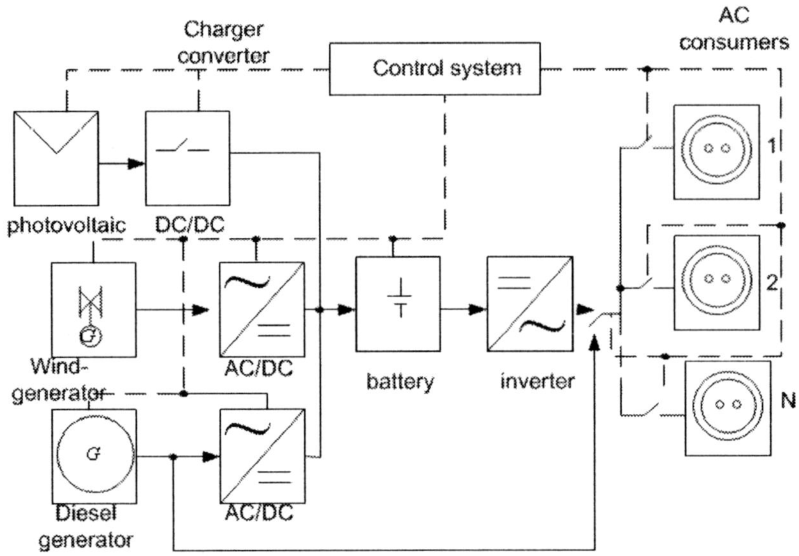

One of the channels in Figure 19 repeats the structure shown in Figure 9. The modules in the second channels are of a great interest. The generated DC energy from the photovoltaic is used to supply a electrolyze module and then to produce hydrogen. I.e. this is a solar installation for a hydrogen generation. The hydrogen is a basic source for the next usage. Using fuel cells, electrical energy is generated. As it is seen, in this structure also heat energy is generated. Moreover, mechanical energy is generated using motor supplied by gas.

Figure 20 shows another system structure. This system allows different AC consumers preliminarily separated into groups to be supplied. This is assumed to be made via the switches in the output.

The energy source may also be air with a boosted pressure. An example of its usage in a structure of double conversion UPS is shown in Figure 21. The system is interesting because here harmful waste products do not exist. One of the channels uses compressed air as a source for energy storage. In this channel, bidirectional conversion of mechanical into pneumatic energy exists. Also, heat energy connected with the processes of compression and expansion of air may be used. At the presence of supply source, energy is transferred from DC wire into both directions – towards kinetic battery and towards turbine, which motions a compressor used to accumulate compressed air into the reservoirs.

Bidirectional power electronic converters are included in both channels. It is obvious, that at a drop of a voltage of the supply network, kinetic energy from the battery or energy from compressed air may be used to supply the consumers.

In 2008 the ideas of distributed electrical energy generation and ecological use of energy sources unified the researchers in Future Renewable Electric Energy Delivery and Management (FREDM) System Center. The center research program purposes are orientated towards Energy Internet (Huang, 2008).

Figure 21. Usage of air as an energy source

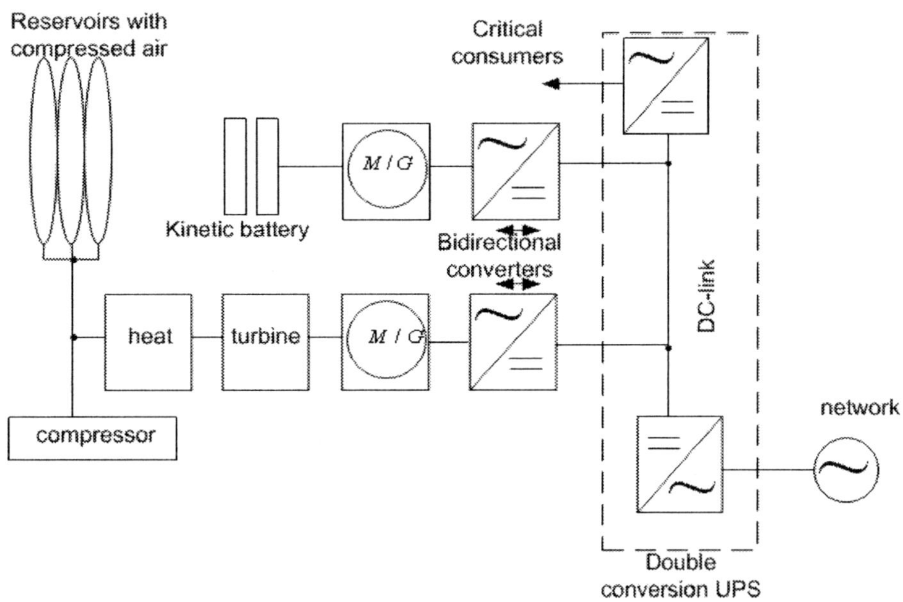

CONVERSION OF WATER ENERGY

Different trends in the study and practical applications of the use of water energy exist. The use of the potential energy of the water at big Hydro Power Plants is the most widespread. Other researches are connected with the use of the energy of the waves, of the high and low tides, etc. At the current part, only the implementation of power electronic converters in small Hydro Power Plants that use energy of water sources of small volume will be studied. Such an implementation is of interest for a future stage development according to the distributed and ecological electrical energy generation. The general requirements towards the electrical energy converters are mainly connected with the decrease of the volume and weight, which are a precondition for mobility, and also with an increase of reliability and stability to climate influences. The system for energy conversion may be with or without galvanic isolation from the public electrical network.

Figure 22 displays different implementations using microturbines putting into motion electrical generators. The version shown in Figure 22.a is without galvanic isolation and that is its disadvantage. Its advantages are connected with the decrease of the number of the power electronic converters – a rectifier and inverter. The energy indicators of the rectifier, such as power factor and harmonic spectrum of the input current, are of great importance for the power of the electrical machine to be more fully used. Rectifiers with transistors and input current modulation according to a certain law (see Chapter 4) are appropriate.

The next two versions use a high frequency transformer to make a galvanic isolation and to reduce the size and weight. The borderline version of the connection of low frequency transformer between the output of the inverter and the network is also possible, but it is not shown and therefore it is not studied here. The requirements for the rectifier converting the energy of the electrical machine remain for the topology shown in Figure 22.b. This version uses DC/DC with galvanic isolation – see Chapter 6.

Figure 22. Structural schemas of conversion of microturbine energy

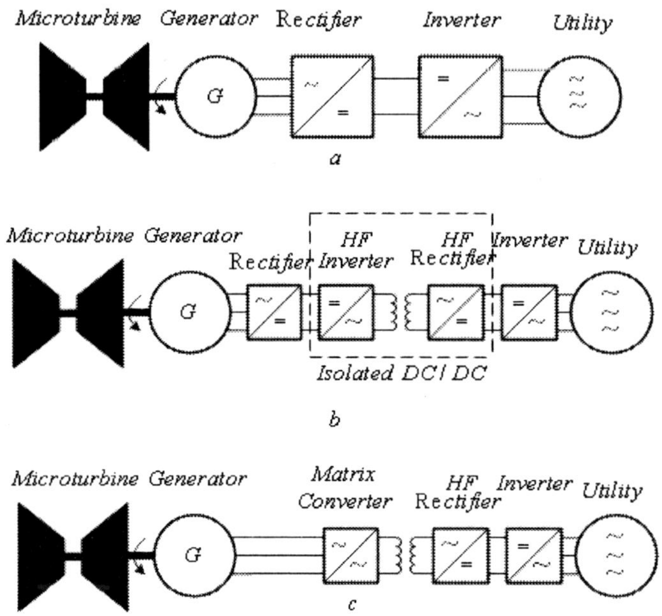

The version shown in Figure 22.c is of interest. A matrix converter (see Chapter 5), combining the functions of a rectifier and inverter and having a possibility to obtain high power factor in respect to the source of the AC energy, is used. This version may be advanced with connection of a matrix conversion of single-phase to three-phase voltage also in the secondary side.

REFERENCES

Aglietti, G., Redi, S., Tatnal, A., & Markvart, T. (2009). Harnessing high altitude solar power. *IEEE Transactions on Energy Conversion*, *24*(2), 442–451. doi:10.1109/TEC.2009.2016026

Blaabjerg, F., Chen, Z., & Kjaer, S. B. (2004). Power electronics as efficient interface in dispersed power generation systems. *IEEE Transactions on Power Electronics*, *19*(5), 1184–1194. doi:10.1109/TPEL.2004.833453

Caisheng, W., & Nehrir, M. (2008). Power Management of a Stand-Alone Wind/Photovoltaic/Fuel Cell Energy System. *IEEE Transactions on Energy Conversion*, *23*(3), 957–967. doi:10.1109/TEC.2007.914200

Carrasco, J. M., Franquelo, L. G., Bialasiewicz, J. T., Galvan, E., Guisado, R. C. P., & Prats, M. A. M. (2006). Power – Electronic systems for the grid integration of renewable energy sources: A survey. *IEEE Transactions on Industrial Electronics*, *53*(4), 1002–1016. doi:10.1109/TIE.2006.878356

Cheng, Z., & Blaabjerg, F. (2006). Wind energy –The world's fastest growing energy source. *IEEE Power Electronics Society Newsletter*, *18*(3), 15–19.

Enslin, J. H. R. (2003). Interconnection of distributed power inverters with the distribution network. *IEEE Power Electronics Newsletter, 15*(4), 7–10.

Gipe, P. (2006). *Wind power: Renewable energy for home, farm and business*. Chelsea Green Publishing.

Ho, B. M. T., & Chung, H. S. H. (2005). An integrated inverter with maximum power tracking for grid – connected PV systems. *IEEE Transactions on Power Electronics, 20*(4), 953–962. doi:10.1109/TPEL.2005.850906

Holm, R., Polinder, H., Ferreira, J. A., Van Gelder, P., & Dill, R. (2002, February). *A comparison on energy storage technologies as energy buffer in renewable energy sources with respect to power capability.* Paper presented at the IEEE Young Researchers Symposium in electrical Power Engineering. Leuven. Belgium.

Huang, A., Heydt, G., Dale, S., Zheng, J., & Crou, M. (2008). Energy Internet – Future renewable electric energy delivery and management (FREEDM) systems. *IEEE Power Electronics Society Newsletter, 20*(4), 8–9.

Jacquot, J. E. (2007). *42.8% efficiency: A new record for solar cells.* Retrieved from http://www.treehuger.com

Kayikci, M., & Milanovic, J. (2007). Reactive power control strategies for DFIG-based plants. *IEEE Transactions on Energy Conversion, 22*(2), 389–396. doi:10.1109/TEC.2006.874215

King, R. R., Law, D. C., & Edmontson, K. M. fetzer, C.M., Kinsey, G.S., Yoon, H., Krut, D.D., Ermer, J.H., Sherif, R.A., & Raram, N.H. (2007). Advances in high – efficiency III-V multijunction solar cell. *Advanced in Optoelectronics. Vol.207, Article ID 29523.* Hundawi Publishing Corporation.

Masters, G. (2004). *Renewable and Efficient Electric Power Systems*. John Wiley & Sons.

Petrone, G., Spagnuolo, G., Teodorescu, R., Veerachary, M., & Vitelli, M. (2008). Reliability issues in photovoltaic power processing systems. *IEEE Transactions on Industrial Electronics, 55*(7), 2569–2580. doi:10.1109/TIE.2008.924016

Petterson, A., Harneforst, L., & Thiringer, T. (2005). Evaluation of current control methods for wind turbines using doubly – fed induction machines. *IEEE Transactions on Power Electronics, 20*(1), 227–235. doi:10.1109/TPEL.2004.839785

Prasad, B., Jain, S., & Agarwal, V. (2008). Universal single-state grid-connected inverter. *IEEE Transactions on Energy Conversion, 23*(1), 128–137. doi:10.1109/TEC.2007.905066

Simoes, M. G., Chakraborty, S., & Wood, R. (2006). Induction generators for small wind energy systems. *IEEE Power Electronics Society Newsletter, 18*(3), 19–23.

Stiebler, M. (2008). *Wind energy systems for electric power generation*. Springer.

Teodorescu, R., & Blaaberg, F. (2005). Photovoltaic systems are with power electronics. *IEEE Power Electronics Society Newsletter, 17*(4), 10–13.

Chapter 10
Uninterruptible Power Supply Systems

INTRODUCTION

Recently, there has been a sharp increase in a number of so-called critical equipment of electrical power. Both separated units and complex objects, whose normal operation is strongly influenced by the parameters of electrical power, are included in this equipment.

The critical equipment includes the following objects:

- Personal computers (PC) and computer systems
- Electrical equipment for continuous technological processes
- Hospitals and other medical objects with modern furnishing
- Communicational system and equipment
- Security services and alarm systems
- Land aircraft equipment
- Fire-security
- Military objects

DOI: 10.4018/978-1-61520-647-6.ch010

At contemporary practice, the normal operation of critical equipment is secured by separated parts of equipment or whole systems. This equipment is fed by electrical energy delivered from public distribution electrical systems with a common level of power quality and with special methods and means for improving the electrical supply quality to a level required for a the normal operation of the critical equipment. These special methods and means applied in devices are received popularity as "UPS" (Uninterruptible Power Supplies).

This name is originally connected only with a definite structure of the supplies, but along with the UPS development it is enhanced - Uninterruptible Power Supply Systems (Griffith, 1989: Emadi, 2005: Gurrero, 2007).

Users more often associate the installation of UPS to a necessity to secure an operation of a responsible consumer at a drop off of the source voltage. Different changes in the quality of the source voltage are possible in practice. More often these changes are:

1. A drop off for a long time – Figure 1.
2. A drop off for a half-period or period of the source voltage – Figure 2.
3. A decrease of the source voltage value below a permissible one – Figure 3.
4. An increase of the source voltage value above a permissible one – Figure 4.
5. A change of the source frequency – Figure 5.
6. Transient overvoltages – Figure 6.

Figure 1. A drop off of the source voltage for a long period of time

Figure 2. A drop off of the source voltage for a half- period or a period

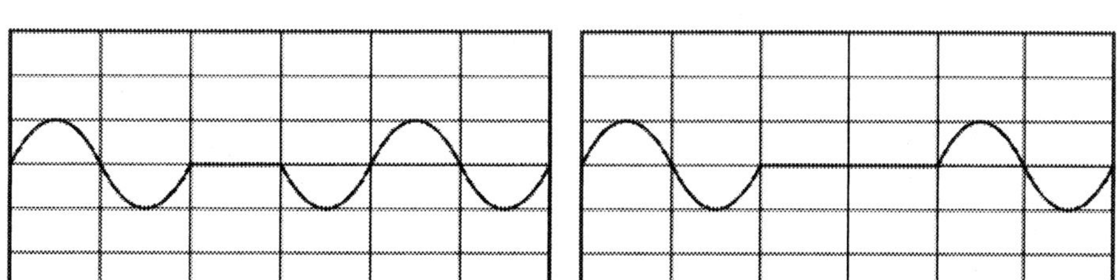

Figure 3. A decrease of the source voltage value below a permissible one

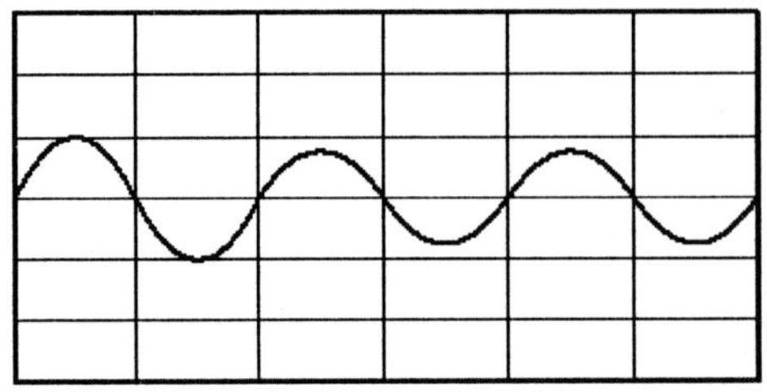

Figure 4. An increase of the source voltage value above a permissible one

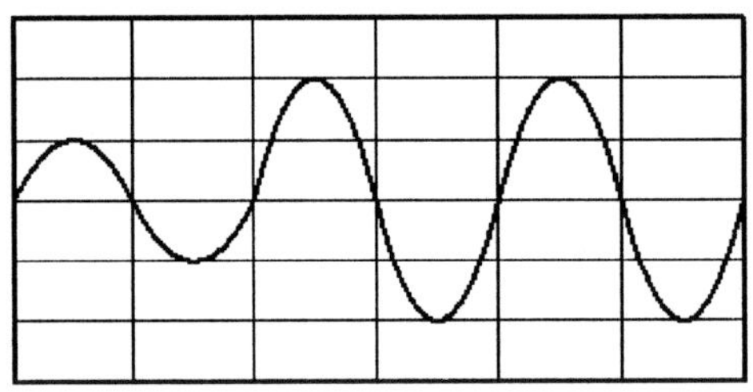

7. Transient decrease of the voltage value – Figure 7

8. Non-periodical disturbances put over the source voltage waveform – Figure 8.

9. An increase of non-sinusoidal coefficient of the voltage waveform above the permissible one – Figure 9.

Figure 5. A change of the source frequency

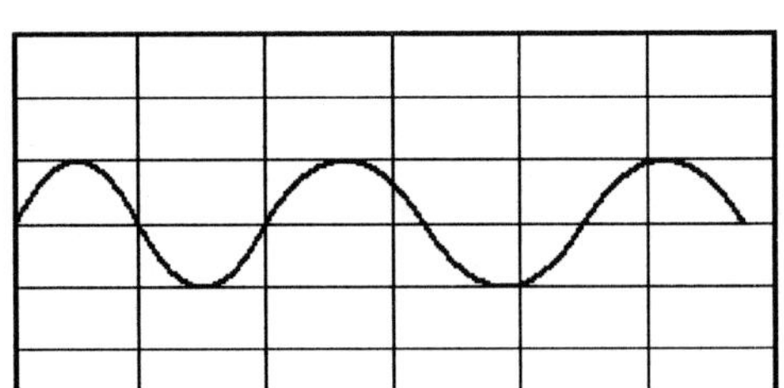

Figure 6. Transient overvoltages

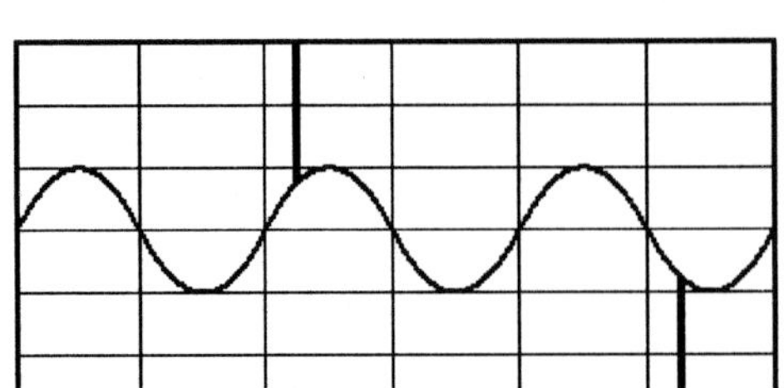

Of course, no all of the abovementioned disturbances in the quality of the source voltage affects in the same way different consumers (Rasmussen, 2003). The affects depends on the consumer type - whether it uses directly AC electrical power or power is preliminarily transferred into DC one in a separated source block (as it is in PCs), also, whether this transmission is connected also with a stabilizing of output voltages of this block, etc. Besides, it is worthy to be known that there are also other ways to protect the consumer from a part of these changes. For example, stabilizers of AC voltages exist to eliminate number 3 and 4; some companies produce filters protecting the consumer to avoid distributions with number 6, 7, 8 and 9. It has not to look at UPS as a universal means to protects a consumer from all possible changes because it will be clarified further down that the protection possibilities depends on the structure and the way of implementing the UPS itself. Some of the systems secure supply at a drop off of the electrical source but they do not resist other changes. Uninterruptible power supplies, whose output voltage has a square waveform (more often of a low power to supply PC), are yet offered at the Market.

Figure 7. Transient decrease of the voltage value

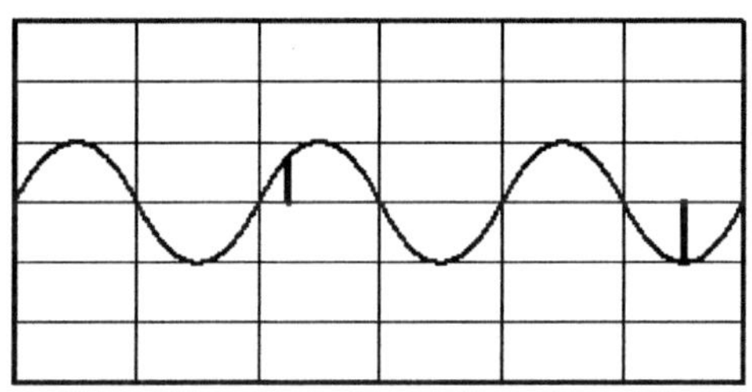

Figure 8. Non-periodical disturbances put over the source voltage waveform

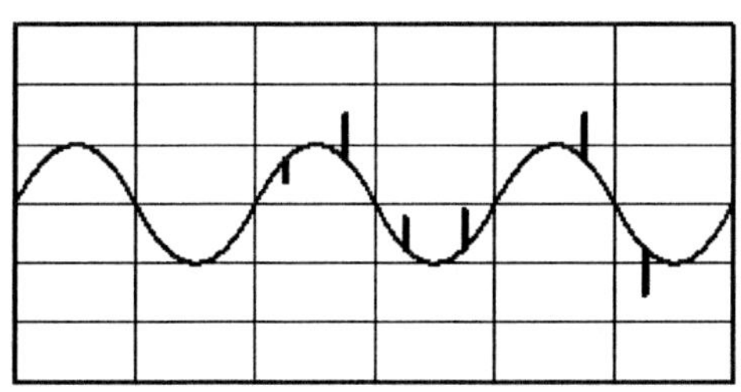

Figure 9. An increase of non-sinusoidal coefficient of the waveform above the permissible one

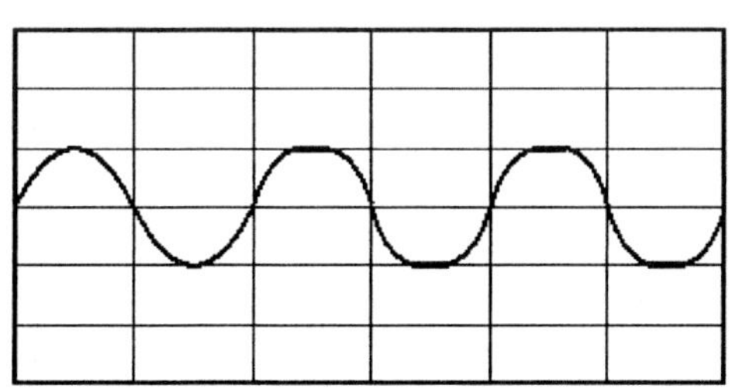

Figure 10. Centralized topology (Source: Powerware. Easy UPS solutions for IT systems.pp.4 – 5)

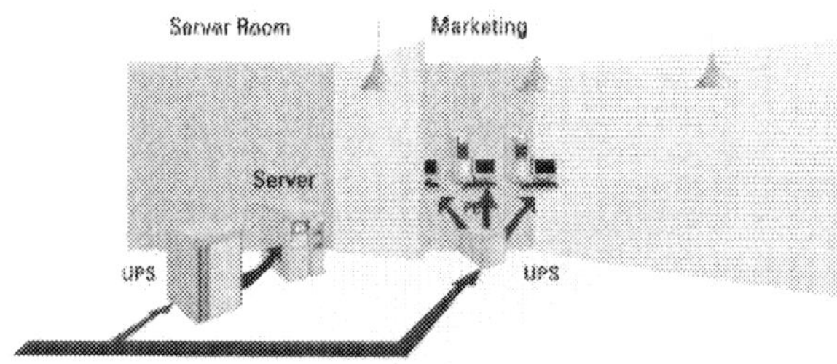

Centralised UPS systems:

And yet, structures to implement systems capable of eliminating all changes in the quality of the source voltage, which would be unpleasantly affects a responsible consumer, exist.

When it is necessary to secure uninterruptible power supply for a separated responsible consumer, UPS intended only for this consumer and with technical data compatible with the specialty of the consumer is used. Such cases are, for example, these of electrical supply of a PC or separated telecommunication station.

Uninterruptible power supply for several different consumers may be secured in three ways – centralized, distributed or combined ways. Figure 10 shows the centralized topology for an uninterruptible power supply. A special feature for this version is that all servers are supplied from a single UPS, as well as all computers, for example, at marketing room are secured also from a single UPS. Here, reliability is lower than in the other topologies but the system price is also lower.

Figure 11 illustrates the distributed topology of uninterruptible power supply. In it, each server and each computer in the marketing and selling rooms are supplied from a separated UPS-system. This topology is characterized by a higher reliability but also very often by a higher price.

The third topology of mixed solution for uninterruptible power supply is shown in Figure 12. Regarding the servers this topology is centralized one and regarding the computers – distributed one. Of course, the opposite variant is also possible – the servers – distributed, and the computers – centralized. From the point of view of reliability and price this topology is at the middle of the previous two.

Figure 11. Distributed topology (Source: Powerware. Easy UPS solutions for IT systems.pp.4 – 5)

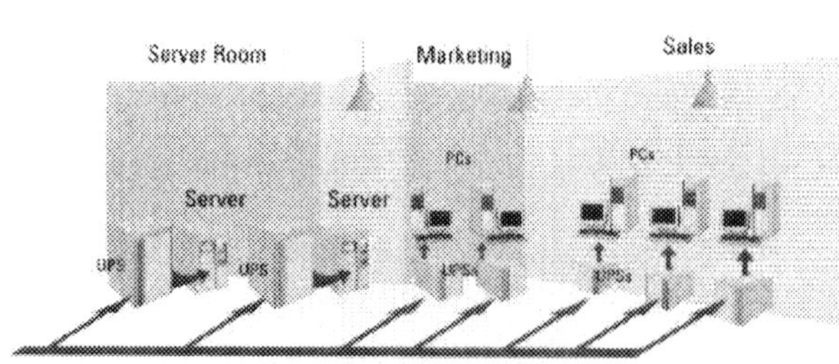

Distributed UPS systems:

Figure 12. Mixed topology (Source: Powerware. Easy UPS solutions for IT systems.pp.4 – 5)

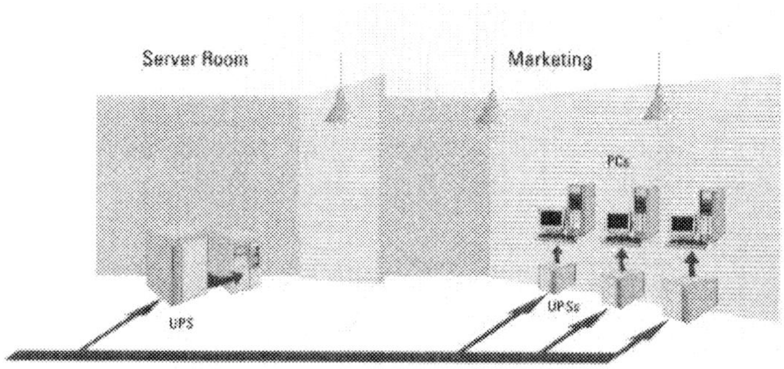

Based on data provided by (Powerware, 2008), the following shares are seen in the already installed UPS in IT: 45% - central topology; 35% - distributed topology and 25% - mixed one.

The question about uninterruptible power supply of individual homes is very often put in practice. Proprietors, designers and design performers are usually considered the aspect of UPS after all the work upon the electrical installations of the building has been finished. In the author's opinion based also in his experience, settling of this issue has to be a part of an initial design for the electrical installation to be designed in an appropriate way and also to be located rooms or spaces for the elements of the supply. An idea about UPS for an individual home is shown in Figure 13.

As it is seen, all the electrical power consumers at the home are separated in three circuits, and also, their tentative powers are shown.

In the first and the most responsible circuits, for example, the following equipment is connected: lightening, security system, audio and video systems, computer, and other informational and communicational systems. Thus, the requirements for the parameters of UPS1 are considered to this equipment.

Figure 13. An idea about uninterruptible power supply for an individual home

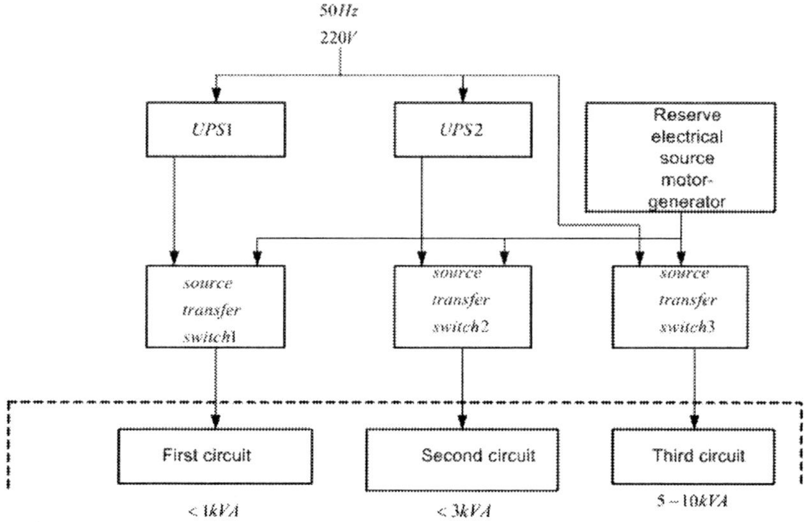

For example, the longest duration of operation may be required after a drop off of the source voltage, also zero switching time and thoroughly separation of the disturbances in the source voltage may be required. The first system of UPS may be of double conversion type, which will be examined further down.

In the second circuit all household appliances may be connected, such as fridge, cooker, laundry, etc. Obviously, the requirements for UPS2 may be lowered. UPS may be of another type. After a voltage drop off, the household appliances continue to operate till completion of their functions – for example, food preparation, etc.

Heating and air-conditioning are included in the third circuit. It is initially supplied from the source voltage and after the drop off of the voltage the circuit may stop to operate for a short time. After this time, a reserve electrical supply has to be switched on – diesel aggregate, and then sing the source transfer switch 3, the heating and air-conditioning to be supplied again.

Using the source transfer switches 1 and 2, the supply of the other two circuits can also be passed to the diesel aggregate at the condition that the source voltage is yet missing and the capabilities of the two UPS are exhausted.

Control of all elements of the supply of the home is using a common microcomputer system and with parameters, which are set and may change by the proprietor.

Topics about different energy sources, others than the network, are examined in Chapter 9.

BASIC SCHEMAS AND THEIR INDICATORS

According to the accumulator and the means of energy conversions, the systems of uninterruptible power supplies are in general classified as (Karve, 2005; Cottuli, 2008):

- Static ones
- Dynamical ones

In static UPS, accumulation of energy is made in electrochemical accumulators (secondary sources) and also the required conversions of electrical power are performed by semi-conduction converters, which do not contain moving parts and are classified as static ones.

Dynamic UPS are realized using converters with moving parts. Energy in them is stored for a short time as kinetic energy into moving flywheels. The required energy conversions are performed in electrical machine. Long-term operation of the dynamic UPS at the drop off of the electrical supply is secured by motors with inner combustion.

Technical solutions of implementation of UPS have been developing together with the development of the consumers requiring such a supply. From the beginning of 1970, so-called on-line UPS shown in Figure 14 have found implementation in securing big calculation machines.

Here, the inverter is fed by the rectifier with DC energy and operates continuously feeding the load. The rectifier secures also energy to charge the battery. Static bypass may transfer the supply of the consumer directly to the AC input (to the supply network) without an interrupt in the case of failure in the main channel or in the processes of initial switching on of the consumer, when it consumes a high current. It is seen that the structure does not exactly correspond to its name. Therefore, some authors make more precise the terminology of on-line as inverter, which feeds continuously the load and connected in series with the network.

Figure 14. Structure of on-line UPS

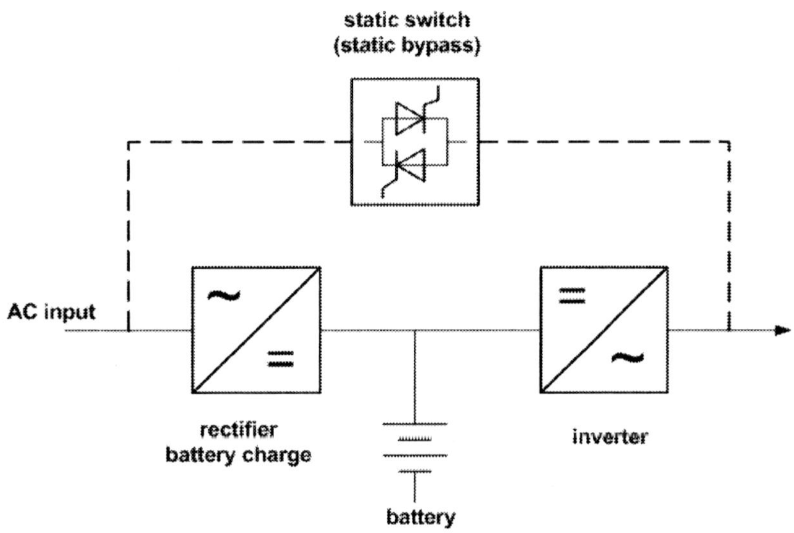

At the beginning of 1980, so-called off-line UPS shown in Figure 15 have been developed in opposition to on-line ones.

The term off-line is used rather as an opposite one to on-line name of the previous topology because here the term off-line should be understood it as parallel connection of the inverter to the network. The inverter does not operate continuously, yet it is in a standby mode. The consumer feeds from the inverter only in the case when the voltage of the AC input leaves the permission range in particular when

Figure 15. Off-line UPS topology

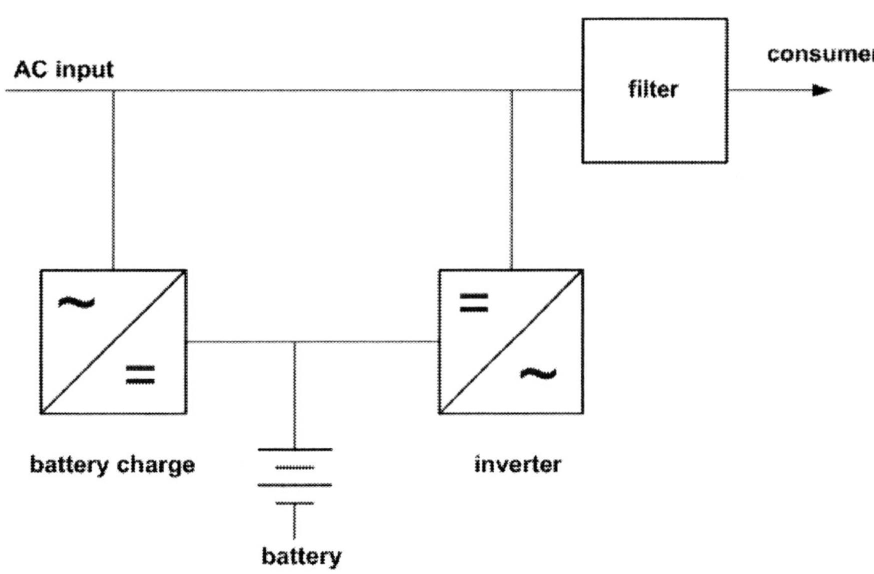

it misses. Functions of the block marked as filter include not only filtering but also, in some systems, stabilizing the value of the voltage passed to the load. Here, the term off-line also does not reflect the nature of the UPS system.

Development of mini and micro computers, as well as telecommunication systems, has leaded to invention of new schematic solutions in the systems securing their uninterruptible supply. The new standard of International Electrotechnical Commission IEC 62040-3 and its European equivalent ENV 50091-3 currently define three topologies of UPS – systems (Karve, 2000; Solter, 2002; Beaudet, 2005):

- Passive standby
- Line interactive
- Double conversion

The passive standby version shown in Figure 16 reminds a lot of off-line topology. Its features are as follows:

Mode when the source voltage presences: Load is continuously fed by the AC network through a filter, which may have also stabilizing functions. The standard does not completely clarify the stabilizing functions of the block but it mentions that a ferro-resonant stabilizer may be included in it. The inverter is in passive-standby mode. The battery is in charging mode.

Mode when the source voltage misses: When the source voltage leaves the permissible limits, the consumer is switched to the supply of the inverter output. The switching has to be performed in less than 10ms using a switch. This mode continues either to the moment of battery discharge to a permissible level or to the moment of the appearance of the source voltage within the defined limits.

Figure 16. Passive stand-by topology

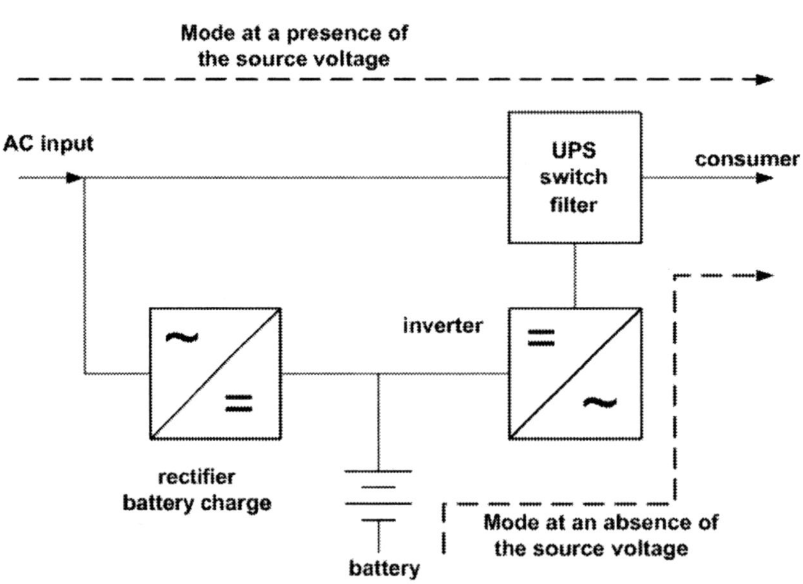

Figure 17. Line interactive topology

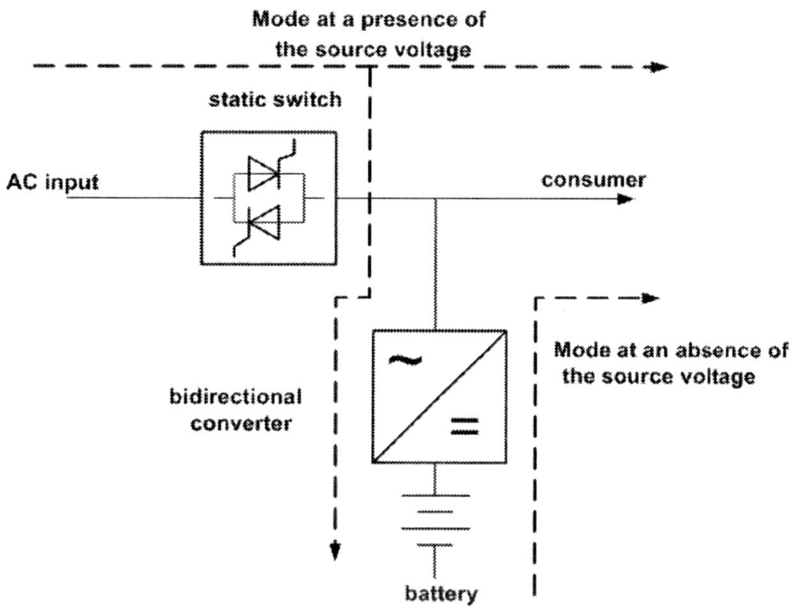

Advantages of this topology are: simplified structure, low price, small volume and weight. Disadvantages are: dependence of the voltage supplying the load in normal mode on the disturbances of the source network; sometimes the switching time appears to be long for the consumer; frequency of the source voltage supplying the consumer can not be regulated. The topology is applied in low powers below 2kVA. It can not be used as a frequency converter for transition between networks with different voltage standards.

Presence of bidirectional converter is typical for line interactive configuration shown in Figure 17.

The standard defines three possible modes of operation for this topology, as follows:

Mode at a presence of the source voltage: The consumer feeds from the source voltage, while the battery is charged through the bidirectional converter operating as AC/DC converter;

Mode at an absence of the source voltage: When the source voltage leaves its permissible limits, the consumer feeds from the battery using the converter operating as DC/AC converter. The static switch breaks off the connection to the network to isolate the converter output from it.

Bypass mode: Some systems of this type can be provided with a possibility for a manual transferring of the supply of the consumer directly only to the source network in case of a system failure.

Advantages of the line interactive configurations are: approximately the same price as this of the second topology and lower than the first topology price. Disadvantages are: there is no thorough isolation of the consumer from the source network and the consumer is affected by the disturbances of the source voltage; there is no possibility to regulate the frequency of the voltage supplying the consumer. This topology implementations are in average power consumers to 10 kVA, which do not require source voltage with frequency different from the source one.

A special feature of double conversion topology shown in Figure18 is the supplying of the consumer always from the inverter output.

For double conversion topology, the standard defines three modes of operation:

Mode at a presence of the source voltage: The consumer is continuously fed from the inverter output using a conversion of AC source voltage first time into DC by the rectifier and second time into AC. The battery is in charging mode.

Mode at an absence of the source voltage: When the source voltage leaves its permissible limits, the consumer feeds again from the output of the inverter receiving energy from the battery.

Bypass mode: This mode is secured automatically by the static switch in several cases. The first is in the case of failure in the main channel, the second case is in transient increase of the consumer current (initial turning on or overloading during operation), the third case is in depletion of battery capacity.

Almost each system of this type is provided with a possibility for a manual bypass in the case of a failure in the rest part of the system.

Advantages of the double conversion are: complete isolation of the load from the disturbances in the source voltage in normal mode; wide range of a change of the source voltage by a precise stabilizing of the supply voltage to the consumer; precise regulation of the frequency of the output voltage and a possibility to use the topology as a frequency converter; there are no transient processes at switching from the network supply to the inverter supply or in contrary. Disadvantages are: high price, in some cases reduced efficiency coefficient due to the double conversion of energy. The UPS, implemented using this topology, are almost every time used in high powers above 10kVA up to several MVA and at the need of frequency conversion.

Figure 18. Double conversion topology

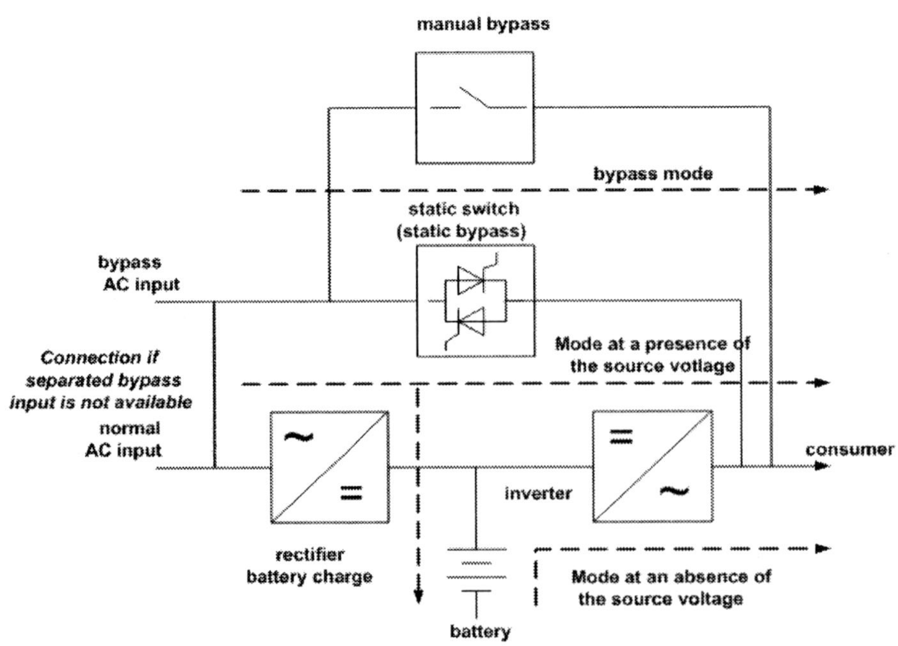

After the examination of the standard three topologies has been made, it appears that in some of them elements and functions of the sooner used terms on-line and off-line may be found. Nevertheless, these terms are not recommended to be used because they are considered not to reflect the nature of the performed functions and capabilities of the corresponding system.

Besides these basic blocks determining the name of the topology, UPS systems also contain: systems for control, regulation and diagnostic of the converters, systems for an appropriate temperature mode, interface and software, system to communicate with the staff performing prevention measurements or repairing, systems and software for telecommunicational connections with the producer. So, as a whole, the UPS system is a complex device that is usually distinguished with a very high reliability. In spite of all, failures are possible in the system, which implies the necessity to take additional measurements in extremely responsible consumers, such as hospitals and hospital rooms, control of vehicles, etc.

Figure 19 displays a dynamic UPS where DC motor 6 motions electrical generator 5. DC voltage to supply the motor is generated from the source network by rectifier 8. When the source voltage drops off, the DC voltage is maintained by the battery 7, which is charged from the source network 1 using rectifier 8.

When the source voltage drops off, consumers 9 will continue to receive supply from the generator without an interrupt. In this mode switch 2 is also provided. The switch turns off when the drop off of the source voltage appears not to pass energy to other consumers. Manual bypass is also shown in the figure with non-marked circuit-breaker. It is used when service procedures or prevention measurement of the UPS are taken.

Choosing an appropriate uninterruptible power supply for a consumer, technical data and parameters given at the producers manuals and datasheets have to be known very well. This knowledge is necessary for one to be able to choose the most appropriate system for ones purpose without unnecessary given more money and also to avoid some misunderstandings and conflicts. The technical data are usually divided into groups.

Figure 19. Dynamic UPS system

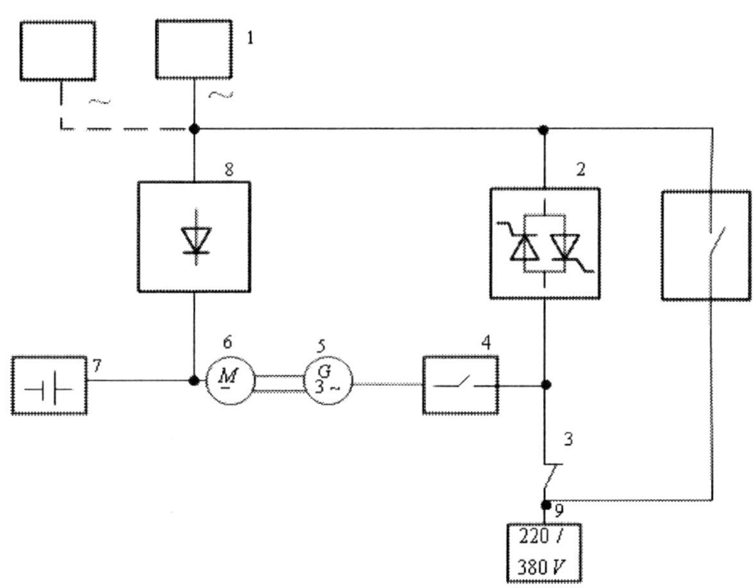

Technical data concerning the output:

- **Output power:** It is desirable that the producer shows two values of this parameter – active power in Watts and total power in VA. For example, active power 24kW, total power 30kVA. Load power factor is also usually shown and the type of load – more often load power factor from 1 to 0.8 and inductive type of load.
- **Output voltage:** If the system is a three-phase system, both phase-to-phase and phase voltage values are indicated. For example, 380/220V, 400/230V, 415/240V. Sometimes, only the value of phase-to-phase value may be shown – for example, 380/400/415V. Some producers make their configurations according to the client's choice. Whether a neutral wire is taken out or not has to be also indicated.
- **Accuracy of maintaining the output voltage:** It is usually shown in static mode – for example, ±1.5%, and the mode is indicated. Sometimes a dynamic parameter may be shown – transient voltage stability – for example, < 10% under a 100% change of load. The best producers indicate also transient time – for example, transient time 2 mS.
- **Accuracy of maintaining the phase angle among the three voltages:** It is given only by several companies – for example, phase angle accuracy ± 3°.
- **Total harmonic distortion THD of the output voltage:** It is desirable to be indicated also in what type of load the THD is listed. For example, THD < 2% in linear load case and THD <5% in non-linear load.
- **Frequency of the output voltage:** It is usually 50/60 Hz with probability to choose. Some producers also indicates whether a synchronization to the mains exists.
- **Accuracy of maintaining the output frequency:** A value under a full loading of the output is obligatory to be indicated – for example, ±0.5 Hz or in percentages – for example, ±1%. Also, the value in free running may be shown – for example, ± 0.05%.
- **Output overloading possibility:** This datum is not given by all producers but it is good for the consumer to take it in mind when choosing the system regarding its load. For example, possible overload of 150% for 30 seconds, 125% for 10 minutes and 110% for 1 hour.
- **Possibility for parallel operation of the output with other systems:** It is given very rarely by single producers and it is mentioned the kind of other systems with which parallel operation is possible.

Technical data concerning the input:

- **Value of the input voltage:** In three-phase system, both phase-to-phase and phase voltages are shown – 380/220V, 400/230V, 415/240V. Only the phase-to-phase voltage may be indicated – 380/400/415 V. The type of the source network is obligatory to be shown – three wired (three-phase) system or four wired (three phase and a neutral one) system.
- **Range of change of the input voltage:** It may be shown for the phase-to-phase voltage in values – for example, 340 – 440 V. It may be indicated in percentage – for example, ± 20% from the value of the input voltage.
- **Frequency of the source voltage:** It may be given in two ways – as a value of the frequency and as a percentage of its change (for example, 50 Hz ±10% or 60 Hz ±5%) or as a range of it change (for example 48-62 Hz).

- **Power factor:** It is extremely important parameter, which determines the behavior of the whole uninterruptible power system in respect to the source network (Blooming, 2008). This parameter however is not shown by all producers. Its minimum value is usually indicated – for example, > 0.96. Some companies show lower value, for example, 0.7, and they offer additional purchase of an active power filter to correct this value up to 0.99.
- **Total harmonic distortion of input current THDI:** It is shown by very few producers, for example, THDI < 3%.

• Technical data for batteries:

- **Battery type:** It is shown by almost all producers.
- **Expiring period of the batteries:** It is usually 5 or 10 years with a possibility to be chosen.
- **Recovery battery time after a discharge:** It is usually given in dependence on the discharge time – for example, 10 times of the discharged time.
- **Number of cycles charge/discharge:** For example, 500 cycles. Not all producers indicate this parameter.

Common technical data for the whole system:

- **Type of the system of uninterruptible power supply (technical decision):** Above examined topologies, as double conversion, etc., are had in mind. This parameter also is not indicated by all companies.
- **Transient time at a drop off or at an appearance of the source voltage:** It is usually less than 10ms – for example, 2ms. In some technical decisions, the experts know that this time is 0 – for example, in double conversion topology, therefore, if the producer is shown the type of the system at the manual he is not obliged to specify also and this time.
- **Backup time:** This is an extremely important parameter for the user. It is given in dependence on the battery type – 5 or 10 year expiring period. Almost all producers indicate a value under a 100% loading of the output, for example, 10-12 minutes and 10 year battery. Only few producers specify the load type – linear or non-linear and its power factor.
- **Common efficiency coefficient:** It is usually shown in full loading of the output – for example, 93%. Several values dependent on the output loading may be given in very few occasions.
- **Level of acoustical noise of the system:** The distance may also be shown – for example, 60 dB at 1 m. It is usually given for a full loading of the output, but it also may be shown as a range dependant on the loading – for example, 50 – 65 dB.
- **Ambient temperature:** Almost all producers give it as a range – for example, 0 – 45 ° C.
- **Air humidity of the environment:** For example, 15 - 80% usually without condensation.
- **Size and weight:** The three sizes and the common weight are usually indicated. They may also be given as separate data for the battery cabinet if it is separated.
- **Correspondence to international standards:** The standard numbers, in accordance with which the experiments have been performed and to which the correspondence is found, are obligatory to be shown. Usually, these are two types of standards – regarding the technical safeness and the electromagnetic compatibility.

- **Technical data concerning communication and software:** Types of communicational interfaces are given – for example, RS 232, also data and possibilities of the software are shown. Usually this is specialized production software. It advantages for the user are listed, also its possibility to operates with different operational systems, etc.

As it is seen, summary analysis of all technical data, comparison among different producers and the decision of the choice are a complex topic, which is very often impossible to be resolved without a consultation to a specialist in the field of Power Electronics. Besides, not all data presented in the advertising production materials and additional queries and specifying are necessary. In all cases, the possibilities service measurements are also analyzed – presence of an official distributor, time of removing the failure, warranty, etc. The producers almost never show data concerning reliability parameters such as average time to failure, etc. It is possible to make an additional analysis and comparison of data for already installed and operating systems at other users for a particular company. Here, the subjects of price and other economical indicators are not discussed, but additionally to the technical data they are also important at taking a final decision.

Each of the above described technical parameters is connected to a particular technical device (basic block) from the UPS system and additional data for some of the parameters are given further down at the description of the specialties of the separated blocks.

The contemporary researched in the field of UPS systems are orientated towards the following directions:

1. Structure and power schematic, for example, (Kamran, 1998; Vazquez, 2002; Ming, 2003).
2. Control and regulation, for example, (Mattaveli, 2005; Pai, 2006).
3. The energy source, for example, (Tao, 2008).

METHODS TO INCREASE THE RELIABILITY

Source transfer switch (STS) is a device, which switches critical AC consumer between two independent sources on one another in dependence on their transient state or on a command passed by the user. The two sources are very often two undependable UPS systems – see Figure 20 (MGE UPS Systems, 2008).

Figure 20. Increase of the reliability using two UPS and STS. Source: MGE UPS Systems. Pulsar STS. Source transfer switch.

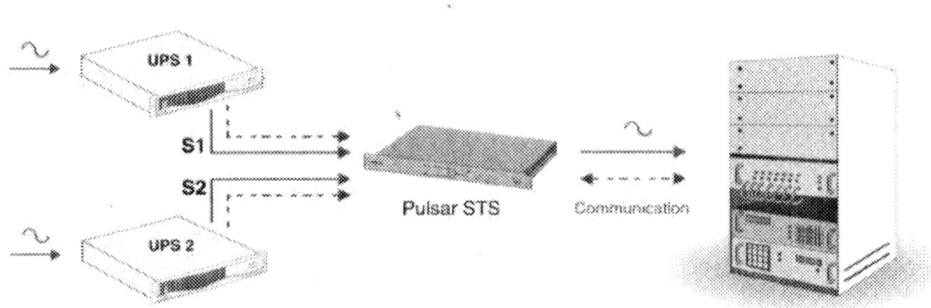

The two systems and the consumer have a connection by STS both regarding the source voltage and the data exchange. At failure in one of the systems, the supply is automatically passed to the other system and the necessary information for the failure state is passed. Besides the nominal current, which the STS may switch, a basic parameter is transient time. For example, STS produced by MGE France, so-called Pulsar STS, guarantee currents of 10, 15 and 16 A at transient time of 6ms.

Important function of the control system is to monitor the transient values of the voltages of both sources, which are usually non-synchronized and to choose the most appropriate moment for switching, which has to happen at minimum time and minimum disturbance in the waveform of the supply voltage for the consumer. The best moment to perform the switching is at the moments when the load current decreases to 0, thus the turning off of the finally conducting triac is guaranteed, see Figure 21.

Since sometimes STS are offered undependably from UPS system, it is necessary to know switch parameters and characteristics.

Technical data concerning the input:

- **Input voltage:** More often this is an output voltage of two undependable one from other UPS systems. They usually guarantee high stability of the voltage value, besides, for this switch parameter higher range is permissible – for example, 220/230/240 V ±10%.
- **Frequency of the input voltage:** It is usually 50/60 Hz
- **Number of the inputs and wire standards:** For example, 2 corresponding to IEC 320 C13.
- **Nominal input current:** For example 16A.

Technical data concerning the output:

- **Number of the outputs and wire standards:** For example, 6 corresponding to IEC 320 C13
- **Nominal output current:** For example 10A.
- **Transient switching time:** It is obligatory to be lower than the value of the half-period of the input voltage. The period of 60Hz is 8.33ms. The producer may be shown 10ms, for example. Thus,

Figure 21. Contact-free switching between two sources using triacs

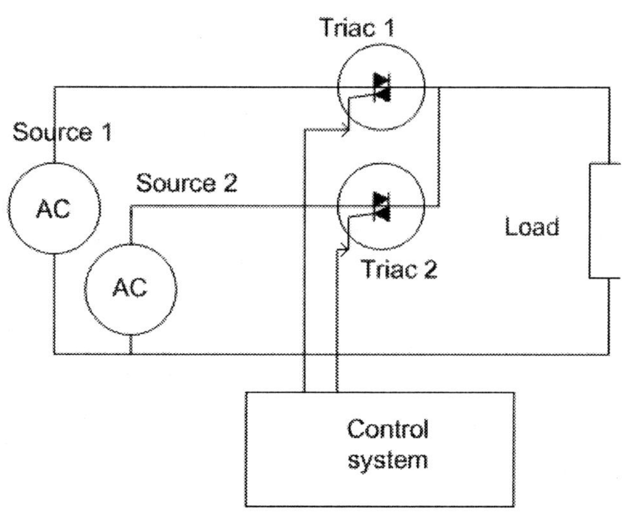

an analysis has to be made whether the consumer supplying by this switch may carry a drop off of the source voltage for this time.

Common technical data:

- **Sizes and weight:** The three sizes and weight are indicated - for example, 400x60x100 mm and 10 kg.
- **Correspondence to international standards:** The standards numbers are obligatory to be given. They are usually regarding the technical safeness and the electromagnetic compatibility – for example, regarding the electromagnetic compatibility IEC 1000-4, and regarding the technical safeness EN 50091-1.

Different schematic versions are often provided to increase the reliability. The first version is depicted in Figure 22. The main requirement to both separated UPS systems is their capability of parallel operation and also each of them has to be capable of delivering the total power required by the load.

In a normal state the power is divided between the two systems, and when a failure occurs in one of them, the power is delivered only by the other. The possibility of so-called centralized bypass is also provided.

In high power some producers design the UPS system itself according to a module principle with possibility for parallel operation of the separate modules and centralized bypass. Figure 23 depicts this version. A possibility for automatic turning on and off of particular modules dependent on the power loading and the module states is often existed in this version.

Another version to increase the reliability is shown in Figure 24. It is for two UPS systems. The system contains STS.

Figure 22. First schematic version to increase the reliability

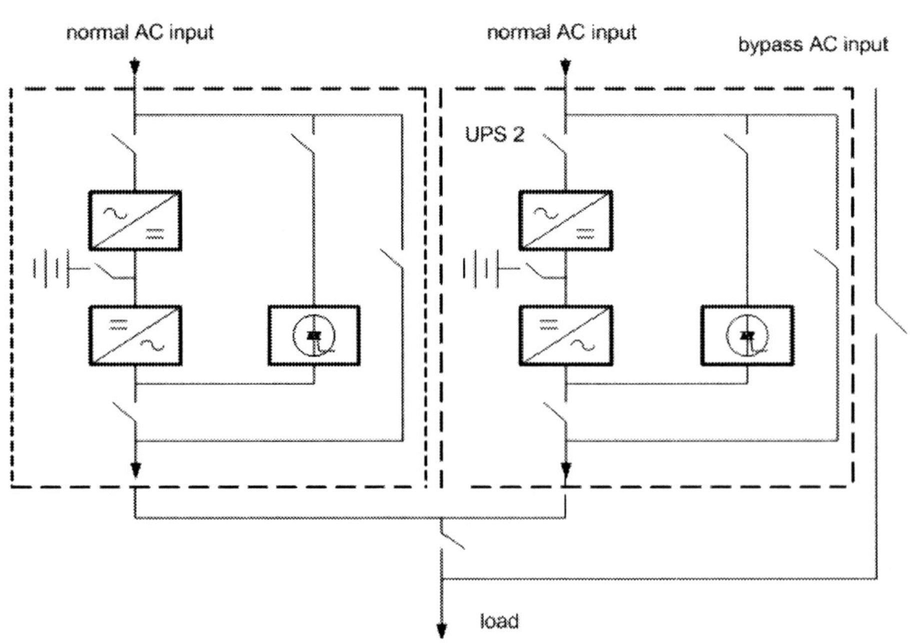

Figure 23. Module principle of UPS applied to increase the reliability

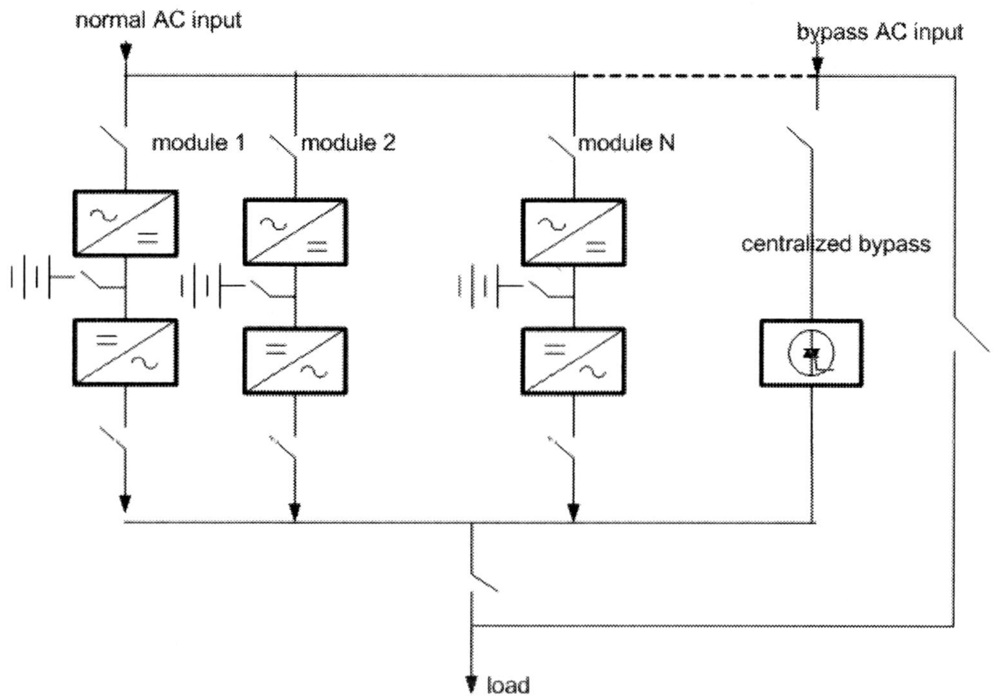

Figure 24. Version to increase the reliability for two UPS systems

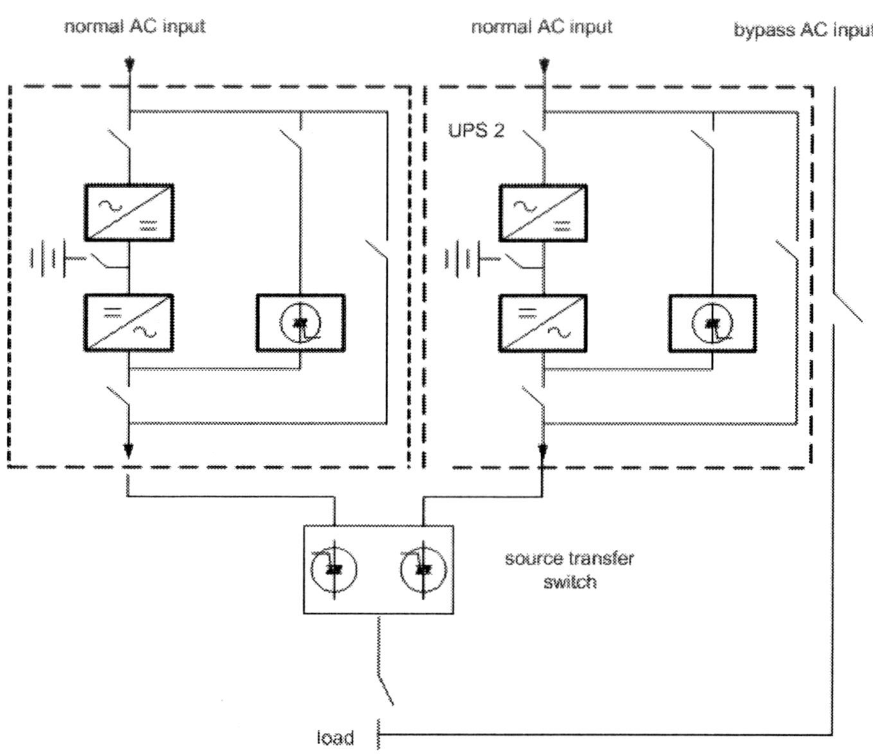

At each moment only one of the systems operates, the other is in standby mode. In failure in one of the systems the supply of the load is passed to the other one, and the passing is made through the STS. Advantage of the version is the fact that the two systems are not loaded simultaneously. It has to be mentioned that this version may be extended for a boosted number of systems and loads using different combinations among them and using several STS.

In all versions examined till now, the time during which the load will be supplied at a drop off of the source voltage depends on two factors. On one hand, it is the state of the battery – charge degree and capacity; on the other hand, it is power which the load consumed at that moment. It is clear that this time is limited by these two factors. For example, if the load consumes power equal to the nominal one of the UPS system and the battery is fully charged, the time during which supply for the load can be maintained after the drop off of the source voltage varies in different producers and it is often in the range of 5-20 minutes. If this does not meet the requirements of the consumer, he may use another UPS system with higher power than his load will consumed consulting the producer what will be the increase of the time in this case, of course, on the account of a higher price. Another option is the consumer to ask for an installation of another battery with an appropriate capacity for the case, of course, if the producer gives this opportunity, which is cheaper than the previous version.

At very responsible loads a group motor- generator supplied by a liquid fuel (so-called reserve supply) is added to the blocks of UPS systems described up to here. In this case the supply for the load relies on from the battery using DC/AC converter for a comparatively short time during which the motor-generator group begins to operate after the source voltage drops off and if there is no yet source voltage after the depletion of the capability of the battery then automatically the supply of the load is transferred to the generator. Thus a reserve electrical supply is realized. An example of its structure is shown in Figure 25.

Figure 25. A structural schema for a reserve electrical supply

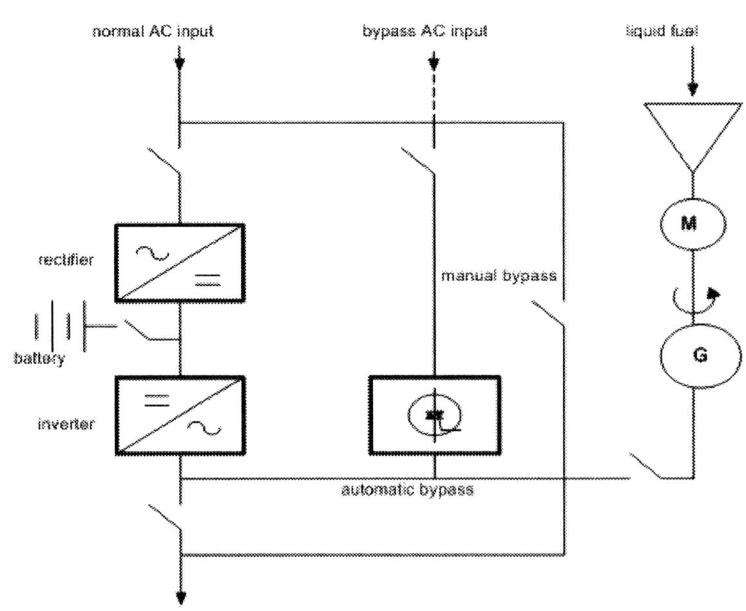

COMMUNICATION BETWEEN UPS SYSTEMS AND DIFFERENT SYSTEMS

It has been clarified at the above made study that the UPS systems are complex devices which are assigned responsible functions connected to the supply of critical equipment. This imposes the need of providing special means to monitor the operation of the devices and at failure the reaction time and the time to eliminate the failure to be as short as possible. The possibilities, which are provided in this aspect by the producers, very often play a significant role in the choice which the consumer may be made among the UPS systems offered at the Market. These possibilities determine in a high degree the speed to eliminate the failures – obligation that the producers take at the contract of warranty.

Recently, a basic share in control systems of power electronic converters used in UPS systems is taken from digital signal processors. The processors themselves have series peripheral interfaces or/and series communicational interfaces with possibilities of synchronous or asynchronous data exchange and thus there embedment in hardware and software systems meant for communication is simplified. During the design of the control systems, possibilities for self-diagnostic of the separated blocks are provided, so, the data to be accessible to the staff. The communication means can be generally divided into two groups:

- Means for monitoring, indication, imitation and control at the point of installation of the UPS system
- Means for monitoring and control from a distance

Liquid-crystal displays for indication and control used by all producers and most frequently controlled by a common microprocessors are from the first group of the communication means. A structure of the system with indication, which blocks operate at that moment is usually shown at the displays, as also different functional buttons to draw in or out data at the display are provided.

This permits:

- Indication of the currently operating blocks by lightening their symbols at the block schema at the display
- Indication of voltage, current and frequency of input/output/battery
- Indication of all alarms for operational disturbance and their history
- Indication of remaining operational time at an absence of the source voltage dependent on the state of the battery, transient load and temperature
- Indication of remaining time of use of the battery dependent on its usage history
- Enable or disable of sound indication
- Configuration of the above mentioned function dependent on the requirements of the consumer – language, indicated parameters, date, time during which the battery is tested, etc.

The consumer also has access to the display. So-called remote mimic and control panel is additionally installed. This panel is a coupling for connection to a special service PC used by the service engineers and signals of "dry contact" type are brought on the panel, as follows:

- Inverter load
- Overloading

- Test for damaged battery
- Network load
- Failure turned off inverter
- Coming stop
- Turned off rectifier
- Turned off battery charge
- Damage in the bypass system, etc.

These are only part of the most important signals, which allow to service engineer to make a quick diagnostic of the failure and to take a decision how to continue its elimination.

The second group of communication means for a distance monitoring and control is realized by specially made centers for this purpose by the producers. In these centers, a continuous flow of information for the state of the most responsible objects arrives using the embedded modems into UPS system through telephone cable or through Internet. This allows monitoring the parameters required for a proper operation, passing the state of a particular system using so-called "reports" at a set period of time. According to the data, eventual malfunctions of the system can be foreseen and both the consumer and the serves group to be informed preliminarily. Of course, the producers have necessary software, shaped in separate parts with different producer names, for means of the two groups. Information about software can be found in advertising materials of the companies.

If the UPS system is used for computer systems and nets, it has to be known that means for connection to the computers themselves are provided. Corresponding software, which may be integrated in different operational system such as WINDOWS[1], UNIX[2], etc., is offered.

REFERENCES

Beaudet, J. P., Fiorina, G. N., & Pinon, O. (2005).*UPS topologies and standards.* France: MGE UPS Systems.

Blooming, T. M. (2008). *Power factor as it relates to UPS products.*Cleveland: Eaton Corporation. Retrieved from http://www.powerware.com

Cottuli, C., & Christin, J.-F. (2008). *Comparison of static and rotary UPS.* American Power Conversion.

Emadi, A., Nasiri, A., & Bekiarov, S. B. (2005). *Uninterruptible power supplies and active filters.* Florida: CRC Press.

Griffith, D. C. (1989). *Uninterruptible power supplies.* New York: Marcel Dekker Inc.

Gurrero, J. M., De Vicuna, L. G., & Uceda, G. (2007). Uninterruptible power supply systems provide protection. *IEEE Industrial Electronics Magazine, 1*(1), 28–38. doi:10.1109/MIE.2007.357184

Kamran, F. (1998). A novel on-line UPS with universal filtering capabilities. *IEEE Transactions on Power Electronics, 13*(3), 410–418. doi:10.1109/63.668099

Karve, S. (2000). Three of a kind [UPS topologies, IEC standard]. *IEE Review*, *46*(2), 27–31. doi:10.1049/ir:20000204

Karve, S. (2005). *Static or rotary? - That is the question.* France: MGE UPS Systems.

Mattaveli, P. (2005). An improved deadbeat control for UPS using disturbance observer. *IEEE Transactions on Industrial Electronics*, *52*(1), 206–211. doi:10.1109/TIE.2004.837912

MGE UPS Systems. *Pulsar STS. Source transfer switch.* France/Cedex. Retrieved from http://www.mgeups.com

Ming, T. T., & Chia, H. L. (2003). Design and implementation of a cost-effective quasi line-interactive UPS with novel topology. *IEEE Transactions on Power Electronics*, *18*(4), 958–965. doi:10.1109/TPEL.2003.813760

Pai, F. S., & Huang, S. J. (2006). A novel design of line – Interactive uninterruptible power supplies without load current sensors. *IEEE Transactions on Power Electronics*, *21*(1), 202–210.

Powerware. (n.d.). *Easy UP solutions for IT systems* (pp. 4-5).

Rasmussen, N. (2003). *The different types of UPS systems*. USA: American Power Conversion.

Solter, W. (2002). *A new international UPS classification by IEC 62040-3*. Paper presented at Telecommunications Energy Conference, 2002.INTELEC.24th Annual International (pp.541-545).

Tao, H., Duarte, J. L., & Hendrix, M. A. M. (2008). Line-interactive UPS using a fuel cell as the primary source. *IEEE Transactions on Industrial Electronics*, *55*(8), 3012–3021. doi:10.1109/TIE.2008.918472

Vazquez, N., Aguilar, C., Arau, G., Caceres, R. O., Barbi, I., & Gallegos, G. A. (2002). A novel uninterruptible power supply system with active power factor correction. *IEEE Transactions on Power Electronics*, *17*(3), 405–412. doi:10.1109/TPEL.2002.1004248

ENDNOTES

1. WINDOWS is a registered trademark of Microsoft Corporation in the United States and other countries.
2. UNIX is a registered trademark of The Open Group.

Chapter 11
Other Applications of Converters and Systems of Converters

INDUSTRIAL APPLICATIONS

The industrial application of power electronic converters and systems of converters are varied (Motorola, 1991, 1993).

Power electronic converters of electrical energy for electrical engineering technologies have a wide spread in industry.

Electrical engineering technology is a technology to perform a particular process using electrical energy. The following positive effects are obtained using electrical energy in electrical engineering technology compared to the usage of other energy sources:

- Reduce of the total expense and prime cost of used energy
- Decrease of the technology process time
- Decrease of the harmful emissions in the environment
- Improving of the conditions of work of the staff, etc.

DOI: 10.4018/978-1-61520-647-6.ch011

Metal heating with a purpose of its melting is more often made into furnaces using coal or natural gas according to the heat exchange principle.

The high-frequency induction heating (HFIH) use is increasingly expanded with the development of the power electronic converters. Besides at the melting, the HFIH is also used at surface hardening, welding, forging, hot printing, etc.

In 1831, Michael Faraday (1791 – 1867) discovered electromagnetic induction. It is used in all induction heating (IH) applied system. Electromagnetic induction refers to the phenomenon by which electrical current is generated in a closed circuit by the fluctuation of current in another circuit placed next to it. AC current flowing through a circuit affects the magnetic movement of a secondary circuit located near it. This is the basic principle of the induction heating, which is an applied form of Faraday's discovery. Heat loss, occurring in the process of electromagnetic induction, could be turned into productive heat energy in an electric heating system.

Absence of any physical contact to heating devices in IH excludes unpleasant electrical accidents. High energy density is achieved by generating sufficient heat energy within a relatively short period of time.

Furthermore, induction heating is a combination of electromagnetic induction, the skin effect, and the principle of heat transfer. Briefly, the IH refers to the generation of heat energy by the current and eddy current created on the surface of a conductive object when it is places in the magnetic field formed around a coil, where the AC current flows through.

The principle of the IH is illustrated in Figure 1. It shows inductor 1, where a high-frequency current flows through, and detail 2 put inside the inductor. The system might be studied as a transformer with a secondary winding where induction current I_2 flows through. The current density decreases according to an exponential law into the inside of the detail shown in Figure 2, according the relationship:

Figure 1. Illustration of IH principle

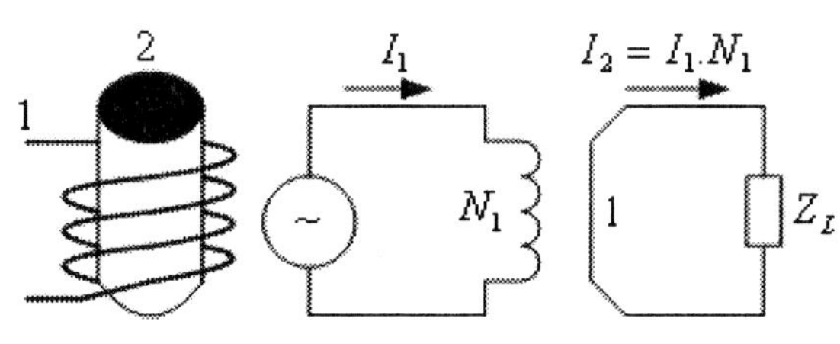

Figure 2. System of inductor – detail and its equivalent schematic

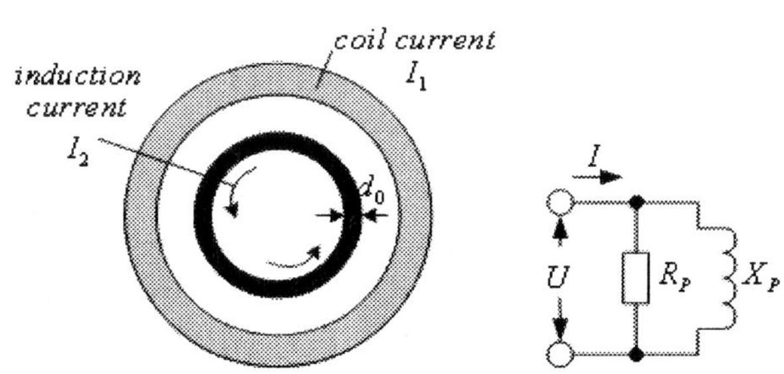

$$I_X = I_0 . e^{-x/d_0},\qquad (11.1)$$

where I_0 is the current density on the detail surface, I_X is the current at x distance inside the detail, d_0 - a constant dependent on the properties of the detail material and the current frequency (Fairchild Semiconductor, 2000):

$$d_0 = k\sqrt{\frac{\rho}{\mu.\omega}},\qquad (11.2)$$

where ρ is specific resistance of the material, k - coefficient, μ - magnetic permeability of the material, ω- angular frequency.

The dependence of the current density on the deepness of the penetration is shown in Figure 3.

The flowing of the current I_X through the detail is connected with active power in the detail. This power heats the detail and transfers into heat.

The system inductor – detail has a parallel equivalent schematic shown in Figure 2 including parameters - R_p, X_p, which change their values during the heating process. For example, R_p may change 1.5 – 2.5 times; X_p - 1.3-1.6 times. The change ranges are different for non-magnetic and ferromagnetic materials.

Frequency dependent on the electrical engineering tasks of IH is required to be about from 50 Hz to several MHz.

At earlier stage, electrical-mechanical converters have been used for the IH purposes. Nowadays, using power electronic converters permits to make economies of electrical energy within the range of 15-40% dependent on the loading.

Figure 4 depicts basic blocks incorporated a frequency converter for IH. The converter derives its name - "frequency converter" from the fact that the output voltage and current frequencies are higher (HFAC) than this of the mains (LFAC). For orientation, the inverter is said to be thyristor current inverter within the frequency range of 500 Hz to 4 kHz, thyristor resonant inverter up to 10 kHz, and transistor resonant inverter for higher frequencies (see Chapter 7). Valve generators are used at megahertz range.

In some cases, the rectifier may be uncontrolled and thereby the regulation of output power is performed using the inverter control system. If the output power has to change within a wide range, controlled rectifier is applied. Its regulation may be combined with the inverter regulation.

Figure 3. Dependence of the current density on the penetration deepness

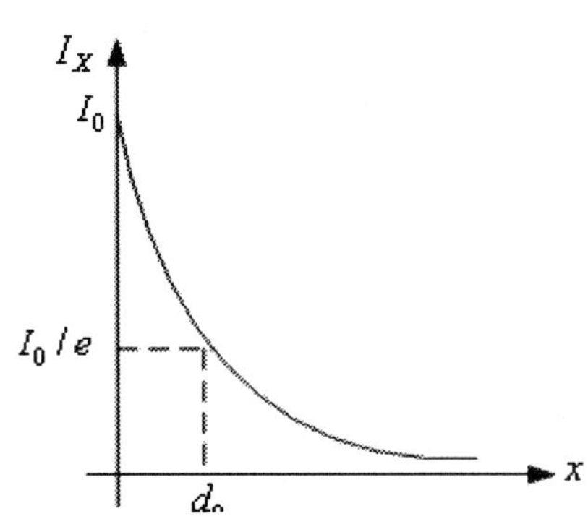

Figure 4. Structural schema of a frequency converter for IH

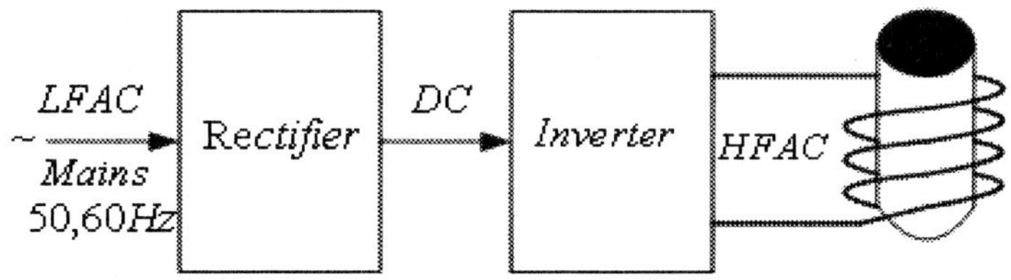

Using of a matrix converter for HFIH purposes as it is shown in Figure 5 is also possible (Antchev, 2007). In this case, a power factor of the entire technology installation in respect to the mains very closed to 1 may be achieved (see Chapter 5).

For a thyristor frequency converter with 1250 kW used for metal melting at furnaces to 5 tons, the following typical data are of interest. Its operational frequency change during the process is from 0.3 kHz to 6 kHz. Also, converter efficiency coefficient is above 90% at a consumption of cooling air about 15 000 m^3/h.

Electric Welding

At electric arc welding, power required for the melting is obtained producing an arc between two electrodes, one of them is usually the metal detail, which has to be welded. The welding in the air atmosphere is the most widely spread.

Volt-ampere characteristic of the process depends on the type of the welding process. Typical characteristic is a strongly dropping one as it is shown in Figure 6. Before the arc creation, the voltage is equal

Figure 5. Use of high-frequency matrix converter for IH

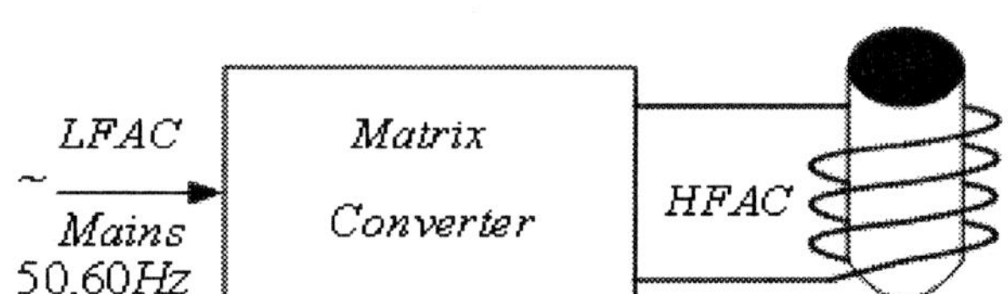

Figure 6. Typical volt-ampere characteristic of a welding aggregate

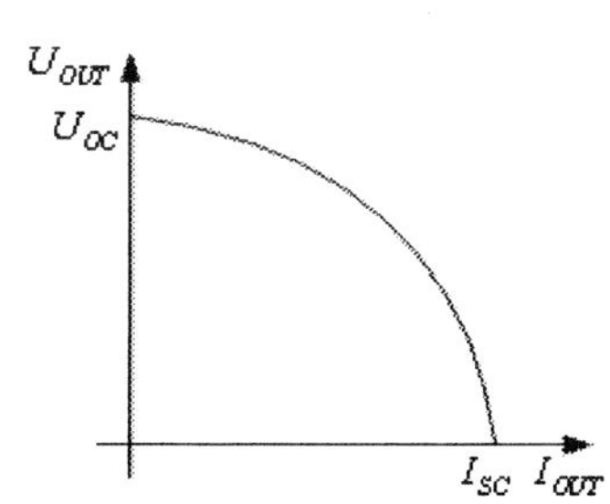

Figure 7. Structural schematics with a low-frequency transformer

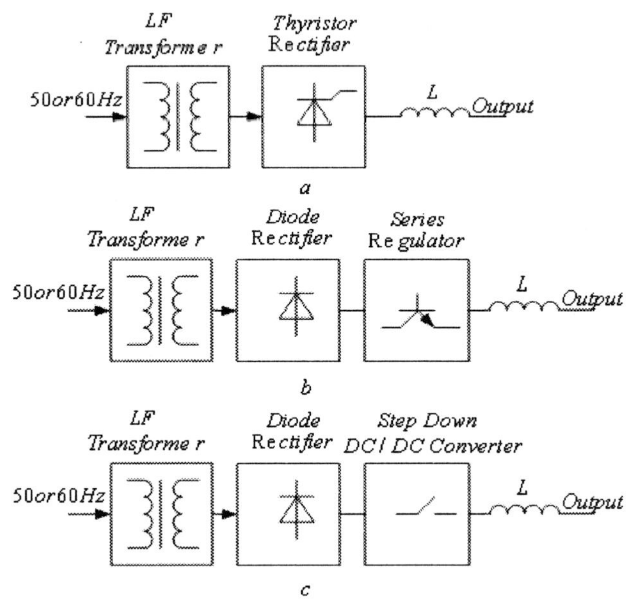

to the voltage of an open - circuit U_{OC}. After the arc creation, the current is limited to the value of the short-circuit current I_{SC}. The welding aggregates usually have a possibility to regulate this current.

Typical values of the idle motion voltage and short-circuit current are 50V and 500A. Moreover, the fluctuations of the output current have to be of minimum value after the arc creation. Therefore, a smoothing inductance is usually connected to the aggregate output. Furthermore, a galvanic isolation from the mains is required to secure safeness for the staff. The isolation is made via high-frequency or low-frequency transformers.

Figure 7 depicts structural schematics of a welding aggregate using a low-frequency transformer. In the schema shown in Figure 7.a, an appropriate control of the thyristors of the rectifier secures the character of the output characteristic. DC voltage from an uncontrolled rectifier is generated in Figure 7.b, and then the regulation is performed using a series transistor operating in active mode. This topology has the lowest efficiency coefficient. The variant in Figure 7.c is implemented using a pulse buck DC/DC converter. Its control system secures the type of the output characteristic. All topologies using a low-frequency transformer have increased volume and sizes.

Figure 8 shows a variant using high-frequency transformer. DC/DC converter is used. It may be with PWM or resonant type. Its efficient coefficient may reach 90% at small sizes and volume.

Figure 8. Schema using high-frequency transformer

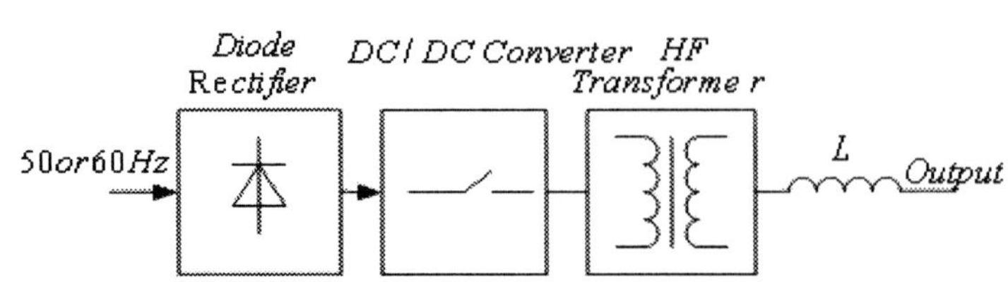

Converters with DC output electrical energy are more and more applied as a supply for aggregates with electrical arc in a vacuum or in gas atmosphere. Vacuum and plasma arcs are applied in electro-metallurgy, chemistry, etc. The application at electro-metallurgy is at welding, electro-arc cutting. At chemistry, the application is to generate acetylene and some other compounds. The requirements for the supplies with electro arc in a vacuum or with a plasma arc are equal in their principles. A stability of arc burning, a possibility for slightly regulation of the current and its stabilization are required. DC source for the electrical welding purpose has to be output short-circuit proved, because the short-circuiting of the output is a normal operational mode in the process of arc lightening. The output voltages of the equipment for plasma welding are within the range of 100-200V with output currents up to 1500A. Also, equipment for micro welding exists whose output current does not exceed 1A.

The devices for plasma welding and cutting have several advantages over the aggregates for electrical arc welding. The plasma arc, burning in the atmosphere of specially fanned gas, is more stable and also has less section than the electrical arc in the air. These advantages permit to obtain, for example, smooth surface at welding and more precise accuracy at cutting. These properties are extremely important at micro-welding. A block schema of aggregate for plasma welding is shown in Figure 9.

The symbols in the figures are as follows:

1 - cathode, 2 - anode, 3 – controlled rectifier, 4 – control regulation system, 5 – current transformer, 6 – auxiliary high-frequency high-voltage block, 7 – detail to be operated with.

Simultaneously with the main arc, which burns between the anode and cathode and which is directed towards the detail, an auxiliary arc, which serves to stabilize the main arc, exist. The rectifier control system monitors the current using the current transformer. Also, the control system stabilizes the current and limits it to a short-circuit mode. Thus, plasma welding or cutting is performed. The voltage is about 400V; the arc current is about 500A.

Vacuum Electrical Arc and Plasma Heating Furnaces

DC or three-phase AC voltage may be used to supply vacuum electrical arc and plasma heating furnaces. At DC voltage supply, the desired arc form, as well as to perform stabilization of its burning, is easily obtained.

A three-phase controlled rectifier is usually applied at voltages within the range of 400-1000V (Chapter 4). If low voltages at high current values are required, the schematic shown in Figure 10 is used. It includes a control rectifier with thyristors in the transformer primary winding.

Figure 9. Schematic of aggregate for plasma

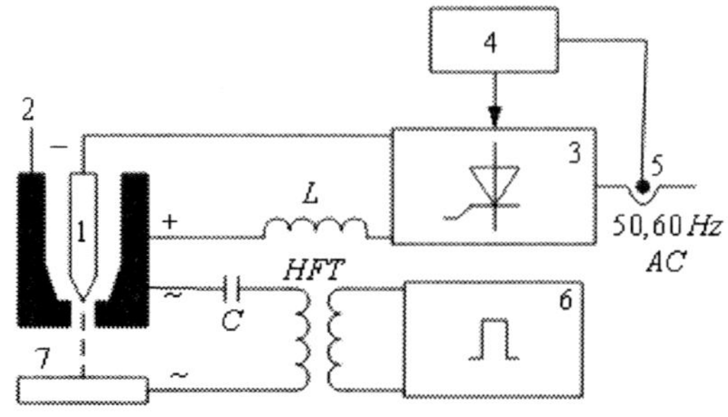

Figure 10. Schema for supply of arc and plasma furnaces

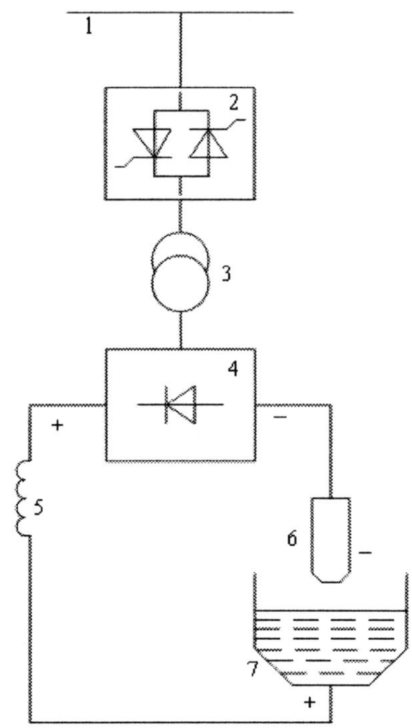

The symbols in the figure are as follows:

1 – three-phase supply network, 2 – three-phase thyristor AC regulator, 3 – buck transformer, 4 – uncontrolled rectifier, 5- smoothing inductance, 6 - cathode, 7 – anode. Blocks 2, 3, and 4, form so-called rectifier with thyristors in the transformer primary winding.

Resistive Electrical Furnaces

Single-phase and three-phase thyristor AC regulators (see Chapter 5) are applicable for temperature regulation in the resistive electrical furnaces. They are applied in the processes of drying, glass melting, hardening, diffusion, etc. It has to be mentioned that the heat constant of such furnaces is much higher than the period of the supply voltage. Thereby, the regulation method of continuous cycling selection (see Chapter 5) is used. Exemplary structural schema used to regulate the temperature of a resistive electrical furnace is shown in Figure 11.

The symbols in the figures are as follows:

1 – furnace, 2 – heater, 3 – thyristor AC regulator, 4 – temperature sensor, 5 – reference temperature, 6 – device for comparison, 7 – electron regulator, 8 – control system for the thyristor AC regulator.

The regulation is performed over a define time interval – about seconds. At a part of this interval, the control system delivered firing pulses for the thyristors at supply voltage zero crossing, i.e. during that subinterval the thyristors are permanently on and the whole supply voltage is applied to the heater. During the rest of the time interval, the thyristors are permanently off. Therefore, this electron regulator is ON-OFF type. This method is also applicable in control of a three-phase resistive electrical furnace

Figure 11. Structural schema of resistive electrical furnace

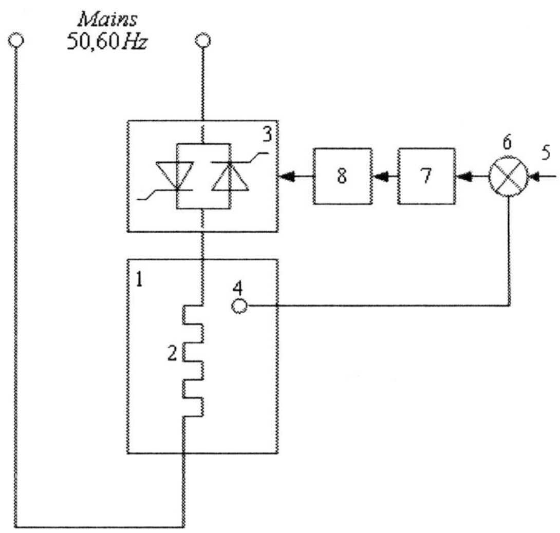

with a three-phase AC regulator. Regulation at different furnace areas using different AC regulators is sometimes performed.

Electrolysis

Electrolysis is a process, which requires DC electrical energy. The electrolysis is made in electrolysis tanks, connected in series with each other. The voltage of a tank, as well as the requirement of the voltage and current of the DC supply source, is different, dependent on the type of the electrolysis – see Table 1.

Recently, controlled thyristor rectifier is used as a DC supply source. However, the requirements of the rectifier control and regulation are specific for the electrolysis. This is conditioned of the different character of the rectifier and group of series connected tanks output characteristics – see Figure 12.

These characteristics are as follows:

1. Rectifier output characteristic
2. Electrolysis of water and water solutions
3. Electrolysis of chlorine

Table 1.

Type of the electrolysis	Working tank voltage,V	Rectifier current, кA	Voltage of a series of tanks,V
Water and water solutions	1,9-2,6	to 10	160
Chlorine	4,0-4,8	to 570	510
Aluminum	4,6-5,0	to 150	800
Copper	0,2-0,4	to 30	200

Figure 12. Output characteristics

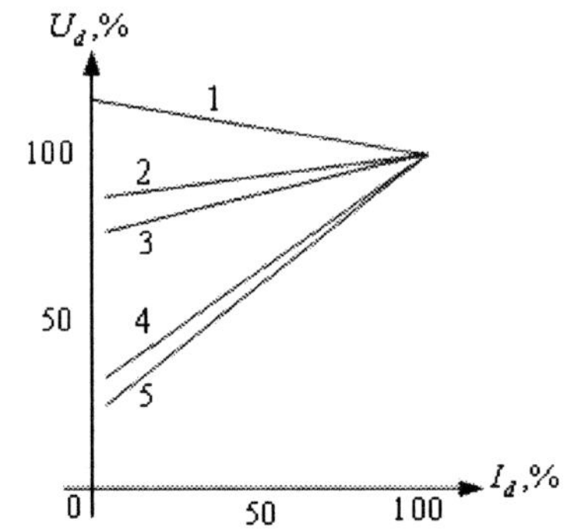

4. Electrolysis of aluminum
5. Electrolysis of copper

At a specific range of current change, the slope of the characteristic for the corresponding type of electrolysis determines the range of the rectifier output voltage change. Regulation of the output voltage is required to maintain an optimal value of the current through the series tanks. The current value determines the speed and quality of the electrolysis process. At a new electrolyser, at the beginning of its use, a voltage with lower value than the nominal one is necessary to be applied for an appropriate forming of the tank. Moreover, an eventual increase of the tank number is provided. Thus, the range of the rectifier output voltage regulation is extended.

As it is seen from Table 1, the electrolysis requires very high electrical energy consumption. This determines the enhanced requirements for the rectifier efficiency coefficient, which reaches 95-98%. The requirements of reliability and diagnostics are especially enhanced. In some cases, additional rectifier modules may be provided. These modules connect automatically to secure the required current and not to disturb the quality of the production in the cases of failure of the main rectifier.

A possibility for an appropriate rectifier is the rectifier with thyristors in the transformer primary side – see blocks 2, 3, and 4 in Figure 10. This rectifier permits the regulation of the AC voltage supplying the diode rectifier to be made using the control system of the three-phase AC regulator. The resistance of the series tanks in the solution electrolysis changes in small ranges after the nominal voltage of the tank is set. Thereby, a requirement for the control system after setting of nominal voltage to compensate only the change of the supply network voltage is imposed.

Since the power is very high, a special attention is paid to the indicators in respect to the supply network of the whole electrolysis installation.

Purifying of Waste Gases

To purify smoke and gases of hard particles and dust in cement plants, in chemical and metallurgy industry, etc., high-voltage electro filters are used. In the filter chamber, an uniformly electrical field is produced. The field is between an electrode, which is incorporated on the center axis of the chamber and to which a negative voltage is applied, and an outer grounded surface. In practice, this surface consists of a big quantity of lamellas with a specific sizes lying at a distance from one another. In the filter chamber, acceleration of the ions and electrons available in the air is performed at an influence of a strong field. At an impact of the present hard particles in the waste gas with the electrons, the electrons and the hard particles are precipitated over the chamber walls. The requirements for the filter voltage value are particular. At an increase of a permissible level, a break due to a discharge arisen is possible. Thus, a sharp decrease of the filter voltage and an increase of the current happen. Furthermore, the flow of the gas is not constant, i.e. to maintain the desired voltage in the chamber, a permanent regulation is required.

Figure 13 displays a structural schema of supply source for the filter.

The symbols are as follows:

1. Supply network
2. Thyristor AC regulator
3. Boost transformer
4. Uncontrolled rectifier;
5. Current-limited inductance
6. Negative electrode
7. Chamber

Figure 13. Supply of an electro filter

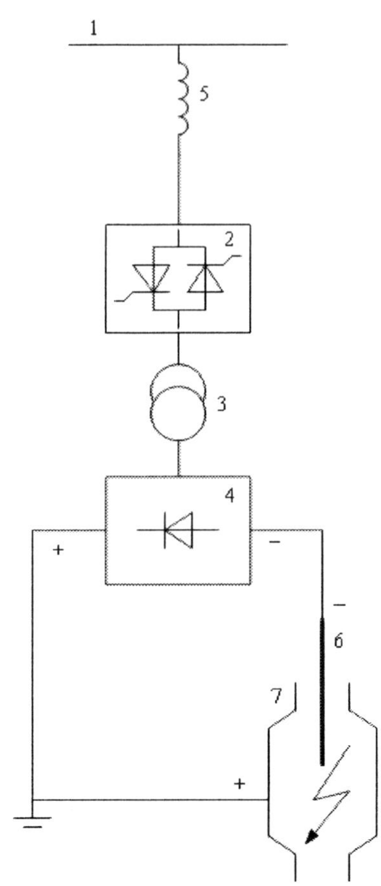

Typical values of the chamber voltage are 25-75 kV at a current through the chamber 100mA to 4A.

Single-phase version of the supply source is usually preferred because of its lower price.

Some specific requirements are imposed to the system for control, regulation and protection. For example, short-circuits at a short time breaks in the chamber are possible at operation, besides, the converter has to maintain its efficiency. The current is limited by the inductance 5 using the control system of the thyristor AC regulator which stops the firing pulses for the thyristors, for example, the pulses are stopped for a period of the source voltage.

Other applications of the power electronic converters of electrical energy for electro-technological purposes are known, such as:

Electromagnetic stirring: A system of AC regulators is used; the system operates in ultra-low frequency. As a result of moving electromagnetic field an agitation, for example, of cast iron and its purifying is performed.

Electro-resistive welding: The aggregate consists of a welding transformer in whose primary side a thyristor AC regulator is connected. The regulator control system regulates the welding current. At the secondary side of the welding transformer, the current reaches 10-80kA at voltage 6-10V.

TRANSPORT APPLICATIONS

The basic part of the electrical transportation uses DC power with different values of the voltage. For example, in trams and trolley cars the voltage value is 600V, in subways – 750 or 1500V, in railway transport – 1500 or 3000V. The energy conversion is made in so-called traction substations. Figure 14 displays a structural schema for energy conversion in such a substation.

The symbols in the figures are as follows:

1. High-voltage AC transmission line
2. Commutation AC installation
3. Buck transformer
4. Diode rectifying blocks
5. System for control, protection and blocking
6. Inductance for limiting the short-circuit current
7. Commutation DC installation
8. Transportation vehicle

As a rectifiers, three-phase bridge schematics with m=6 or m=12 are used because of the low ripples of the output voltage and of the increased power factor in respect to the AC mains.

Specific requirements as follows are imposed to these converters:

- Securing of the required value of the output voltage with a permissible margin and of the desired output power
- Working efficiency at strongly changeable loads – from idle running to maximum power including exceeding of its permissible level. Appearance of short-circuit in the output due to a break of the isolation of the contact network or in the transportation vehicle is sometimes possible

Figure 14. Structural schema for conversion in a traction substation

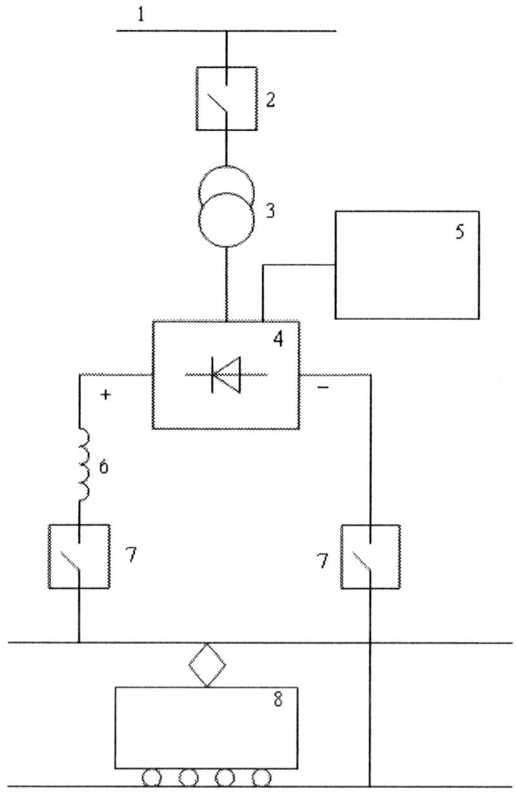

- High reliability and simplicity of the service. Substations operate in most of the cases without serving staff. The time to eliminate the failure has to be minimized. Thereby reserve rectifying modules, which should turn on at a failure event in the main modules, are usually provided in the converter itself. The turn on should be made via system for diagnosis and control
- Very reliable commutation and protection installations. The separate protection should have selective operation and a determined priority in such a way that a truly broken parts, which can be change easily and quickly, to be turned off

The rectifiers are designed and implemented in some cases with a possibility to carry the whole current of a short-circuit in the output for a long time, for example, till the turning on of the protection commutation installation.

The implementation of electrical motors in the modern cars increases (Venkateswara, 2004). The source of electrical energy is a battery, fuel cell or ultracapacitor (Lukic, 2008).

Operational features and characteristics of the fuel cells are studied in Chapter 2. System, implemented on the basis of the fuel cells, contains different blocks – to store hydrogen, oxygen compressor, for water processing, for processing of heat energy, for processing unnecessary gases and for producing electrical energy. In this chart, the basic attention is paid to the conversion of the produced electrical energy. Because of the specific features of producing of this energy, a necessity for its accumulation and storing for further use arises. For the time being, the accumulator batteries are the main mean to accomplish this necessity. Also, researches are made into use of the ultracapacitors.

Figure 15 systemizes the principle solutions to produce electrical energy from fuel cells (Pera, 2006, 2007). In all schematics, a diode connected in series with the fuel cell group is used to prevent reverse direction current, which may damage the cells. The first two topologies are the simplest ones.

Figure 15.a shows a supply of a DC consumer. Its supply voltage is generated from the main DC/DC converter. A bidirectional DC/DC converter is connected to the supply voltage of the load. This converter charges a battery during the operation of the fuel cell system. When the system does not operate, the DC/DC converter converts the battery voltage to supply the consumer.

Figure 15.b depicts a similar case to this of Figure 15.a with the difference that the load is AC. Of course, an inverter is included in the schematic. The supply of the inverter is generated either from the fuel cells or from the battery. A bidirectional DC/DC converter is also required.

Figure 15. Conversion of energy from the fuel cells

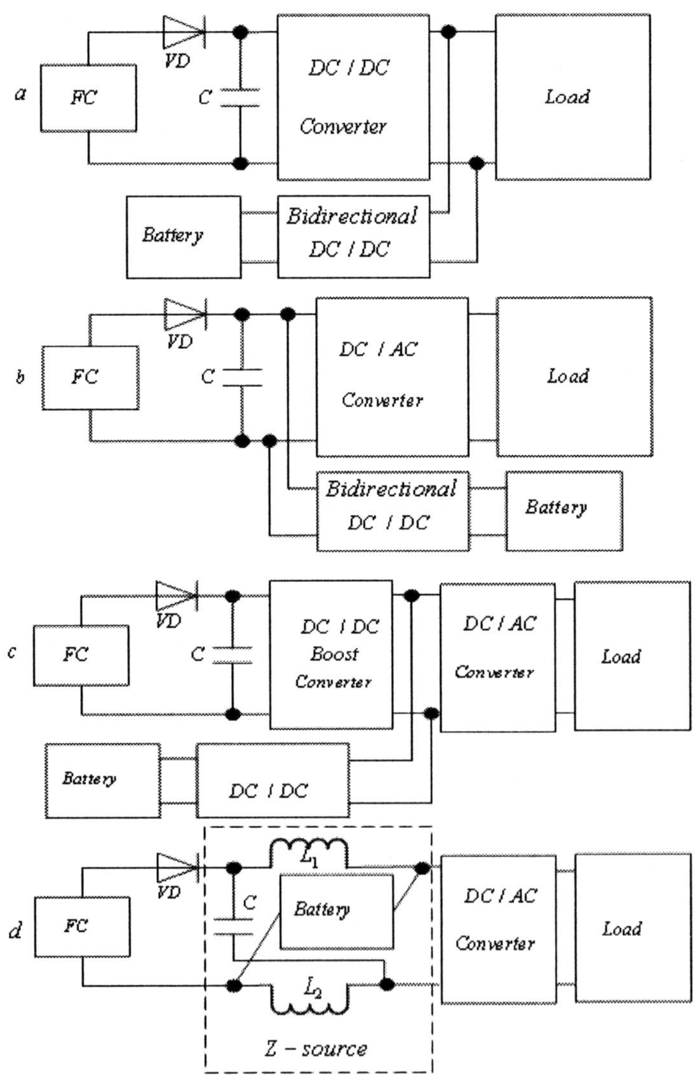

Topology shown in Figure 15.c uses an intermediate DC conversion using boost converter. The variants used for its implementations are push-pull converter, half-bridge or bridge converters, moreover, the galvanic isolation in the converter is via a high-frequency transformer.

The topology shown in Figure 15.d is of great interest. A Z-source inverter is used in this schematic (Peng, 2003). In this inverter, the effective value of its supply voltage is higher than the output voltage generated from the fuel cells.

The use of different resonant DC/DC converters with ZCS or ZVS (see Chapter 6) with improved energy indicators is possible (Fahimi, 2007).

The basic methods to charge accumulator batteries and different types of the output characteristics of the charge sources have been studied in Chapter 2. The charge of the accumulator batteries in their producing is usually made in a big group, where in they are connected in series with or in parallel to each other in so-called charge stations. In this case, a DC source with a higher power and with a particular type of its output characteristic, working under immobile conditions, is required. As such a source, a three-phase controlled rectifier is most often used. It should have a possibility to operate as a dependable inverter at a specific type of the accumulator batteries. During the technological process at a definite intervals, discharge of the accumulator batteries, the energy from it is delivered on the supply network, is also performed. Besides the desired voltage and current feedbacks used to form the output characteristic at a charge, also, monitoring of the battery temperature is provided. The implementation, based on microprocessors and microcontrollers, of the control and regulation system of these rectifiers permits to set important parameters of the technologic process, such as maximum charge voltage, value of the charge current, type of the characteristic, temperature, etc., in the dialog mode with the user.

Charge of an accumulator battery put in a transportation vehicle, which uses electrical energy for its motion, such as trams, trolley cars, electrical railway, is very often necessary to be secured. The voltage of the contact network is a source for this charge. The voltage is produced in the substation – Figure 14.

Figure 16. Conversion of the contact network energy for charging an accumulator battery

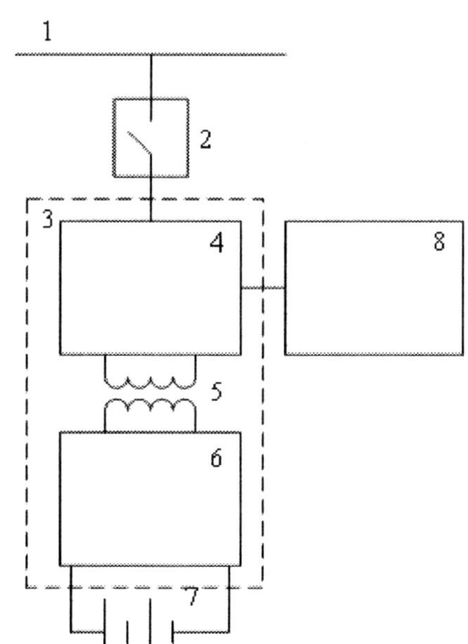

Moreover, the charge converter in this case is in the transportation vehicle. The requirements for converter constructive implementation are enhanced. A variant of such converter is shown in Figure 16.

The symbols in the figure are as follows:

1. DC contact network
2. Protection elements
3. DC/DC converter with galvanic isolation
4. High-frequency inverter;
5. High-frequency transformer
6. High-frequency rectifier
7. Accumulator battery
8. Control and regulation system

The conversion is made with galvanic isolation at a higher frequency. The purpose is to reduce the size and volume of the converter. Half-bridge and bridge electrical schematics, studied in Chapter 6, of DC/DC converter are used. The control and regulation system monitors the output voltage and output current of the converter. The accumulator battery usually operates in buffer mode. A part of the converter output current is used to supply DC consumers of the transportation vehicle.

As it has been mentioned in Chapter 2 at the study of the ultracapacitor features, they might be used as a source of energy with a high value for a comparatively short time. For example, such energy is required at a start of very powerful motors (16 cylinder motors), in trains, ships, etc. In this mode the current value reaches 1000-3000 A. A hybrid car is a similar case where the electrical motor might be supplied in combination from an accumulator battery and an ultra capacitor.

The requirements for the charging devices of ultracapacitors are connected with the necessity that the charge should be made for a short possible time without a disturbance of the limitations of the capacitor maximum voltage. DC/DC converter without galvanic isolation (see Chapter 6) in different regulation modes has found widest application as a charge converter. Figure 17 (Maxwell Technologies, 2005) shows buck converter consisting of a transistor switch, diode and inductance in the mode of constant

Figure 17. Charging with a buck converter at a constant current and voltage limitation

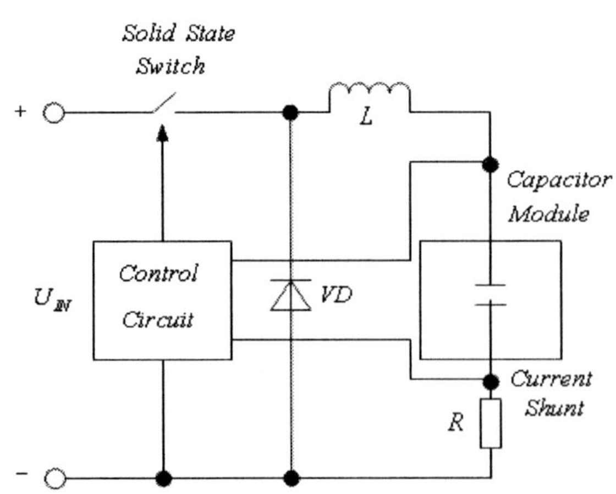

Figure 18. Charge with a buck converter and constant power

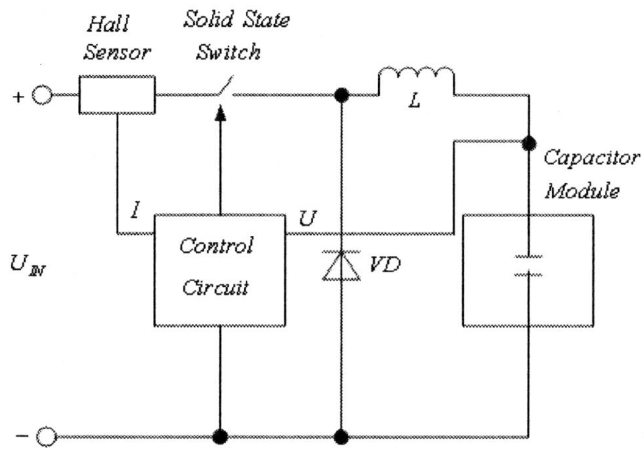

current with a voltage limitation. The charge current is monitored through the feedback from the by-passed resistor. Simultaneously, the ultracapacitor voltage is also monitored.

Figure 18 illustrates transistor switch current and ultracapacitor voltage monitoring. The schematic is recommended at a constant power charge (Maxwell Technologies, 2005).

It is seen from the comparison made in Figure 19, that the constant power charge method has an approximately twice short charging time (Maxwell Technologies, 2005).

At a charge from the low voltage supply network, a transistor flyback converter with an appropriate regulation might be used (Maxwell Technologies, 2005).

There are also several additional applications of the power electronic converters in transportation:

- **In the motion systems of electric trucks:** The energy source is usually an accumulator battery while control and regulation of a DC electrical motor is required. Thereby, DC/DC converter implemented on the basis of power transistors (see Chapter 6) is connected between the source and motor;

Figure 19. Dependence of the charge time on the method used

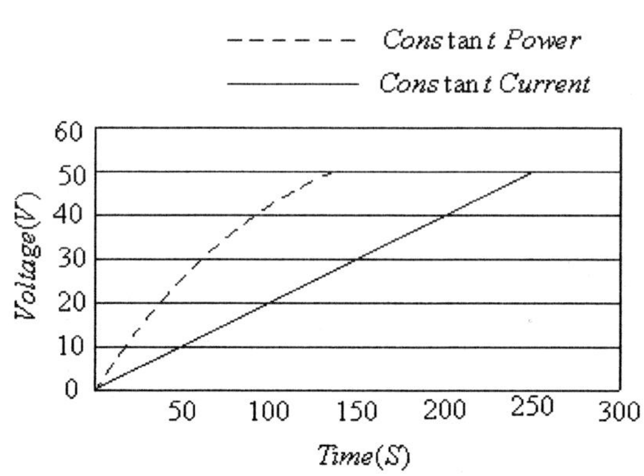

- **In the motion systems of transmanipulators:** Outlined trend is the use of asynchronous electrical motors supplied from a three-phase bridge inverter controlled according to a method described in Chapter 7. The inverter is supplied from a rectifier connected to the mains. Recently, a bidirectional converter (see Chapter 4) is used as rectifier. Thus, a possibility to return energy on the supply network in specific modes of load motion on the transmanipulators exists. The bidirectional converter most often controlled in accordance with a frequency control has the widest application. During comparatively slower processes of load motion (from several seconds to tenth seconds), control and regulation of the entire electrical energy converter is possible to be on the basis of programmable controllers, produced by many firms and also used in the field of the automatic control and regulation.
- **In additional electrical source systems on railway carriages:** Inverters for electro luminescent lighting; inverters to generate voltage with a value of the supply network with a purpose to supply low-powered everyday electrical devices.

HOME APPLIANCES

Figure 20 depicts the energy expenses for house holding (Iowa Energy Center, 2008).

Figure 20. Energy expenses for house holding. (Source: Iowa Energy Center, Home series booklets. Book four: Major home appliances)

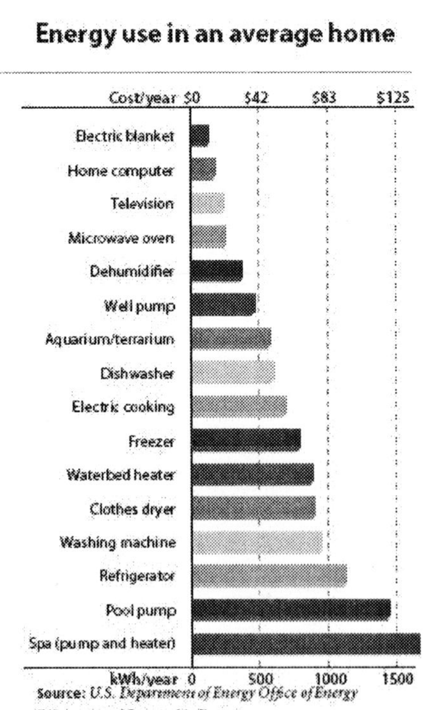

Figure 21. Distribution of energy in incandescent and fluorescent lamps

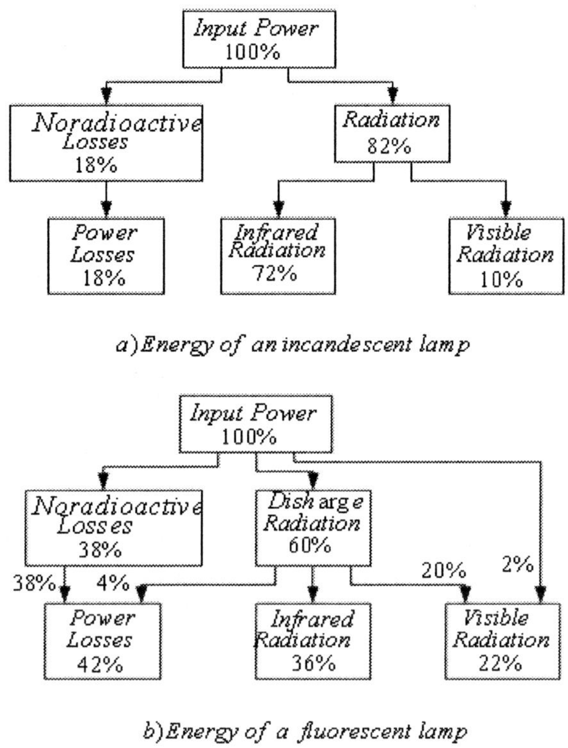

Lighting

Fluorescent lamps have a wide spreading for the lightning needs. The principle advantages of the fluorescent lamps are long durability, high light production, variety of colored temperatures, and high index of color transmission. At same input power, they produce almost two times higher lighting power compared to the incandescent lamps – Figure 21.

To fire a fluorescent lamp (appearance of discharge between the electrodes) is necessary to increase the voltage applied to the lamp. After the lightening, the voltage decreases, and afterwards, the current through the lamp is limited to a value which is necessary to maintain the discharge. One of the possible lightening ways is using so-called electromagnetic starting-regulation apparatus. It consists of a starter and inductance. The inductivity of the inductance connected in series with the lamp limits the current through the lamp to the desired value. Figure 22 shows the connection of an electromagnetic ballast and the waveforms of the voltage across and the current through the lamp. Due to the square waveform of the voltage, the power factor in respect to the supply network decreases and also high harmonics in the source current are available. Compensating capacitors are used to compensate the power factor. Moreover, due to the pulse waveform of the lamp voltage with steep fronts, electromagnetic radiation on the environment is increased.

Disadvantages of the method are the high losses across the inductance – about 10%, possibility of blinking and noise during starting, appearance of flicker effect, and, finally, partially fault lamp might continue operating at bad quality of the light parameters.

Figure 22. Operation of a fluorescent lamp with electromagnetic ballast at network frequency

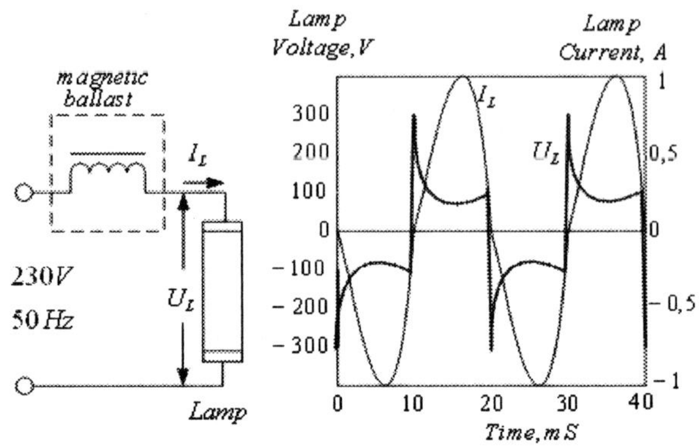

The operation of a fluorescent lamp at higher frequency has several advantages, as follows:

- An established fact is that at operation at higher frequencies, the light production of the fluorescent lamp is increased – Figure 23. The increase after 30 kHz is negligible
- The losses across the system lamp – ballast are lower. Thus, the high power omitted is lower. The operating expenses connected with air conditioning are decreased
- Owing to the less loading of luminofour and electrodes, the live duration of the lamp operating at high frequency with electronic ballast increases with 30% to 50%
- The starting is quick, noiseless and without blinking. The operation duration of the lamp is enhanced at an initial warming up of the electrodes
- In a case of failure at the lamp, it turns off automatically from the electronic ballast
- The sensibility to the ripples of the network voltage values and network frequency is brought to minimum

So-called electronic ballasts for fluorescent lamps are used at operation in high frequency. Figure 24 depicts the way of connection of the ballast and the waveforms of the current and voltage of the

Figure 23. Increase of the light production at operation at high frequencies

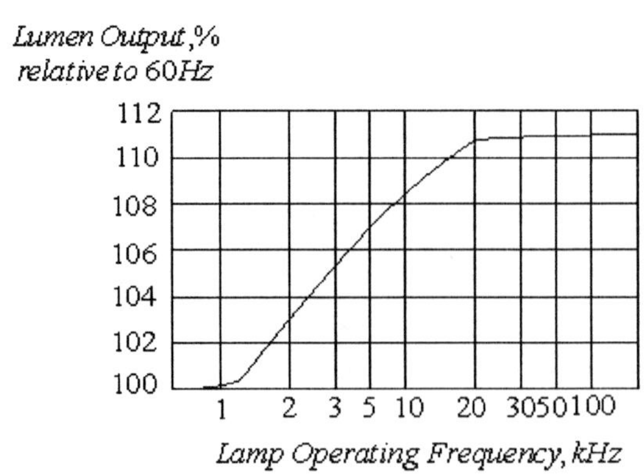

Figure 24. Operation of a fluorescent lamp with electronic ballast at high frequency

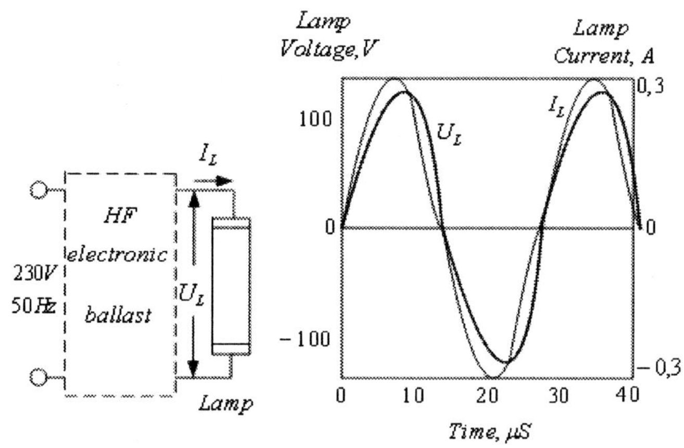

lamp. The current waveform is close to the sinusoidal one. Thus, the electromagnetic disturbances in the environment are decreased.

Here, transistor resonant inverters, more often half-bridge schema and rarely bridge schema, are used as basic power schemas. Figure 25 illustrates the electronic ballast basic block – uncontrolled rectifier and transistor resonant inverter (Texas Instruments, 1994). To increase the efficiency coefficient of the transistor inverter, its operation above the resonance frequency is provided. ZVS characterizes this operational mode (see Chapter 7). Figure 25 does not show the control system of the transistor inverter. The control system might be implemented with a self-excitation or with undependable excitation. Special integral circuits are produced for the control systems of transistor inverters both with a self-excitation and with undependable excitation.

The figure does not also show the electromagnetic interference (EMI) filter connected between the mains and the electronic ballast. The EMI filter purpose is to limit the disturbances, which may enter on the mains from the ballast, within the radiofrequency range.

Very often in the schema shown in Figure 25, elements for passive or active power factor corrections are included. The researches in the lighting field are towards mixed schematics of the converters combining power factor correction and high-frequency control of the fluorescent lamp.

Figure 25. Basic elements of an electronic ballast for fluorescent lamps

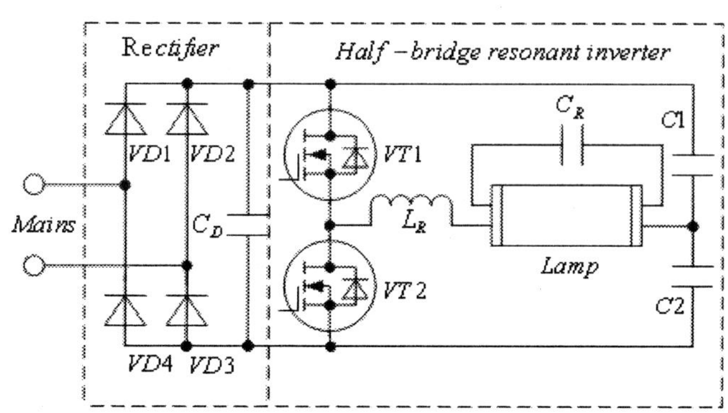

Warming, Cooling and Air Purifying Appliances

Two approaches are used for the needs of warming cooling and air purifying - room air conditioning and central air conditioning. The first approach is more often met in practice. Control of the electrical motor incorporated in the apparatus is recently controlled by transistor inverters. Figure 26 illustrates a structure used to explain the operation of the room air conditioner (National Resources Canada's Office of Energy Efficiency, 2004).

A room air conditioner extracts heat from the space that is being cooled and conveys the heat outside this space. A fan circulates room air through the evaporator, which contains low-pressure refrigerant. Evaporation of the refrigerant cools the tubes and fins, extracting heat on the evaporator outer surface. Then, the cooler, drier air is returned to the room. The gaseous refrigerant leaving the evaporator is drawn into the compressor where mechanical compression raises its temperature and pressure. The hot, high-pressure refrigerant passes through the condenser, where it loses heat to outdoor air (which is blown over it with a second fan) and condenses. This high-pressure liquid refrigerant passes through a restriction and into the low-pressure side of the circuit, and the entire process is repeated over and over again.

The inverter control of air control systems has several advantages compared to the rest air conditioners. Because of the slowly start up of the motor in the inverter air conditioning, the main advantage is less power use to initially start up. Reaching the set temperature of the room, the motor will start slowing down, thus using less electricity. And as an additional benefit, the temperature in the room will

Figure 26. Illustration of the operation of the room air conditioner (Source: National Resources Canada's Office of Energy Efficiency, 2004, EnerGuide)

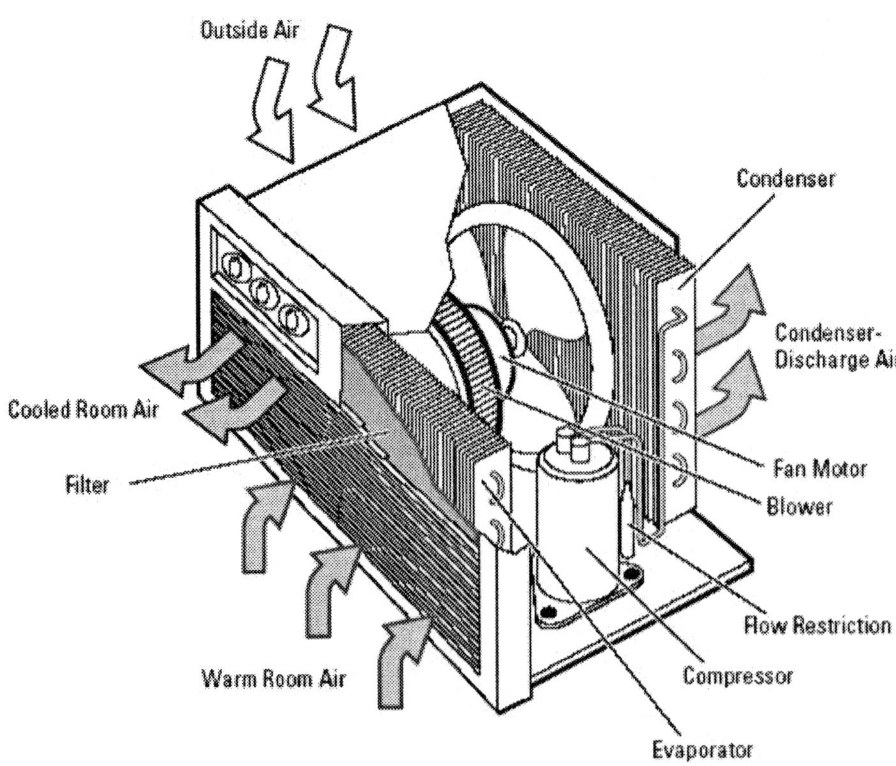

remain more constant. Therefore, the motor does not have to completely stop and because of this less electricity is used removing the need to try to overcome inertia trying to make the motor initiate from a complete stop.

Briefly, the advantages of the inverter technology implemented in the air conditioner are;

- Faster reach of the set temperature
- Start up time is reduced by 1/3
- The power consumption is reduced by 30%
- Avoids cycling of the compressor meaning that there are no voltage peaks
- Temperature fluctuations are also reduced

Household Electrical Appliances

The induction principle described at the beginning of this Chapter finds also implementation in the induction kitchen hot-plates. Figure 27 depicts energy conversion in the separate blocks in induction hot-plates. The conversion is made during the following stages:

First, an AC/DC conversion is made using a rectifier. Then, the generated DC current is connected to a high-frequency switching circuit to administer high-frequency current to the heating coil. According to the Ampere's Law, a high-frequency magnetic field is created around the heated coil. If a conductive

Figure 27. Energy conversion in the induction hot-plate

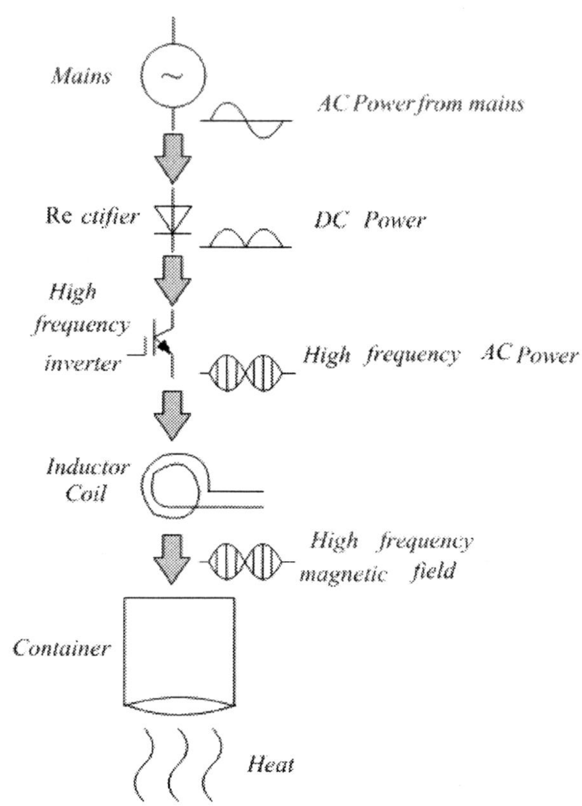

object, e.g. the container of a rice cooker is put inside the magnetic field, then induced voltage and an eddy current are created on the skin depth of the container as a result of the skin effect and Faraday's Law. This generates heat energy on the surface of the container. Rice is cooked by using this heat energy.

A half-bridge resonant inverter – Figure 28, and quasi-resonant inverter – Figure 29, are more often used in the induction hot-plates as inverter schemas (Fairchild Semiconductor, 2000).

The half-bridge resonant inverter switches at a high frequency. Zero voltage/current turn-on switching is enabled by turning the MOS while the diode is in turn on period. The resonant circuit includes resonant inductance L_R and resonant capacitance C_R. The capacitors C_1 and C_2 are the lossless turn-off snubbers for the switches *VT*1 and *VT*2.

The quasi-resonant inverter shown in Figure 29 executes high-frequency switching. By turning on the MOS while its antiparallel diode is in turn-on state, it is possible to do a turn on switching with the voltage and current remaining at zero. The resonant circuit is composed of resonant inductance L_R and resonant capacitance C_R.

Power electronic converters also find place in the following electrical house appliances – microwave ovens, refrigerators, washing machines, etc. DC/DC converters designed on the basis of high-frequency transistor resonant inverters with galvanic isolation (see Chapter 7) are usually implemented in the source blocks of plasma displays.

Figure 28. Use of a half-bridge resonant inverter

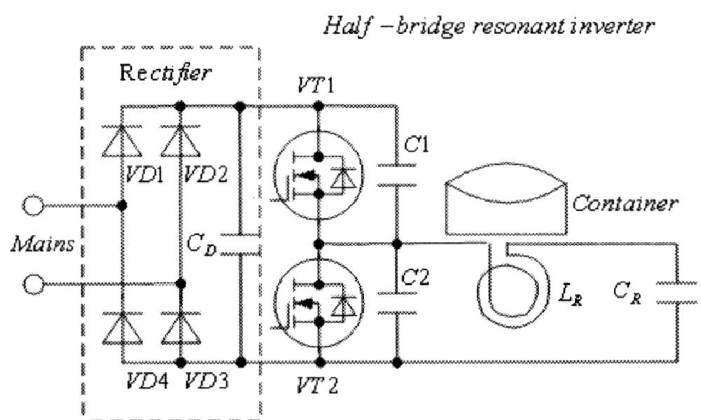

Figure 29. Use of a quasi-resonant inverter

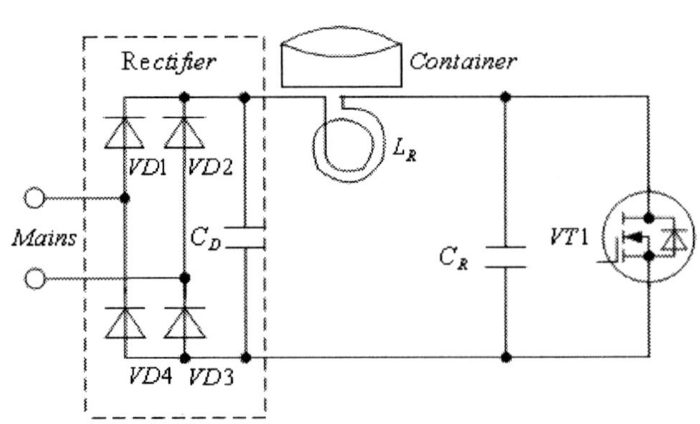

ELEVATORS

Figure 30 displays the implementation principles for control of elevator electrical motors.

In the three topologies, the control of the asynchronous electrical motor is realized via three-phase transistor inverter. The control methods of the inverters are different in the different producers and they are – frequency, vector, field orientated methods, etc. In the first variant shown in Figure 30, energy at stopping is dissipated on resistor via turning on of an additional transistor. The input rectifier in this variant is usually uncontrolled. The second topology requires bidirectional AC/DC converter used to returned energy at stopping on the mains. In the third case shown, this energy is used to charge a block of super capacitors and it may be used later as a source of the inverter at a drop off of the supply network voltage.

APPLICATIONS IN COMMUNICATION

Supply of Telecommunication Centrums

Besides the basic technical parameters of a supply source, an additional requirement for a high reliability is imposed on the supply source of telecommunication centrums because of the necessity of continuous operation of the centrums. Figure 31 shows a way to meet this requirement.

Figure 30. Converters for elevator electrical motor control

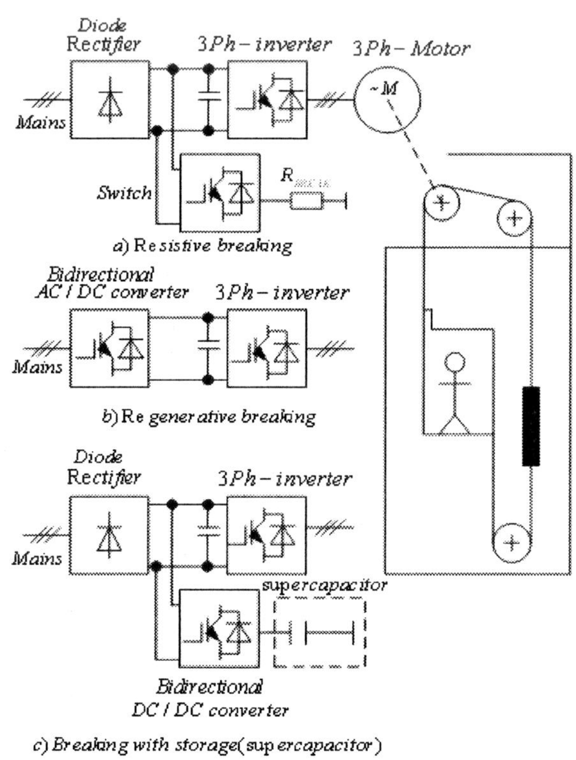

Figure 31. Supply from two undependable lines

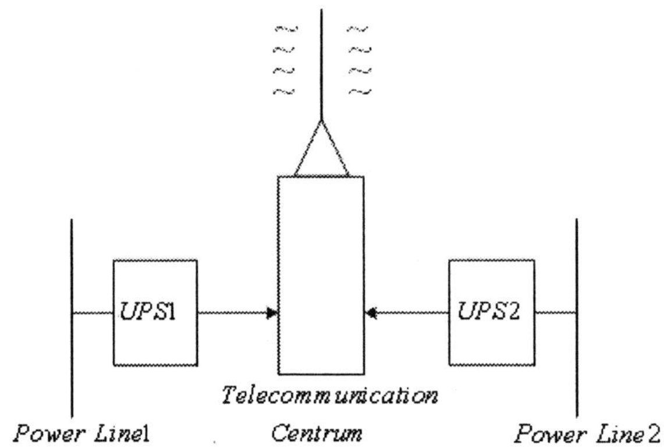

As it is seen, the supply is secured from two undependable lines. For each line an UPS system (see Chapter 10) is provided. Each of the systems is with a desirable power to supply independently the telecommunication centrum.

Supply of Telecommunication Equipment

Two approaches, which are often combined, are imposed at securing the supply of separated telecommunication equipment, for example, telecommunication cells. The first is with separate AC/DC converters as the one shown in Figure 32.

The converter usually may operate at 110/230V 50/60 Hz single-phase supply network and it has capability to supply 3-4 consumers. These may be 3-4 modems, which supply voltages is most often 48V. The consumption of the each modem is about 2-3 A. Each single AC/DC converter is provided with a possibility to be incorporated in the same case as the modem. Also, indication of the input and output states is provided.

The second approach uses an AC/DC converter, which is embedded with accumulator batteries in a separate special box, and thus forming so-called Power and Battery Backup System. In this way, uninterruption of the electrical supply is secured. The accumulator batteries operate in buffer mode, i.e. the supply of DC consumers is secured. The outputs of such a system are 4-5.

Figure 32. Single AC/DC converter

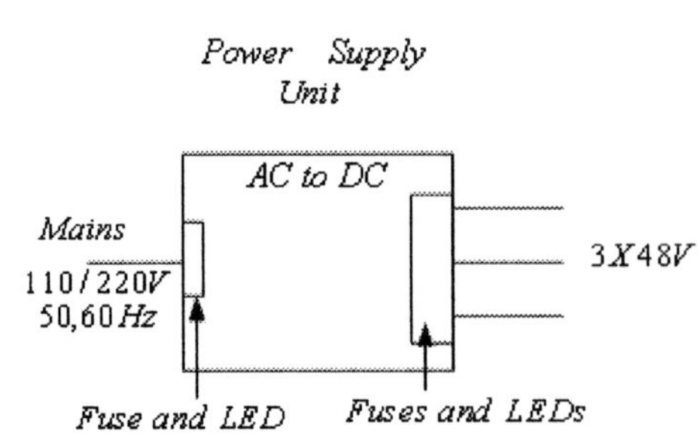

The rectifiers usually operate in parallel according their outputs. If one of them breaks the other has the possibility to take the whole loading. Some producers have special PC software, which helps to test a system as one shown in Figure 33, for example, it helps to check the state of the batteries.

As an additional apparatus to this system, so-called DC Distribution Unit is designed. Its input connects to a primary DC voltage – 24, 48 or 60V – Figure 34. The outputs are usually up to 5 each having appropriate fuses and indication. The power of each output is about 100W. The DC Distribution Unit is provided with possibility to be embedded in the main case as it is in single DC/DC converter. Therefore, the connection wire length and type are of great importance - both from the source L_{SUPPLY} and to each of the consumers $L_{CONSUMER}$. Specialized methods provided by different producers exist. These methods give as graphics the dependence of the source wire length on the consumed power and also give a recommendation about the cable type. The purpose is the supply voltage for the consumer always to be above one minimum value. For example, for 24V the minimum value is 20.4V, for 48V – 40.8V.

MEDICAL APPLICATIONS

In the medicine field, the following power electronic converters find the widest spread place as an electrical source for different part of electro-medical equipment: AC/DC converters, DC/DC converters, Battery Chargers. Their schematic solutions do not differ from the already studied in the book in the corresponding chapters. The general trend here is towards getting small volume and weight. It has to be mentioned that the power of a converter for medical needs only in specific cases is above 500W. The spread of low-powered converters with tentative power below 100W predominates.

In this chart, more specific requirements taking in consideration the operation conditions connected with the equipment for medicine needs that these converters have to meet will only be studied. Quality vendors has to provide a certification file for the power supply. The file has to provide basic performance and safety measurements for the product, including leakage current and electromagnetic interference (EMI) requirements, constructional features, and thermal data, etc.

Figure 33. Structure of Power and battery backup system

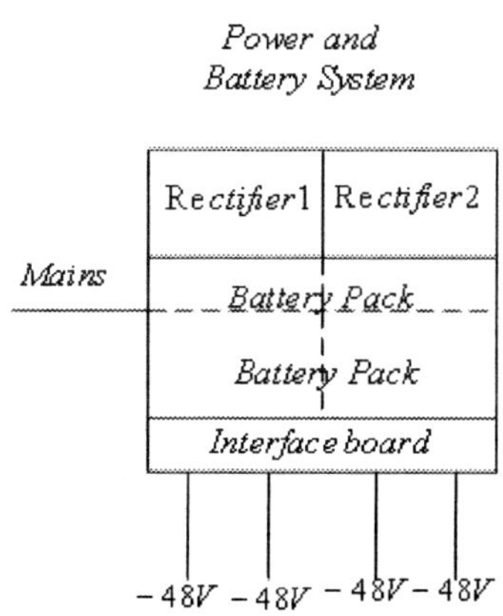

Figure 34. Structural schema of a DC distribution unit

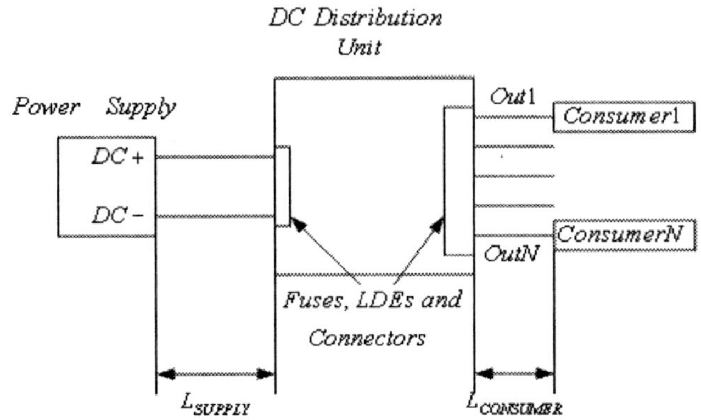

The special requirements of medical equipment are put in international standards. IEC60601-1 standards contain the safety standards for medical power supplies for most of the World, including Europe and North America. IEC states for the International Electrotechnical Committee. UL, CSA and EN standards are derived from the IEC standards. The present version of OEC60601-1 is the second edition (originated in 1988). A 3rd edition of the IEC60601-1 (originated in 2005) is under review by power supply manufacturers and global safety certifying agencies for future adaptation.

Thermal Effect Stability

Thermal requirements are generally stricter for medical power supplies and require lower touch temperatures than those required by commercial or industrial safety certification standards. This is due to the risk of patient contact and the vulnerability of patients. Often, special power enclosure techniques are employed in medical products that decrease the touch temperature and thus, the possibility of adverse reaction of patients touching the power supply when operating at full load.

Additional Requirements Regarding the Reliability

Additionally, quality power supply vendors provide documentation not often requested by industrial and consumer-focused companies. The records include failure mode effects analysis (FMEA), mean time between failures (MTBF), accelerated life testing (ALT), and other supporting documentation.

IPXStandards

IPX standards limit the amount of liquid or solid ingress that can penetrate a power supply. Although the standards were not developed specifically for the medical device industry, medical product will often request IPX1 (Protected against falling water Equivalent to 3-5 mm rainfall per minute for a duration of 10 minutes. Unit is placed in its normal operating position) certification for external power supplies. Medical power supplies are frequently operating in environments where liquids may drip on the power supply. If the dripping liquids ingress the power supply, the power supply may fail to operate or a serious safety situation may arise.Achieving compliance with IPX1 often poses difficult challenges for power supply manufacturers, as it requires custom enclosures and cable assemblies.

Spacing

Spacing can be broken down into two types: creepage, the shortest path between two conductive components along an insulating surface, and clearance, the shortest distance through air. Power supplies that incorporate correct creepage and clearance distances will provide an acceptable degree of electrical safety. Spacing in power supplies is influenced by factors such as the type of insulation, working voltage, pollution degree, input over-voltage, the insulator tracking index, and even operating altitude. Spacing in medical power supplies is influenced by the reference voltage as measured across the insulation during normal operation.

Insulation Voltage

The voltage that the insulation under consideration is being subjected to during normal operation is referred to as the working voltage and the reference voltage in medical applications. This voltage is normally measured when the equipment is operating at its rated input or at the upper point of the rated input voltage under normal use. In switch-mode power supplies, the voltage depends on the rated operating input, power circuit topology used, and the PFC circuit, with the highest voltage normally found between the primary and secondary side of the power transformer. For clearance measurements in medical power supplies, the peak or dc and rms working voltages are required, and if the dc value is used, the peak value of any superimposed ripple is included. For creepage measurements, the required working voltage is the rms or dc value, and if the dc value is used any superimposed ripple and short-term disturbances such as transients are not taken into account. For medical power supplies, clearance and creepage measurements require rms or dc reference voltage.

Other differences from standard power supplies include reinforced or double insulation. This allows operating from 240V AC mains, for example, and so must withstand a dielectric test at AC 4kV for medical applications. Power supplies that are approved to less than AC 4kV may be used in medical applications as part of a reinforced barrier, provided that the insulation within the power supply is regarded as a lesser basic or supplementary barrier. In this case, additional isolation must be provided within the end-product medical equipment by the equipment manufacturer to achieve the requirements of reinforced barrier between the AC mains supply and patient. The 3rd edition of the IEC60601-1 separates the requirements for the patient and operator whereas the 2nd edition treated them as equal.

Pollution Degree

Spacing considered during the design stage may be compromised during the life of the power supply due to pollution in the operating environment that could enter into the power supply circuitry either through the cooling system or through enclosure slots, vents, and louvers used for cooling.

Transients

Because input overvoltage transients are also considered in determining the clearance in the standards, such events divided into four categories, with the higher numbers representing higher surge voltages. The most common category and used in UL 60950-1, Category II involves local level and covers appliances and most office equipment. Category III involves distribution level such as building wiring and other fixed installations with smaller transient overvoltages than overvoltage category IV, which involves primary distribution systems such as overhead lines.

Immunity

Immunity testing for the product's resistance to surge, burst, and electrostatic discharge should also be documented as well as power line harmonics for power supplies over 50 W. These immunity and emissions tests are very important for reliable operation of, not only the power supply, but also all of the other electronic equipment found in a typical hospital or clinic

Leakage Current

Leakage current in commercial and medical power supplies is determined by where and how it is measured, the maximum values allowed, and where the current is flowing. Protective conductor current and earth leakage current in medical supplies are the unwanted current that flows through the protective earth conductor going back to earth.

There are Several Specific Requirements Towards Batteries and Battery Charges

Batteries and charging equipment are prevalent in today's hospital settings, and power supply manufacturers must design battery chargers to achieve maximum battery service life. This is of particular importance when considering that a battery may be charged to power something as critical as a portable heart defibrillator.

Proper selection of a battery charger will ensure reliable operation of battery-powered equipment. The type of charger selected should be based on the frequency of use of the battery-operated equipment, the type of battery being charged, and the service life expectations for the equipment, among other factors. Sophisticated power supply vendors can design the exact battery charger to meet your battery requirements and they can provide innovative mechanical solutions for cradling the battery-operated device while it is charging.

REFERENCES

Antchev, M. H., & Kanov, G. Tz. (2007). *Study of single-phase high frequency matrix converter for induction heating applications.* Paper presented at the 14-th Symposium Power Electronics Ee2007, Novi Sad, Serbia.

Bermann, M., Norton, D., Skinner, A., & Taylor, R. (March 2009), What to look for in a medical power supply. *Medical Design, 33*. Retrieved from http://www.medicaldesign.com

Fahimi, B. (2007). High efficiency and compact DC/DC converter for high power fuel cell systems. *IEEE Power Electronics Society Newsletter*, *19*(3), 14–19.

Fairchild Semiconductor. (2000, July). *Induction heating system topology review application note 9012*. Retrieved 2008 from http://www.fairchildsemi.com

Iowa Energy Center. (n.d.). *Home series booklets. Book four: Major home appliances*. Retrieved from http://www.energy.iastate.edu

Lukic, S. M., Cao, J., Bansal, R. C., Rodriquez, F., & Emadi, A. (2008). Energy storage systems for automotive applications. *IEEE Transactions on Industrial Electronics*, *55*(6), 2258–2267. doi:10.1109/TIE.2008.918390

Maxwell Technologies. (2005, December). *Charging of ultracapacitors*. Application Note 1008981. Retrieved 2008 from http://www.maxwell.com

Motorola (1991). *Industrial control applications*. UK: Tavistock Press (Bedford) Ltd.

Motorola (1993). *Power applications manual*. UK: Tavistock Press (Bedford) Ltd.

National Resources Canada's Office of Energy Efficiency. (2004). *Air conditioning your home*. Canada/Ottawa. Retrieved 2008 from http://dsp-psd.pwgsc.gc.ca

Peng, F. Z. (2003). Z-source inverter. *IEEE Transactions on Industrial Electronics*, *9*(2), 504–510.

Pera, M.C., Candusso, D., Hissel, D., & Kauffmann, J.M. (2006). Fuel cell power generators: a great idea but a stubborn reality. *IEEE Industrial electronics Society Newsletter, 55*(3), 11-13.

Pera, M.C., Candusso, D., Hissel, D., & Kauffmann, J.M. (2007). Power generation by fuel cells. *Industrial electronics magazine, 1*(3), 28-37.

Texas Instruments. (1994). *Lighting products. Data manual & applications report.* UK

Venkateswara, A. S. (2004). Introducing power electronics in Ford hybrid Escape vehicle. *IEEE Power Electronics Society Newsletter*, *16*(3), 13–15.

Chapter 12
State-of-the-Art Review on Power Electronics

During 1991-2004 years, International Workshop on the Future of Electronic Power Processing and Conversion (FEPPCON) held four times with the support of IEEE Power Electronics Society. The main purpose is to outline the possibilities of development in different fields during the next period as a result of the discussions. At the last meeting held in Italy, the tendency to increase the role of Power Electronics during the next 25-30 years in the processes of energy conversion has been confirmed (Blaabjerg, 2005). Special attention is paid to the role of Power Electronics at a system energy conversion level, because it is not in the position to dictate the trends in the development of this level. Nevertheless, without the power electronics tools, future serious achievements in power processing are impossible. Therefore, the power electronics implantation at a system level at a system energy conversion level is an issue of the efforts of the specialists in this field (Agrawal, 2001). This, of course, imposes also some changes and adaptation towards Power electronics role also in the process of schooling of specialists paying attention mainly to multidisciplinary of Power Electronics. For example, a necessity of further tuition in electrochemistry, mechanics, physics (especially electromagnetic and thermal processes), etc, is outlined. Remote access to complex power electronics laboratory equipment and the possibility of remotely driving experiments and measurement is represented (Rodriguez, 2009). Power Electronics takes a significant part in the following systems: Generic Systems; Energy Storage; Power Systems, including Alternative Energy Supply; Automotive Systems (Bose, 2009).

DOI: 10.4018/978-1-61520-647-6.ch012

In this chapter, some of the latest achievement in the three basic levels in the Power Electronics – components, circuits and systems, are consecutively studied. To prepare and write this chapter, mainly survey papers in the different fields are used. In these papers the reader may find a lot of additional references in certain filed of interest.

CONTEMPORARY STUDY IN THE FIELD OF COMPONENTS FOR POWER ELECTRONICS

Thyristors (SCRs) dominated as power switches in the period 1958-1975 years. Their implementation in controlled rectifiers causes problems connected with the worsen power factor. Because of that, the rectifiers with phase control are displaced step by step by these with PWM based on fully-controlled power switches – see Chapter 4. The capabilities of a standard thyristor regarding the commutation of high voltages and currents at low frequencies outline its dominant use in the coming future in the following fields – high voltage DC transmission of energy (HVDC); cycloconverters and some drivers with very high powers. The creation of high-power gate turn-off thyristor (GTO) in 1980 year and its upgrading especially by Japan companies make preconditions for the displacement of a standard thyristor from some of the power electronics converters with high powers. The possible variations of GTO with or without blocking capability in the reverse bias, as well as the borders reached (6kV, 6kA), make its implementation in multimegawatt converters the most probable one. The problems for wider implementation of GTO-s are connected with its slow turning on and off, complicated structure of the control drivers, the required snubbers group. Thus, the implementation is in the field of low frequencies – to several hundred Hz. In the field of power MOSFET it has to be mentioned the creation of CoolMOS by Infineon Technology, characterized by a high fast reaction and decreased losses compared to the rest MOSFET structures. For the present, the CoolMOS voltage capabilities are to 800V. With the creation of IGBT in 1983, its development during the following period and its parameters, it may be considered that IGBT fully displaces power bi-junction transistor (BJT). Till now, the IGBT voltage and current capabilities reached are 6.5kV and 700A, respectively. Recently, the attention towards integrated gate-commutated thyristor (IGCT) the first one demonstrated by ABB in 1996 is quickened. It is probable to be used in high powers – values for single devices reached are 6.5kV and 4kA. In these powers, the capabilities of IGCT are commensurate to those of IGBT regarding the switching frequency, while in high voltages IGCT has lower voltage across it in ON state compared to IGBT. Step by step, IGBT and IGCT displace GTOs in the whole ranges of frequencies and powers. The static induction transistor (SIT) is a normally ON state device with a high voltage in forward bias. This limits its implementation possibilities but its development will permit its use in high frequencies. Notes of other semi-conductor devices, at an experimental level for the time being as MOS-controlled thyristor, static induction thyristor, MOS turn-off thyristor, etc, exist.

In the field of creation of new power devices, the attention is paid towards so-called wide bandgap semiconductors. The aim is basically to increase breakdown voltage, to achieve higher operational temperatures of the junction, as well as to achieve lower values of the resistance in the ON state. Also, better temperature conductivity is searching for to facilitate the integration of several power devices if it is possible together with control drivers in a common body. The researches are towards material close to silicon (1.12 eV) – GaN (3.44eV); GaP(2.26eV), SiC (2.36-3.25eV) C (diamond) (5.46-5.6eV)

(Hudgins,2003). The high value of electron mobility of diamond – 2200 $\frac{cm^2}{V.s}$ makes it appropriate for creation of power bi-junction devices. The diamond is also characterized by the highest temperature conductivity of the materials mentioned - 600-2000 $\frac{W}{m.K}$. Lower values of the ON state resistance compared to these of the silicon devices are obtained using GaN and SiC. Values are reached regarding the temperature in MOSFET based on SiC to 300° C, while base on diamond to 600° C. The state of the art of some power devices based on SiC at the time being is (Bose,2009): Sckottky barier diodes – 1700V, 10A; P-I-N diodes – 10 kV; BJT – 1.8 kV; GTOs – 2.6 kV; MOSFET – 1.2 kV,10A. The integration of CoolMOS and SiC-diode is of interest to obtain higher switching frequencies and to decrease the losses in the power factor correction schematics – see Chapter 4 (Liang, 2005).

At power switches marker, a tendency has been imposed to offer modules including several devices connected in a certain schematic – buck, boost, phase leg, full bridge etc (Krein,1998). Better known producers of such modules are: Mitsubishi Electric, Fuji, Toshiba, International Rectifier, Advanced Power Technology, Infineon, etc. Thus, the design and implementation of the electrical power converters are significantly facilitated. Integration in a single body of several power devices, connected in a certain circuit, together with their control drivers is a material achievement of microelectronics. Implementation of Intelligent Power Modules leads to solution of serious problems connected with the electrical compatibility in power electronic converters. Along with this, the implementation has driven to new solutions in design of a whole power electronic converter (Mitsubishi Electric, 1998). Yet, the use of Intelligent Power Modules is not wide enough due to several facts – price, reliability, repair possibilities, etc. Probably, in the future, the possibilities of the conception mentioned have to be checked in the use of the above mentioned new materials for power devices. It is possible also another conception to be imposed regarding the requirements of higher insulation voltages. Till now, only integrated control system based on SiC are reported.

At the ambition to facilitate the users at the use of products, some of the producers (International Rectifier, Infineon) propose also models for computer simulation of their new devices. These models are compatible with more known software.

Continuous development is seen also in passive elements which are applicable in Power Electronics – such as transformers, inductances, capacitors, elements protecting from over voltages and over currents, etc. It is worthy to be mentioned that in some power electronic converters magnetic materials determine in significant degree such parameters as sizes, volume, and electromagnetic compatibility. The aim in researches and experiments are to increase the operational frequency of the materials at obtaining low specific losses. Regarding this, the creation of ferrite material N97 by EPCOS with losses to 54% less than the already obtained in the temperature range 60 – 100°C deserves its attention. At design of power high-frequency transformers the application of magnetic alloys is extended. Magnetics (1999) proposes a comparison at the use of following alloys in the transformers for the voltage inverters: orthanol (50% Ni, 50% Iron);Sq. Permalloy (79% Ni, 17% Iron, 4% Mo); 48 Alloy (48% NI, 52% Iron); supermalloy (78% Ni, 17% Iron, 5% Mo) и magnesil(3% Si, 97% Iron). For example, the closest to rectangular form is the hysteresis of orthanol $\left(\frac{B_r}{B_m} = 0.96\right)$. The comparison is made at 2.4 kHz.

At low powers, the ambition to integrate passive components (inductances, capacitors) and semiconductor devices in a single body is a dominant one (van Wyk,2005). Of course, this is yet possible for example in DC/DC converters for low voltages and low consummation such as mobile phones, cameras, computers, etc. For example, with the increase of the operation frequency in the range of 2-10MHz in synchronous buck converter, the values of the inductance and capacitor decreases to 8 and 2.5 times, respectively, (Mathuna,2005). Besides, the losses in the core of thin films type are less compared to the losses in ferrites in the range of inductance change 10-100 mT. The values of saturation inductance in thin films are 1-2T while in ferrites they are 0.3-0.4T. The two facts mentioned determine higher power reached in smaller sizes and high efficiency coefficient. The expansion of the applicable fields of the power integrated in a chip is connected with achievement of higher values of the inductances at enhanced current density, as well as with dissipation of the thermal losses at high powers.

CONTEMPORARY STUDY IN THE FIELD OF CIRCUITS FOR POWER ELECTRONIC CONVERTERS

The survey of some trends in the development of the schemas of the power electronic converters is made further down in two groups – converters of AC energy (AC/DC and AC/AC converters) and converters of DC energy (DC/DC and DC/AC converters).

At AC energy conversion, the use of transistor schemas is extended also towards the field of high powers. The general aim, as it has been studied in Chapter 4, is to improve the harmonic spectrum of the current consumed from the AC source and to improve power factor. A survey of different schemas in AC/DC converters with power factor correction is made from Garcia (2003). Different schemas of converters – boost, flyback, SEPIC, etc, are used. Development exists not only in the methods for active correction, but also in those of the passive correction. In some cases, the aim is with comparatively less elements source current waveform to meet the requirements of the standard EN 61000-3-2 for converters class D, not the current waveform to be a sine wave. In a classic version, studied in Chapter 4, a separate block for correction and a separate block for DC/DC conversion are used. To decrease the number of the components, as well as to improve the indicators, combine schemas both for power factor correction and DC/DC conversion with galvanic isolation are developed in this field. This combination is probable in comparatively low powers in the time being. In these powers, also resonant converters are used with the purpose to increase the efficiency of the power factor correction schema (Liu, 2009), (O'Sullivan, 2009). The use of resonant modes in several levels has started. A comparison of versions of three-level resonant AC/DC converters is made by Agamy (2009). The schemas of bidirectional transistor converters studied in Chapter 4 are investigated at different types of control. For example, to trace the reference sine wave, one-cycle control in modified version is used (Ghodke, 2008). In three-phase converters, one of the study directions is different versions of VIENNA–rectifiers (Qiao, 2003), (Youssef, 2008). Study to improve the harmonic spectrum of the current from network continues also in three-phase rectifiers with fully-controlled power devices and inductive type of load. The use of space vector modulation for this purpose is of interest (Lopes,2009). It has to be mentioned, that although on a small scale, improvements and modifications in classical thyristor controlled rectifiers are also studied. An interesting combination of two three-phase bridge thyristor rectifiers with power IGBTs in their DC circuits are studied in (Villablanca, 2007). At equivalent m =12 (see Chapter 4) the power factor is increased. Besides the worsen power factor of the classical schemas of the control thyristor rectifier with phase control, another disadvantage at the use of standard regulators (mostly PI) is their time delay at a reaction of control influence of the automatic control system, as well as the oscillation type of the transient processes at aiming the fastest reaction possible. Obtaining the fastest reaction combined with fast transient process is very important in the HVDC transmission systems. Good indicators in this regard are got from (Jeong, 2007) at the use of predictive control.

At conversion of AC into AC energies, the attention in researches and designs is paid to matrix converters. Depending on the proportion of the frequency of the output AC voltage and this of the network voltage, converters with increased or decreased output frequency are studied. More covered are those with decrease of the output frequency with probability for its regulation. The matrix converter implementation is towards frequency control of AC electro motors. Versions of single-phase matrix converter are known (Kulekcioglu, 2004), but the main attention is towards the three-phase converters – for example use of predictive control to decrease the losses and to increase the efficiency coefficient (Vargas, 2009). Use of sigma-delta modulation in single-phase to three-phase matrix converter enables to improve the harmonic spectrum of the output voltage in comparison to those of pulse width modulation (Hirota, 2008). A new direction in researches is the use of Z-Source converters to make direct AC/AC conversion in single-phase or three-phase version (Fang, 2005), (Peng, 2007). In Chapter 5, a single-phase matrix converter, which output frequency is higher than the network frequency, has been studied. Such converters may be used for electrotechnical purposes. It is possible to transfer their output voltage using high frequency transformer and further transformation into DC (rectifier) or into multi-phase (again matrix converter). In Chapter 3 a computer simulation of three-phase to single-phase matrix converter has been studied. This principle may be apply in implementation of dimmable lighting system using high frequency output transformer and appropriate control (Sabahi, 2009).

To convert DC into DC energy, a main trend is to use soft commutation (Mohan, 2003). This is the basic way to increase operational frequency, to reduce the volume and weight, as well as to reduce electromagnetic interference. One of the first researches is connected with the implementation of ZVC in a bridge DC/DC converter with galvanic isolation using so-called "phase shift control" (Andreycak, 1993). For the control purpose, UNITRODE has designed a special integral circuit – UC3875. The use of the circuit in a bit different controller is proposed by Lee (2008). The difference is the use of voltage-doubler-type rectifier in the secondary side. The lack of an inductance in the secondary side leads to ease conditions of operation of the rectifier diodes and higher efficiency. This control method evolves in time and one of its new versions is studied by Bai (2008). The power schema is bidirectional DC/DC converter, which consists of two bridge converters examined in Chapter 6. The control method is called "dual phase shift control" and consists of a phase shift between the primary and secondary voltages of the isolation transformer and a phase shift between the gate signals of the diagonal switches of each H-bridge. Two modifications of half-bridge converter studied in Chapter 6 are presented by Deblecker (2008). Additional transistors to achieve ZCS and to increase efficiency are used. The two modifications are compared in their use as auxiliary railway supply. It has to be mentioned that the approach to include additional power devices and elements in the standard schemas of DC/DC converters (with hard commutation) aiming to obtain soft commutation is a widely met one. It is used in the galvanic isolated schemas – more often flyback, half-bridge and full-bridge. The last mentioned is a separated direction in DC/DC converter deveopment. Recently, an example of this is use of a separate cell for commutation (Wang, 2008). Another research direction is to gain resonant modes and soft switching using passive elements in the primary side of the high-frequency transformer connected in a special way. Martin-Ramos (2008) proposes a study of hybrid series parallel resonant topology, PRC-LCC, suitable for implementation of high output voltages. So-called topology LCL-T RC in half-bridge converter is widely investigated because of the fact that such a converter operating at resonant frequency has properties of a current source. This makes it suitable for a technological applications connected with operation at high output voltage and periodical discharges in the secondary side. A design taking in consideration the capacity of the transformer windings is presented by Borage (2009). By the discussion made in Chapter 6, it is seen that the operation of DC/DC converters at frequency control is connected with large component stresses due to high peak currents and voltages. Therefore, the researches of soft switching in PWM converters are developing. Conventional forward and flyback converters with one transistor have the disadvantage to have over-voltages due to the dissipation inductance of the transformer and the recovery time of the rectifier diodes. To reduce these over-voltages and to obtain ZVS in such converters, additional active and passive elements forming so-called "active clamp" are included. Different configurations of such converters exist but as an example of the recently proposed ones it may be examined the one proposed by Choi (2009) boost type active clamp and by Lee (2008) dual series-resonant active clamp. To increase the output power, a parallel connection of the separate DC/DC converters operating with different phases in time forming so-called multiphase converters is used. Such a connection is used at first in converters with hard switching and without galvanic isolation more often buck converter. The basic problem is connected with the currents given by the separated converters. The standard approach to solve the problem is connected with the measurement of the total current and currents from the separate converters and control in the separate converters. Recently, new technology has been developed such as sensorless current sharing (Quhouq, 2008). The method is applicable in low powers and requires digital control. The proposed method of using digital filtering (Kelly, 2009) is based on such a control. Parallel connection of the outputs is also examined in the resonant converters (Lin, 2008). Connected with the use of renewable energy sources, another direction of the research of DC/DC converters has developed. It is very often required the converter to have a possibility to operate using several sources of energy – solar array, fuel cell, commercial ac line. This research direction is connected with so-called multi-input converters. Different schematics for their implementation are generalized by Liu (2009). It has to be mentioned that there are also other way to solve the problem connected with the currents given by the separated converters – operation of different DC/DC converters connected to a common DC line, which way is also studied (Karlson, 2003)

At conversion of DC into DC energies at higher powers, recently, it has been seen a quicken interest towards three-phase converters using a three-phase high-frequency transformer and a three-phase rectifier. This allows to reduce the volume, weight and price of the whole system owing to the better use of the reactive components of the converter. The increased frequency of the ripples of the output voltage compared to the operating frequency of the converter facilities also the implementation of the output filter. Based on the operational principle of push-pull converter (see Chapter 6) a three-phase version of the conversion at feeding from a current source is designed (Andersen, 2009). Three-phase DC/DC converter with galvanic isolation consisting of a three-phase bridge high-frequency inverter in the primary side and a three-phase bridge rectifier in the secondary side is studied by Cha (2008). The converter is fed by a current source and it uses ZVS and active clamp through additional transistor in the primary side.

The ambition to improve the harmonic spectrum of the output voltage of voltage inverters when DC energy is converted into AC energy has brought the most serious development in multilevel inverters regarding the schematic. Skvarenina (2002) examines different schematic versions - neutral point clamped diodes, flying capacitors, multicell inverters. Problems of the versions containing diodes are connected with balancing of the voltages across the capacitors that divide the input DC voltage in a certain ratio. Versions with flying capacitors require galvanic isolation of the feeding sources. Different methods to balance voltages across the capacitors are studies, as an example the one proposed in Khajehoddin (2008). The most widely spread control method is sinusoidal pulse width modulation (SPWM). Space vector modulation (SVM) method is used more seldom. One of the research trends in these inverters are namely the modulation methods. Liu (2009) investigates step modulation in cascade multilevel inverters. For the investigation purpose, an algorithm to calculate the switching angles in real time is proposed. Study of new hybrid PWM (HPWM), as well as a comparison between HPWM and SVM in three-phase inverters with limitation diode is examined by Zaragoza (2009). As a result of the use of HPWM, the voltages across the power devices are reduced. Leon (2009) proposes study of feed-forward SVM in two-cell cascade multilevel inverter. The feed-forward SVM makes possible optimization between the values of the supply voltages of the two transistors of the bridge. Investigation of predictive control in multilevel inverters is discussed by Perez (2008). There is development not only regarding the control methods but also regarding the power schemas of multilevel inverters. Different schemas appear under the continuous ambition to increase the number of modulation levels. For example, multilevel inverter using fully-controlled devices with bidirectional conductivity and their reduced number is proposed (Babaei, 2008). Using the "duality" principle, Bai (2008) proposes synthesis of multilevel inverters at feeding using current sources. Also, implementation of multilevel inverters based on several cells of Z-source inverter is known (Loh, 2008). (Ceballos, 2008) studies a new schematic of a three-phase inverter containing three main legs (standard neutral point clamped topology) which are connected to the output phases, and a fourth leg with flying-capacitor structure. An interesting schema of a single-phase inverter using a bidirectional power switch and five level of modulation is examined by Selvaraj (2009). The development of resonant transistor inverters is mainly connected with their application in the field of lightening. The application is characterized with comparatively lower powers. It has to be mentioned that large amount of schematics of such converters exists therefore here only the basic trends in their research will be listed: control of the operation of the resonant inverter in such a way that a high power factor of the uncontrolled rectifier connected to the network to be obtained; an appropriate control of the frequency or of the source voltage to obtain dimming-effect; making equal the currents trough several shunt connected lamps; different methods to warm up the electrodes and to start the inverter with the ambition to prolong the expiring period of the lamps, etc. For the standard voltage inverter schemas investigations of comparatively known modulation methods such as selective elimination of harmonics exist (Khaligh, 2008). Sliding mode PWM is studied by Yan (2008). Wavelet Modulation of Single- Phase Inverters is described and compared with PWM and random PWM (Saleh, 2009). A bit-stream PWM is also investigated (Patel, 2009).With the expansion of the renewable energy source applications, a new trend of investigation of the standard voltage inverters has been develop – the inverter application when they are connected to the network - grid connected inverters. One of the serious problems to be solved is connected with islanding detection. This is a whole direction of research into such inverters and it develops with a quick rate. The islanding of a grid connected inverters occur when a utility is disconnected from the grid line but the grid connected inverter continue to energize energy to the isolated section of line. The islanding control can be achieved through inverters and control system can be designed on the basic of detection of grid voltage, frequency variation, increase in harmonics or measurement of impedance. Different solutions for the problem of the islanding of a grid connected inverters exist. Application of a special programmed logic monitoring the changes of the voltage and frequency in the point of common coupling is of interest (Tunlasakun, 2008). The use of LCL-filter in the output of the inverter is predominant. Special attention is paid to current loop control to obtain the best dynamic parameters of the system inverter-grid (Park, 2008). The application of hysteresis-current control in a system containing two inverters is of interest (Milosevic, 2004). Another research direction is connected with synchronization and accuracy of the measurement of the required variables for inverter operation. The researches are based on the measurements of different variables – voltage, current and their parameters – phase, amplitude, and frequency. The most widely used method is using phase-locked loops (PLL), also, open-loop methods are developing. An example is the results presented by Freijedo (2009). Separate direction of the researches is connected with the insurance of the reliable isolation of the public network, suitable grounding, electromagnetic disturbance protection, protection from over voltages due to the electrostatic discharges, etc. Objects of research are also the problems in the following fields: resetting and reconnection with grid; over-current protection; voltage limits and frequency limits.

CONTEMPORARY STUDY IN THE FIELD OF SYSTEMS OF POWER ELECTRONIC CONVERTERS

In Section 3 the application of systems of power electronic converters in different fields has been scrutinized. In all of them, scientific researches and practical implementations are continuously made (Lai, 2009). In the present section, the attention is paid only to the most quickly developed fields - Uninterruptible Power Supply and converters for Renewable Energy Sources (sun and wind). The author selects them because of their importance in the energy use in the future.

During the past years, it is seen a significant development in the field of Uninterruptible Power Supply (UPS) systems both in the scientific papers and as a new products at the market. The basic system solutions are listed in Chapter 10. With regard to the distributed generation of electrical energy a tendency to remove the conventional single systems of high power and replace them with distributed UPS systems has occurred. This solution is convenient in microgrids. However, it is connected with parallel operation regarding the outputs of the inverters of the separate systems and this problem is more serious than the parallel connection for example of DC sources. If the output inverter voltages are of equal amplitude phase and frequency, then the currents would be disturbed equal in their outputs. Practically, due to the differences of the separated inverters, and also as a result of the influence of the impedance of the AC line, the load current is not equally disturbed and the so-called circulation current among the inverters appears. The problems which have arisen from the circulation current, as well as different versions of active load sharing technique are discussed by Guerrero (2008). This technique requires mutual communication among inverters and is studied in four versions: centralized control (CC), master-slave (MS), average load-sharing (ALS), circular chain control (3C). At CC the total current is divided to the number of the modules and reference signal for the current is generated for each module. In each of them the reference signal is compared to the module current value and the error is processed using the automatic regulation system. At MS a single master module regulates the load voltage, i.e. it operates as a voltage source while slave modules operate as current sources. At ALS every module tracks the average current done by all the modules. Special feature of 3C is generation of a reference signal for the current of each module by the previous one, and all the modules are connected in ring. The output currents may be also made equal by droop control method, in which the mutual communication among the separate inverters is prevented. The active and reactive powers supplied to the AC bus are sensed and averaged and the resulting signals are used to adjust the frequency and amplitude of the UPS inverter output voltage reference. Low (2008) discusses model predictive current control of parallel-connected inverters. At the research into the schemas to implement of separate UPS, one of the directions is the control of the operation of the converter system with the ambition to obtain high power factor in respect to the source network. The rectifiers with an improved power factor studied in Chapter 4 are used but also other solutions are sought. For example, Vazquez (2008) proposes as a first stage towards the source network multifunctional converter that operates as battery charger. Rodriguez (2008) combines the operation of the schematic for power correction with this of the inverter for charging an accumulator battery. Disadvantage of the version is the lack of galvanic isolation between the battery and consumer. This disadvantage presents also in the different versions of the transformerless UPS, created with the aim to reduce the volume and the weight of the equipment (Park, 2008), (Branco, 2000). This direction of investigation continues to develop. Another direction of investigation in the field of single UPS is high-frequency energy conversion using high-frequency transformers – with the aim to reduce the volume and weight but ensuring galvanic isolation (Tao, 2008). Also, the use of the Z-source inverter in UPS, which is not typical application of the inverter, appears (Zhou, 2008). Probably in the future the trend towards the investigation and use of others energy storage elements besides the already gained widespread hermetically closed accumulator batteries in the UPS will be strengthened. These energy storage elements are kinetic batteries, fuel cells, supercapacitors, etc.

Other fields, which develop with a fast rate during the last 5 years, using systems of power electronic converters, are distributed generation of electrical energy and use of the renewable energy sources (Wang, 2009). Basic system solutions are discussed in Chapter 9. Regarding this, the special issue of Masters (2004) is of interest. The reader may get accountant also with the review made by Bialasiewicz (2006) and Balcells (2006). The use of the energy from photovoltaic modules and wind generation has the fastest development for that those two directions will be only discussed. Survey in the field of the use of photovoltaic energy is made by Carrasco (2006) and Spagnuolo (2008). Concerning the combination of converters the following versions are developing: (1) High frequency transformation of the energy of the photovoltaic module with following generation of the current through the input inductance of the bridge schema. Essentially, the bridge schema is in the current inverter mode with switching of the diagonally connected transistors in the moments of the grid voltage zero crossing (Rodriguez, 2008). The use of a three-phase current inverter to implement integrated transformerless converter is also discussed by Sahan (2008). (2) Usage of several DC/DC converters to obtain DC voltage from the module. The version in which the converters are connected in series in respect to their inputs towards the output of the photovoltaic module and also operating with a common output transformer is also possible. The secondary windings of the transformer are connected in series and the voltage from them is rectified through a rectifier (Lee, 2008). It has to be mentioned, that it is also possible a version using separate transformers and rectifying the voltage of each converter. The rectifiers in the secondary side are connected in series regarding their outputs. Sometimes, parallel connection of DC/DC converters regarding their outputs to obtain voltage of the DC line with following transformation using common inverter is also used. (3) Use of multilevel inverters. Detailed discuss of the different topologies is made by Daher (2008). The use of single-phase and tree-phase diode clamped multilevel inverter without output transformers is illustrated by Gonzalez (2008) and Busquets-Monte (2008). New methods for maximum power point tracking (MPPT) are continuously suggested. In these methods either the output voltage and the output current (the power respectively) of the module or the output voltage of the module and the effective value of the output current of the inverter connected to the network are monitored. Enormous number of publications of the different methods exists. Here, only for an orientation for both approaches as an example for the first method Liu (2008) is listed and for the second one – Kwon (2006). Survey in the field of the systems of converters for wind energy use is made by Carrasco (2006) and Chen (2008). The conversion of wind energy is scrutinized in specialize issues by Gipe (2006) and Stiebler (2008). The interest towards the systems using doubly-fed induction generator (DFIG) with a conventional structure with the use of bidirectional transistor converters is quickened. Lu (2008) presents coordinated control of the rotor side converter and grid side converter, which is appropriate at operation during network unbalance. The operation of the converters at DFIG with constant decreased frequency of switching of the power devices of the systems with high power with the aim to reduce the electromagnetic torque and rotor flux ripples is proposed by Abad (2008). This is possible when predictive direct torque control is applied. Improved control algorithms are considered by Yao (2008) and Moursi (2008). Also, applications of current inverters connected to the source network at permanent magnet synchronous generator are studied (Tenca, 2008). Separate direction is the study of applications of static compensators (STATCOM), Static voltage compensators (SVC) and active power filters (APF) in the systems with wind energy conversion. As an example the comparatively study made by Molinas (2008) may be given. The development in the future of the systems of converters applicable in the use of wind energy is very tightly connected with the development and advance in the field of electrical machines. New solutions in the electrical machines will enforce new requirements during the energy conversion processes.

CONTEMPORARY STUDY IN THE FIELD OF THE CONTROL SYSTEMS OF POWER ELECTRONIC CONVERTERS

In the field of the control systems of the electronic energy conversion during the last 20 years a fast development has been seen based on the perfectly new methods and means of control (Kazmierkowski, 2002). Precondition for this is the discreet character itself of the most frequently used power electronic devices. Artificial Intelligence Techniques (AI) – expert system (ES), artificial neural network (ANN), fuzzy logic (FL) and genetic algorithm (GA) have been applied in power electronics. The quicken implantation of the ANN technology has been made from 1990 although the technology dated from 1940. The theory of GA, with wider application after 1970, is based on the principles of genetic (the Darwin's theory of the species evolution). Very often contemporary intelligent control systems combine neuro, fuzzy and GA techniques to increase their adapting. Survey of applications of ANN in the field of Power Electronics is made by Bose (2007). As examples are presented as follows: generator of three-phase system of sine wave voltage, three-phase pulse width modulation at a variable frequency, space-vector modulation in two- and three- level inverters, induction motor vector drive with ANN-Based space vector modulation and Flux Estimation. Control of a three-phase parallel active power filter based on ANN is proposed by Abdeslam (2007). These contemporary techniques require hardware and software insurances and have been leaded to fasten development of micro processor and digital signal processors. The control of the contemporary power electronic converters feeding sophisticated consumers is yet unthinkable without use of embedded systems based on digital signal processors. Specific hardware is field-programmable gate arrays (FPGA), combining software means of programming with a high level of integration gained in the field of microelectronics. Survey of their use in Power Electronics and an example of control of diesel – driven stand-alone synchronous generator is presented by Monmasson (2007). More applications at inverters with PWM, selective elimination of harmonics with regulation of the output voltage, multilevel inverters, power factor correction, etc. are also known. Survey of the FPGA use in implementation of current controllers for AC machine drives is made by Naouar (2007).

Simultaneously with the development of these new methods and means, the investigation and use of the already affirmed ones continue. The affirmed methods are such as: PWM, for example, when Walsh function is used and space-vector modulation, for example in multilevel inverters (Leon, 2009). It is interesting to be mentioned, that sometimes the needs of Power Electronics provoke the creation of the relevant control means. An example is Texas Instruments digital signal processors of series TMS-320F2XXX, whose architecture is designed with the possibility to control three-phase voltage inverters for asynchronous electromotors using so-called "field orientated control" and space vector modulation. Hysteresis regulation at different converters and tracing a reference curve for current or voltage also develops. Examples for this development are hysteresis regulation with constant frequency or maximum frequency limitation (Antchev, 2007), using of fuzzy logic at hysteresis regulation, etc. Adaptive fuzzy controller for wind energy system is investigated (Galdi, 2008).The use of sliding mode control in the control systems of power electronic converters, for example in three-phase bidirectional converter (Shtessel, 2008), has widened significantly. Survey of the use of predictive control in Power Electronics is made by Kennel (2008). Combination of predictive control and space vector modulation is examined by Zeng (2008).

CONTEMPORARY STUDY IN THE FIELD OF COMPUTER SIMULATION OF POWER ELECTRONIC CONVERTERS

In the field of study of power electronics converters through computer simulation, the ambition to make closer the conditions of the simulation to those of the practical equipment is has been recently dominated (Ropp, 2009). Study of impulse source converter through a virtual prototype is examined (Li, 2003). In this virtual prototype, the procedures of design, verification and testing are divided in the following groups: system partitioning, multilevel modeling of device/function block, hierarchical test sequence and multilevel simulation. The prototype enables the following test levels: basic function test, steady stage analysis, small-signal stability analysis, large-signal transient analysis, subsystem interaction test and system interaction test. The contemporary control systems based on digital signal processors perform a significant part of their functions using software. This is precondition for developing of so-called "real - time simulation". Moreover, the hardware part of the control system is practically constructed and produces control signals. The simulation of the power schema may be performed using PC in MATLAB/SIMULINK. Abourida (2009) proposes design of RT-LAB platform for real-time simulation of motor drives, power converters and power systems. In the basis of the platform realization is very powerful hardware: parallel processing, multi-core processors, fast I/O devices, FPGA-based computation. Study only of the control systems applicable in Power Electronics based on combination of software and hardware is suggested in Monti (2003). Software includes virtual test bed (VTB) and MATLAB/SIMULINK, while hardware – dSpace DSP. Probably in the future the approach studied will develop, and its use in simulating the effects such as electromagnetic influences and thermal drift is of a special interest. This statement is based on the increase of reaction of digital signal processors which connected with the presence of software part in the described systems of real - time simulation, would enable to consider the accidental character of those effects.

REFERENCES

Abad, G., Rodriguez, M., & Poza, J. (2008). Two-level vsc based predictive direct torque control of the doubly fed induction machine with reduced torque and flux ripples at low constant switching frequency. *IEEE Transactions on Power Electronics*, *23*(3), 1050–1061. doi:10.1109/TPEL.2008.921160

Abdeslam, D., Wira, P., Merckle, J., Flieller, D., & Chapuis, Y. (2007). A unified artificial neural network architecture for active power filters. *IEEE Transactions on Industrial Electronics*, *54*(4), 61–76. doi:10.1109/TIE.2006.888758

Abourida, S., & Belanger, J. (2009). *Real time platform for the control prototyping and simulation of power electronics and motor drives*. Paper presented at the Third International conference of Modeling, Simulation and Applied Optimization, Sharjah, UAE, January 20-22, 2009.

Agamy, M., & Praveen, J. (2009). Performance Comparison of Single-Stage Three-Level Resonant AC/DC Converter Topologies. *IEEE Transactions on Power Electronics*, *24*(4), 1023–1031. doi:10.1109/TPEL.2008.2010769

Agrawal, J. (2001). *Power electronic systems: Theory and design*. Prentice Hall.

Andersen, R., & Barbi, I. (2009). A three-phase current-fed push-pull DC-DC converter. *IEEE Transactions on Power Electronics*, *24*(2), 358–368. doi:10.1109/TPEL.2008.2007727

Andreycak, B. (1993). Designing a phase shifted zero voltage transition (ZVT) power converter. *Power Supply Design Seminar, Unitrode Corp.* (pp. 3.1-3.15).

Antchev, M., Petkova, M., & Kostov, A. (2007). *Hysteresis - Current control of single - phase shunt active power filter using frequency limitation*. Paper Presented at conf. PES 2007, Clearwater, USA, 2007.

Babaei, E. (2008). A cascade multilevel converter topology with reduced number of switches. *IEEE Transactions on Power Electronics*, *23*(6), 2657–2664. doi:10.1109/TPEL.2008.2005192

Bai, H., & Mi, C. (2008). Eliminate reactive power and increase system efficiency of isolated bidirectional dual-active –bridge DC-DC converters using novel dual-phase-shift control. *IEEE Transactions on Power Electronics*, *23*(6), 2905–2914. doi:10.1109/TPEL.2008.2005103

Bai, Z., & Zhang, Z.Conformation of Multilevel current source converter topologies using the duality principle. *IEEE Transactions on Power Electronics*, *23*(5), 2260–2267.

Balcells, J., & Bialasiewicz, J. (2006). Special section on renewable energy and distributed generation systems-Part II: Control of distributed generation systems, Guest editorial. *IEEE Transactions on Industrial Electronics*, *53*(5), 1394. doi:10.1109/TIE.2006.883720

Basquets-Monte, S., Rocabert, J., Rodriguez, P., Alepuz, S., & Bordonau, J. (2008). Multilevel diode-clamped converter for photovoltaic generators with independent voltage control of each solar array. *IEEE Transactions on Industrial Electronics*, *55*(7), 2713–2723. doi:10.1109/TIE.2008.924011

Bialasiewicz, J., & Balcells, J. (2008). Special Section on Renewable Energy and Distributed Generation Systems-Part I- Renewable Energy Generation and Storage Systems, Guest Editorial. *IEEE Transactions on Industrial Electronics*, *53*(4), 998. doi:10.1109/TIE.2006.878357

Blaabjerg, F., Consoli, A., Ferreira, J., & van Wyk, J. (2005). The Future of Electronic Power Processing and Conversion. *IEEE Transactions on Power Electronics, 20*(3), 715–720. doi:10.1109/TPEL.2005.846516

Blaabjerg, F., Teodorescu, R., Lissere, M., & Timbus, A. (2006). Overview of Control and Grid Synchronization for Distributed Power Generation Systems. *IEEE Transactions on Industrial Electronics, 53*(5), 1398–1409. doi:10.1109/TIE.2006.881997

Borage, M., Nagesh, K., Bhatia, M., & Tiwari, S. (2008). Designing of LCL-T Resonant Converter Including the Effect of Transformer Winding Capacitance. *IEEE Transactions on Industrial Electronics, 56*(5), 1420–1427. doi:10.1109/TIE.2009.2012417

Bose, B. (2007). Neural Network Applications in Power Electronics and Motor Drives – An Introduction and Perspective. *IEEE Transactions on Industrial Electronics, 54*(1), 14–33. doi:10.1109/TIE.2006.888683

Bose, B. (2009). Power Electronics and Motor Drives Recent progress and Perspective. *IEEE Transactions on Industrial Electronics, 56*(2), 581–588. doi:10.1109/TIE.2008.2002726

Bose, B. (2009). The Past, Present, and Future of Power Electronics. *IEEE Industrial Electronics magazine, 3*(2), 7-14.

Branco, C., Cruz, C., Torriso-Bascope, R., & Antunes, F. (2008). A Nonisolated Single-Phase UPS Topology With 110-V/220-V Input-Output Voltage Ratings. *IEEE Transactions on Industrial Electronics, 55*(8), 2974–2983. doi:10.1109/TIE.2008.918478

Carrasco, J., Franquelo, L., Bialasiewicz, J., Galvan, E., Guisado, R., & Prats, M. (2006). Power – Electronic Systems for the Grid Integration of Renewable Energy Sources: A Survey. *IEEE Transactions on Industrial Electronics, 53*(4), 1002–1016. doi:10.1109/TIE.2006.878356

Cha, H., Choi, J., & Enjeti, P. (2008). A Three-Phase Current-Fed DC/DC Converter With Active Clamp for Low-DC Renewable Energy Sources. *IEEE Transactions on Power Electronics, 23*(6), 2784–2793. doi:10.1109/TPEL.2008.2003256

Chen, Z., & Guerrero, J. (2008). Special Section on Power Electronics fir Wind Energy Conversion. *IEEE Transactions on Power Electronics, 23*(3), 1038. doi:10.1109/TPEL.2008.920875

Choi, W., Kwon, J., Lee, J., Jang, H., & Kwon, B. (2009). Single-Stage Soft-Switching Converter With Boost Type of Active Clamp for Wide Input Voltage Range. *IEEE Transactions on Power Electronics, 24*(3), 730–741. doi:10.1109/TPEL.2008.2008307

Daher, S., Schmid, J., & Antunes, F. (2008). Multilevel Inverter Topologies for Stand-Alone PV Systems. *IEEE Transactions on Industrial Electronics, 55*(7), 2703–2712. doi:10.1109/TIE.2008.922601

Deblecker, O., Moretti, A., & Vallee, F. (2008). Comparative Study of Soft-Switched Isolated DC-DC Converters for Auxiliary Railway Supply. *IEEE Transactions on Power Electronics, 23*(5), 2218–2229. doi:10.1109/TPEL.2008.2001879

Emadi, A., Lee, Y., & Rajashekara, K. (2008). Power Electronics and Motor Drives in Electric, Hybrid Electric, and Plug-In Hybrid Electric Vehicles. *IEEE Transactions on Industrial Electronics*, *55*(6), 2237–2245. doi:10.1109/TIE.2008.922768

Fang, X., Qian, Z., & Peng, F. (2005). Single-Phase Z-Source PWM AC-AC Converters. *IEEE Power Electronics Letters*, *3*(4), 121–124. doi:10.1109/LPEL.2005.860453

Freijedo, F., Doval-Gandoy, J., Lopez, O., & Acha, E. (2008). A Generic Open-Loop Algorithm for Three-Phase Grid Voltage/Current Synchronization With Particular reference to Phase, frequency, and Amplitude Estimation. *IEEE Transactions on Power Electronics*, *24*(1), 94–107. doi:10.1109/TPEL.2008.2005580

Funch, J. (2008). Controlled AC Electrical Drives. *IEEE Transactions on Industrial Electronics*, *55*(2), 481–491. doi:10.1109/TIE.2007.911209

Galdi, V., Piccolo, A., & Siano, P. (2008). Designing an Adaptive Fuzzy Controller for maximum Wind Energy Extraction. *IEEE Transactions on Energy Conversion*, *23*(2), 559–569. doi:10.1109/TEC.2007.914164

Garcia, O., Cobos, J., Prieto, R., Alou, P., & Uceda, J. (2003). Single Phase Power Factor Correction: A Survey. *IEEE Transactions on Power Electronics*, *18*(3), 749–755. doi:10.1109/TPEL.2003.810856

Ghodke, D., Chatterjee, K., & Fernandes, B. (2008). Modified One-Cycle Controlled Bidirectional High-Power –Factor AC-to-DC Converter. *IEEE Transactions on Industrial Electronics*, *55*(6), 2459–2472. doi:10.1109/TIE.2008.921671

Gipe, P. (2006). *Wind Power: renewable Energy for Home, Farm and Business*. Chelsea Green Publishing, USA.

Gonzalez, R., Gubia, E., Lopez, J., & Marroyo, L. (2008). Transformerless Single-Phase Multilevel-Based Photovoltaic Inverter. *IEEE Transactions on Industrial Electronics*, *55*(7), 2694–2702. doi:10.1109/TIE.2008.924015

Guerrero, J., Hang, L., & Uceda, J. (2008). Control of Distributed Uninterruptible Power Supply System. *IEEE Transactions on Industrial Electronics*, *55*(8), 2845–2859. doi:10.1109/TIE.2008.924173

Hirota, A., Nagai, S., & Nakaoka, M. (2008). Suppressing Noise Peak Single Phase to Three Phase AC-ac Direct Converter Introducing Delta Sigma Modulation. In *Proc. of PESC 2008* (pp. 3320-3323).

Hudgins, J., Simin, G., Santi, E., & Khan, M. (2003). An Assessment of Wide Bandgap Semiconductors for Power Devices. *IEEE Transactions on Power Electronics*, *18*(3), 907–914. doi:10.1109/TPEL.2003.810840

Jeong, S., & Song, S. (2007). Improvement of Predictive Current Control Performance Using Online Parameter Estimation in Phase Controlled Rectifier. *IEEE Transactions on Power Electronics*, *22*(5), 1820–1825. doi:10.1109/TPEL.2007.904235

Karlsson, P., & Svensson, J. (2003). DC Bus Voltage Control for a Distributed Power System. *IEEE Transactions on Power Electronics*, *18*(6), 1405–1412. doi:10.1109/TPEL.2003.818872

Kazmierkowski, M. (2002). Control in Power Electronics: selected problems. New York: Academic Press.

Kelly, A. (2009). Current Share in Multifase DC-DC Converters using Digital Filtering Techniques. *IEEE Transactions on Power Electronics*, *24*(1), 212–220. doi:10.1109/TPEL.2008.2006752

Kennel, R., Kazmierkovski, M., Antoniewicz, A., Rodriquez, J., & Cortes, P. (2008). *Predictive Control in Power Electronics and Drives*. Paper presented at the 13th International European Power Electronics and Motion Control Conference, 2008, Poznan, Poland.

Khajehoddin, S., Bakhashai, A., & Jain, P. (2008). A Simple Voltage Balancing Scheme for m-Level Diode-Clamped Multilevel Converters Based on a Generalized Current Flow Model. *IEEE Transactions on Power Electronics*, *23*(5), 2248–2259. doi:10.1109/TPEL.2008.2001892

Khaligh, A., Wells, J., Chapmann, P., & Krein, P. (2008). Dead-Time Distortion in Generalized Selective Harmonic Control. *IEEE Transactions on Power Electronics*, *23*(3), 1511–1517. doi:10.1109/TPEL.2008.921162

Kolar, J., Schafmeister, F., Round, S., & Ertl, H. (2007). Novel Three-Phase AC-DC Sparse Matrix Converter. *IEEE Transactions on Power Electronics*, *22*(5), 1649–1661. doi:10.1109/TPEL.2007.904178

Krein, P. (1998). *Elements of Power Electronics*. New York: Oxford University Press.

Kulekcioglu, A., Sunter, S., & Ozdemir, M. (2004). A Single-Phase In Single-Phase Out Direct AC-AC Converter with Variable Voltage and Frequency. In *Proc. of International Aegean Conference on Electrical machines and Power Electronics, Istanbul, Turkey.*

Kwon, J., Nam, K., & Kwon, B. (2008). Photovoltaic Power Conditioning System with Line Connection. *IEEE Transactions on Industrial Electronics*, *53*(4), 1048–1054. doi:10.1109/TIE.2006.878329

Lai, J. (2009, June). Power Conditioning Circuit Topologies. *IEEE Industrial Electronics magazine, 3*(2), 24-34.

Lee, C., Leung, J., Hui, S., & Chung, H. (2003). Circuit-Level Comparison of STATCOM Technologies. *IEEE Transactions on Power Electronics*, *18*(4), 1084–1092. doi:10.1109/TPEL.2003.813771

Lee, J., Kwon, J., Kim, E., & Kwon, B. (2008). Dual Series-resonant Active-Clamp Converter. *IEEE Transactions on Industrial Electronics*, *55*(2), 698–709.

Lee, J., Min, B., Kim, T., Yoo, D., & Yoo, J. (2008). A Novel Topology for Photovoltaic DC/DC Full-Bridge Converter With Flat Efficiency Under Wide PV Module Voltage and Load Range. *IEEE Transactions on Industrial Electronics*, *55*(7), 2655–2663. doi:10.1109/TIE.2008.924165

Lee, W., Kim, C., Moon, G., & Han, S. (2008). A New Phase-Shifted Full-Bridge Converter With Voltage-Doubler–Type Rectifier fir High-Efficiency PDP Sustaining Power Module. *IEEE Transactions on Industrial Electronics*, *55*(6), 2450–2458. doi:10.1109/TIE.2008.921462

Leon, J., Vazquez, S., Watson, A., Franquelo, L., Wheeler, P., & Carrasco, J. (2007). Feed-Forward Space Vector Modulation for Single-Phase Multilevel Cascaded Converters With Any DC Voltage Ratio. *IEEE Transactions on Industrial Electronics*, *56*(2), 315–325. doi:10.1109/TIE.2008.926777

Li, Q., Lee, F., & Wilson, T. (2003). Design Verification and Testing of Power Supply System by Using Virtual Prototype. *IEEE Transactions on Power Electronics*, *18*(3), 733–739. doi:10.1109/TPEL.2003.810845

Liang, Z., Lu, B., van Wyk, D., & Lee, F. (2005). Integrated CoolMOS FET/SiC –Diode Module for High performance Power Switching. *IEEE Transactions on Power Electronics*, *20*(3), 679–686. doi:10.1109/TPEL.2005.846547

Lin, B., Huang, C., & Wan, J. (2008). Analysis, design and Implementation of a Parallel ZVS Converter. *IEEE Transactions on Industrial Electronics*, *55*(4), 1586–1593. doi:10.1109/TIE.2007.910632

Liu, F., Duan, S., Liu, F., Liu, B., & Kang, Y. (2008). A Variable Step Size INC MPPT Method for PV Systems. *IEEE Transactions on Industrial Electronics*, *55*(7), 2622–2628. doi:10.1109/TIE.2008.920550

Liu, Y., & Chang, L. (2009). Single-Stage Soft-Switching AC-DC Converter With Input-Current Shaping for Universal Line Applications. *IEEE Transactions on Industrial Electronics*, *56*(2), 467–479. doi:10.1109/TIE.2008.2004392

Liu, Y., & Chen, Y. (2009). A Systematic Approach to Synthesizing Multi-Input DC-DC Converters . *IEEE Transactions on Power Electronics*, *24*(1), 116–127. doi:10.1109/TPEL.2008.2009170

Liu, Y., Homg, H., & Huang, A. (2009). Real – Time Calculation of Switching Angles Minimizing THD for Multilevel Inverters With Step Modulation. *IEEE Transactions on Industrial Electronics*, *56*(2), 285–293. doi:10.1109/TIE.2008.918461

Loh, P., Gao, F., & Blaabjerg, F. (2008). Topological and Modulation design of Three-level Z-Source Inverters. *IEEE Transactions on Power Electronics*, *23*(5), 2268–2277. doi:10.1109/TPEL.2008.2002452

Lopes, L., & Naguib, M. (2009). Space Vector Modulation for Low Switching Frequency Current Source Converters With Reduced Low-Order Noncharacteristic Harmonics . *IEEE Transactions on Power Electronics*, *24*(4), 903–910. doi:10.1109/TPEL.2008.2011270

Low, K., & Cao, R. (2008). Model Predictive Control of Parallel-Connected Inverters for Uninterruptible Power Supplies. *IEEE Transactions on Industrial Electronics*, *55*(8), 2884–2893. doi:10.1109/TIE.2008.918474

Lu, X. (2008). Coordinated Control of DFIG's Rotor and Grid Side Converters During Network Unbalance. *IEEE Transactions on Power Electronics*, *23*(3), 1041–1049. doi:10.1109/TPEL.2008.921157

Magnetics (1999). *Inverter Transformer Core Design and Material Selection*. Retrieved from http://www.mag-inc.com

Martin-Ramos, J., Pernia, A., Diaz, J., Nuno, F., & Martinez, J. (2008). Power Supply for a High-Voltage Application. *IEEE Transactions on Power Electronics*, *23*(4), 1608–1619. doi:10.1109/TPEL.2008.925153

Masters, G. (2004). *Renewable and Efficient Electric Power Systems*. John Wiley & Sons.

Mathuna, S., O'Donnell, T., Wang, N., & Rinne, K. (2005). Magnetics on Silicon: An Enabling Technology for Power Supply on Chip. *IEEE Transactions on Power Electronics*, *20*(3), 585–592. doi:10.1109/TPEL.2005.846537

Milosevic, M., Allmeling, J., & Andersson, G. (2004). Interaction Between Hysteresis Controlled Inverters used in Distributed Generation Systems. In *Proc. of Power Engineering Society General Meeting* (pp.2187-2192).

Mitsubishi Electric. (1998). *General Considerations for IGBT and Intelligent Power Modules*. Retrieved from http://www.mitsubishielectric.com

Mohan, N., Undeland, T., & Robbins, W. (2003). *Power Electronics: Converters, Applications, and Design* (3rd Ed.) John Wiley & Sons.

Molinas, M., Suul, J., & Undeland, T. (2008). Low Voltage Ride Through of Wind Farms With Cage generators: STATCOM Versus SVC. *IEEE Transactions on Power Electronics*, *23*(3), 1104–1117. doi:10.1109/TPEL.2008.921169

Monmasson, E., & Cirstea, M.FPGA Design Methodology for Industrial Control Systems – A Review. *IEEE Transactions on Industrial Electronics*, *54*(4), 1824–1842.

Monti, A., Santi, E., Dougal, R., & Riva, M. (2003). Rapid Prototyping of Digital Controls for Power Electronics. *IEEE Transactions on Power Electronics*, *18*(3), 915–923. doi:10.1109/TPEL.2003.810864

Moursi, M., Joos, G., & Abbey, C. (2008). A Secondary Voltage Control Strategy for transmission Level Interconnection of Wind Generation. *IEEE Transactions on Power Electronics*, *23*(3), 1178–1190. doi:10.1109/TPEL.2008.921195

Naouar, M., Monmason, E., & Naassani, A., Slama-Belkhodja, & Patin, N. (2007). FPGA-Based Current Controllers for AC Machine Drives- A Review. *IEEE Transactions on Industrial Electronics*, *54*(4), 1907–1925. doi:10.1109/TIE.2007.898302

O'Sullivan, D., Egan, M., & Willers, M. (2009). A Family óf Single-Stage Resonant AC/DC Converters With PFC. *IEEE Transactions on Power Electronics*, *24*(2), 399–408. doi:10.1109/TPEL.2008.2005521

Park, J., Kwon, J., Kim, E., & Kwon, B. (2008). High-Performance Transformerless Online UPS. *IEEE Transactions on Industrial Electronics*, *55*(8), 2943–2953. doi:10.1109/TIE.2008.918606

Park, S., Chen, C., Lai, J., & Moon, S. (2008). Admittance Compensation in Current Loop Control for a Grid-tie LCL Fuel Cell Inverter. *IEEE Transactions on Power Electronics*, *23*(4), 1716–1723. doi:10.1109/TPEL.2008.924828

Patel, N., & Madawala, U. (2009). A Bit-Stream-Based PWM Technique for Sine-Wave Gemeration. *IEEE Transactions on Industrial Electronics*, *56*(7), 2530–2539. doi:10.1109/TIE.2009.2021682

Peng, F., & Fang, X. (2007). Novel Three-Phase Current-Fed Z-Source AC/AC Converter. In *Proc. of PESC 2007* (pp. 2993-2996).

Perez, M., Cortes, P., & Rodriquez, J. (2008). Predictive Control Algorithm Technique for Multilevel Asymetric Cascaded H-Bridge Inverters. *IEEE Transactions on Industrial Electronics*, *55*(12), 4354–4361. doi:10.1109/TIE.2008.2006948

Petrone, G., Spagnuolo, G., Teodorescu, R., Veerachary, M., & Vitelli, M. (2008). Reliability Issues in Photovoltaic Power Processing Systems. *IEEE Transactions on Industrial Electronics*, *55*(7), 2569–2580. doi:10.1109/TIE.2008.924016

Qahoug, J., Huang, L., & Huard, D. (2008). Sensorless Current Sharing Analysis and Scheme For Multiphase Converters. *IEEE Transactions on Power Electronics*, *23*(5), 2237–2247. doi:10.1109/TPEL.2008.2001897

Qiao, C., & Smedley, K. (2003). Three-Phase Unity-Power Factor Star-Connected Switch (VIENNA) Rectifier With Unified Constant-Frequency Integration Control. *IEEE Transactions on Power Electronics*, *18*(4), 952–957. doi:10.1109/TPEL.2003.813759

Rodriguez, C., & Amaratunda, G. (2008). Long-Lifetime Power Inverter for Photovoltaic AC Modules, Guest editorial. *IEEE Transactions on Industrial Electronics*, *55*(7), 2593–2601. doi:10.1109/TIE.2008.922401

Rodriguez, E., Vazquez, N., Hernandez, C., & Correa, J. (2008). A Novel AC UPS With High Power Factor and Fast Dynamic Response. *IEEE Transactions on Industrial Electronics*, *55*(8), 2963–2973. doi:10.1109/TIE.2008.918484

Rodriguez, F., Giron, C., Bueno, E., Hernandez, A., Cobreces, S., & Martin, P. (2009). Remote Laboratory for Experimentation With Multilevel Power Converters. *IEEE Transactions on Industrial Electronics*, *56*(7), 2450–2463. doi:10.1109/TIE.2009.2017493

Rodriquez, J., Dixon, J., Espinoza, J., Pontt, J., & Lezana, P. (2005). PWM Regenerative Rectifiers: State of the Art. *IEEE Transactions on Industrial Electronics*, *52*(1), 5–22. doi:10.1109/TIE.2004.841149

Ropp, M., & Gonzales, S. (2009). Development of a MATLAB/ Simulink Model of a Single-Phase Grid- Connected Photovoltaic System. *IEEE Transactions on Energy Conversion*, *24*(1), 195–202. doi:10.1109/TEC.2008.2003206

Sabahi, M., Hosseini, S., Sharifian, M., Goharrizi, A., & Gharekpetian, G. (2009). A three-Phase Dimmable Lighting System Using a Bidirectional Power Electronic Transformer. *IEEE Transactions on Power Electronics*, *24*(3), 830–837. doi:10.1109/TPEL.2008.2010344

Sahan, B., & Vergara, A. Henze, N. Engler & Zacharias, P. (2008). A Single-State PV Module Integrated Converter Based on a Low-Power Current-Source Inverter. *IEEE Transactions on Industrial Electronics*, *55*(7), 2603–2609. doi:10.1109/TIE.2008.924160

Saleh, S., Mo;oney, C. & Rahman, A. (2009). Development and Testing of Wavelet Modulation for Single-Phase Inverters. *IEEE Transactions on Industrial Electronics*, *56*(7), 2588–2599. doi:10.1109/TIE.2009.2019776

Selvaraj, J., & Rahim, N. (2009). Multilevel Inverter For Grid-Connected PV System Employing Digital PI Controller. *IEEE Transactions on Industrial Electronics*, *56*(1), 149–158. doi:10.1109/TIE.2008.928116

Shtessel, Y., Baev, S., & Biglari, H. (2008). Unity Power Factor Control in Three-Phase AC/DC Boost Converter Using Sliding Modes. *IEEE Transactions on Industrial Electronics*, *55*(11), 3875–3882. doi:10.1109/TIE.2008.2003203

Spagnuolo, G., Petrone, G., Teodoresku, R., Veerachary, M., & Vitelli, M. (2008). Special Section on Photovoltaic Power Processing Systems, Guest editorial. *IEEE Transactions on Industrial Electronics*, *55*(7), 2566. doi:10.1109/TIE.2008.924445

Stiebler, M. (2008). *Wind Energy Systems for Electric Power Generation*. Springer.

Tao, H., Duarte, J., & Hendrix, M. (2008). Line-Interactive UPS Using a Fuel Cell as the Primary Source. *IEEE Transactions on Industrial Electronics*, *55*(8), 3012–3021. doi:10.1109/TIE.2008.918472

Tenca, P., Rockhill, A., Lipo, T., & Tricoli, P. (2008). Current Source Topology for Wind Turbines With decreased mains Current Harmonics, Further Reducible via Functional Minimization. *IEEE Transactions on Power Electronics*, *23*(3), 1143–1155. doi:10.1109/TPEL.2008.921183

Tunlasakun, K. (2008). PLD-s Based Islanding detection for Grid Connected Inverters. *International Journal of Electrical System Science and Engineering*, *1*(4), 236–240.

Van Wyk, J., Lee, F., Liang, Z., Chen, R., Wang, S., & Lu, B. (2005). Integrating Active, Passive and EMI-Filter Functions in Power Electronics Systems: A Case Study of Some technologies. *IEEE Transactions on Power Electronics*, *20*(3), 523–536. doi:10.1109/TPEL.2005.846553

Vargas, R., Amman, U., & Rodriquez, J. (2009). Predictive Approach to Increase Efficiency and Reduce Switching Losses on Matrix Converters. *IEEE Transactions on Power Electronics*, *24*(4), 894–902. doi:10.1109/TPEL.2008.2011907

Vazquez, N., Villegas-Saucillo, J., Hernandez, C., Rodriguez, E., & Arau, J. (2008). Two-State Uninterruptible Power Supply With High Power Factor. *IEEE Transactions on Industrial Electronics*, *55*(8), 2954–2962. doi:10.1109/TIE.2008.918475

Villablanka, M., Nadal, J., & Bravo, M. (2007). A 12-Pulse AC-DC Rectifier With High-Quality Input/Output Waveforms. *IEEE Transactions on Power Electronics*, *22*(5), 1875–1881. doi:10.1109/TPEL.2007.904185

Wang, C. (2008). A Novel ZCS-PWM Flyback Converter With a Simple ZCS-PWM Commutation Cell. *IEEE Transactions on Industrial Electronics*, *55*(2), 749–757. doi:10.1109/TIE.2007.911917

Wang, J., Huang, A., Sung, W., Liu, Y. & Baliga, B. (2009). Smart grid technologies. *IEEE Industrial Electronics magazine, 3*(2), 16-23.

Yan, W., Hu, J., Utkin, V., & Xu, L. (2008). Sliding Mode Pulse-width Modulation. *IEEE Transactions on Power Electronics*, *23*(2), 619–626. doi:10.1109/TPEL.2007.915775

Yao, J., Li, H., Liao, Y., & Chen, Z. (2008). An Improved Control Strategy of Limiting DC-Link Voltage Fluctuation for a Doubly fed Induction Wind Generator. *IEEE Transactions on Power Electronics*, *23*(3), 1205–1213. doi:10.1109/TPEL.2008.921177

Youssef, N., Al-Haddad, K., & Kanaan, H. (2008). Real-time implementation of a discrete nonlinearity conpensating multiloops control technique for a 1,5 –kw three-phase/switch/level Vienna converter. *IEEE Transactions on Industrial Electronics*, *55*(3), 1225–1234. doi:10.1109/TIE.2007.910630

Zaragoza, J., Pou, J., Ceballos, S., Robles, E., Ibanez, P., & Villate, J. (2009). A Comprehensive Study of a Hybrid Modulation Technique for the Neutral-Point-Clamped Converter. *IEEE Transactions on Industrial Electronics*, *56*(2), 294–304. doi:10.1109/TIE.2008.2005132

Zeng, Q., & Chang, L. (2008). An advanced SVPWM-based predictive current controller for three-phase inverters in distributed generation systems. *IEEE Transactions on Industrial Electronics*, *55*(3), 1235–1246. doi:10.1109/TIE.2007.907674

Zhou, Z., Zhang, X., & Shen, W. (2008). Single-phase uninterruptible power supply based on z-source inverter. *IEEE Transactions on Industrial Electronics*, *55*(3), 2997–3004. doi:10.1109/TIE.2008.924202

Appendix: Fourier Analysis and Total Harmonic Distortion (THD) of Waveforms

For the purposes of the harmonic analysis of some functions, characterizing the basic variables of the power electronic converters, at first an auxiliary function shown in Fig.A.1 is examined. This allows using a unified approach appropriate for the basic waveforms of some variables at these converters.

The auxiliary function consists of positive and negative pulses with amplitude 1 and duration *w*. If the beginning of the coordinate system is the middle of a positive pulses, then the coefficients of Fourier series for the positive pulses are:

$$A_0 = \frac{1}{2\pi}\int_{-\frac{w}{2}}^{\frac{w}{2}} 1.d\vartheta = \frac{w}{2.\pi} \quad \text{(A.1)}$$

$$A_k = \frac{1}{\pi}\int_{-\frac{w}{2}}^{\frac{w}{2}} 1.cos\left(k.\vartheta\right).d\,\vartheta = \frac{2}{k.\pi} sin\frac{k.w}{2} \quad \text{(A.2)}$$

Also, the Fourier series for the positive pulses is:

Figure A.1. Auxiliary function waveform

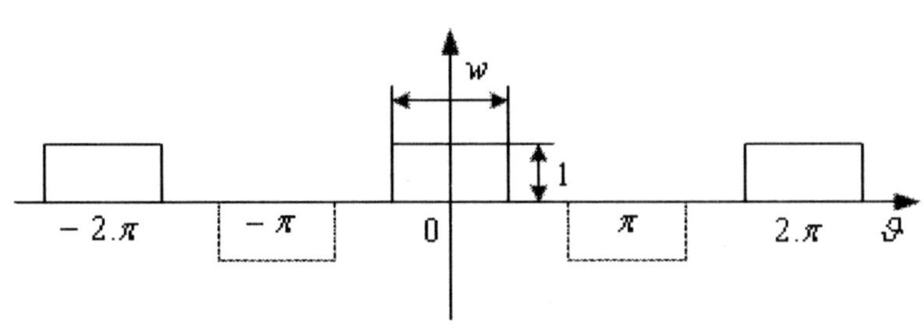

$$F_P = \frac{2}{\pi}\left(\frac{w}{4} + sin\frac{w}{2}cos\vartheta + \frac{1}{2}sin\frac{2.w}{2}cos\,2.\,\vartheta + \frac{1}{3}sin\frac{3.w}{2}cos\,3.\,\vartheta + \frac{1}{4}sin\frac{4.w}{2}cos\,4.\,\vartheta + \frac{1}{5}sin\frac{5.w}{2}cos\,5.\,\vartheta ...\right) \quad (A.3)$$

Analogically, the Fourier series for the negative pulses is found, and, it is:

$$F_P = \frac{2}{\pi}\left(-\frac{w}{4} + sin\frac{w}{2}cos\vartheta - \frac{1}{2}sin\frac{2.w}{2}cos\,2.\,\vartheta + \frac{1}{3}sin\frac{3.w}{2}cos\,3.\,\vartheta - \frac{1}{4}sin\frac{4.w}{2}cos\,4.\,\vartheta + \frac{1}{5}sin\frac{5.w}{2}cos\,5.\,\vartheta ...\right) \quad (A.4)$$

Summing (A.3) and (A.4), the Fourier series for the auxiliary function is found as:

$$F = \frac{4}{\pi}\left(sin\frac{w}{2}cos\vartheta + \frac{1}{3}sin\frac{3.w}{2}cos\,3.\,\vartheta + \frac{1}{5}sin\frac{5.w}{2}cos\,5.\,\vartheta + ...\right) \quad (A.5)$$

Fig.A.2 depicts three waveforms typical for the power electronic converters.

Using equation (A.5), the harmonic content of the three functions may be determined consistently.

For Fig.A.2.a – taking in consideration that the pulse amplitude is F, and their duration is $w = \pi$

$$f = \frac{4.F}{\pi}.\left(cos\vartheta - \frac{1}{3}cos\,3.\,\vartheta + \frac{1}{5}cos\,5.\,\vartheta - ...\right) \quad (A.6)$$

Figure A.2. Waveforms typical for the power electronic converters, a) source current in the rectifiers shown in Chapter 4, Figure 15 and 17; output voltage in single-phase uncontrolled inverter – Chapter 7, Figure 6. b) source current in the rectifiers shown in Chapter 4, Figure 16 and Figure 18; output voltage in single-phase controlled inverter – Chapter 7, Figure 6. c) source current for a phase in the rectifiers shown Chapter 4, Figure 21; phase-to-phase voltage in three-phase inverter – Chapter 7, Figure 18.

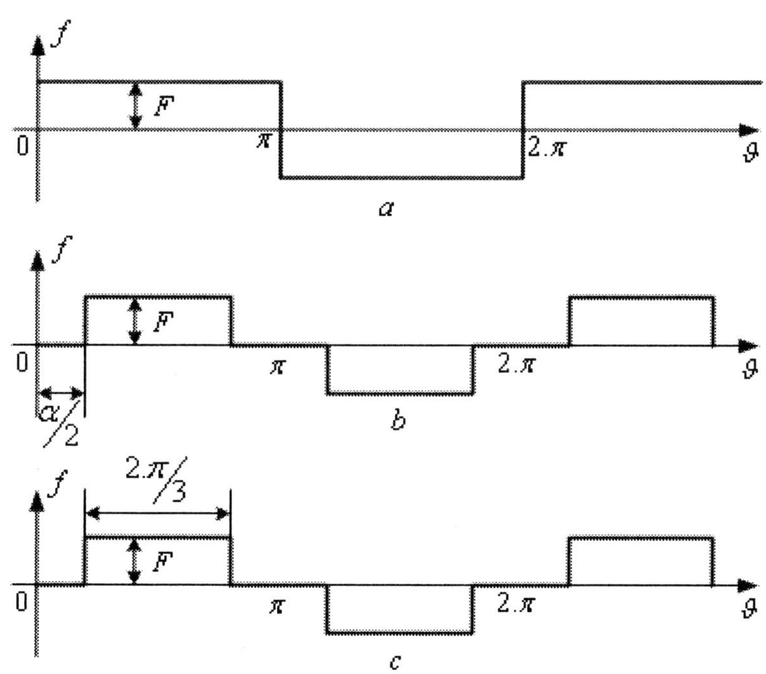

Fig.A.3 shows the distribution of the most important harmonics.

Equation (A.6) shows the relationship between the values of the higher harmonics and this of the first harmonic as:

$$\frac{F_k}{F_1} = \frac{1}{k} \qquad \text{(A.7)}$$

For Fig.A.2.b - taking in consideration that the pulse amplitude is F, and their duration is $w = \pi - \alpha$

$$f = \frac{4.F}{\pi}\left(sin\frac{\pi - \alpha}{2} cos\vartheta + \frac{1}{3} sin\frac{3.(\pi - \alpha)}{2} cos3.\vartheta + \frac{1}{5} sin\frac{5.(\pi - \alpha)}{2} cos5.\vartheta + ...\right) \qquad \text{(A.8)}$$

The numbers of the harmonics are the same as in the previous case but their percentage content is dependent on the firing angle α . The relationship between the value of the kth higher harmonic and this of the first harmonics is:

$$\frac{F_k}{F_1} = \frac{1}{k} . \frac{\sin\frac{k.(\pi - \alpha)}{2}}{\sin\frac{\pi - \alpha}{2}} \qquad \text{(A.9)}$$

For Fig.A.2.c - taking in consideration that the pulse amplitude is F, and their duration is $w = \frac{2.\pi}{3}$

$$f = \frac{2.\sqrt{3}.F}{\pi} . \left(cos\vartheta - \frac{1}{5} cos5.\vartheta + \frac{1}{7} cos7.\vartheta - \frac{1}{11} cos11.\vartheta - \frac{1}{13} cos13. ..\vartheta \right) \qquad \text{(A.10)}$$

For this case, Fig.A.4 depicts the distribution of the most important harmonics.

Figure A.3. Distribution of harmonics of the function shown in Figure A.2.a

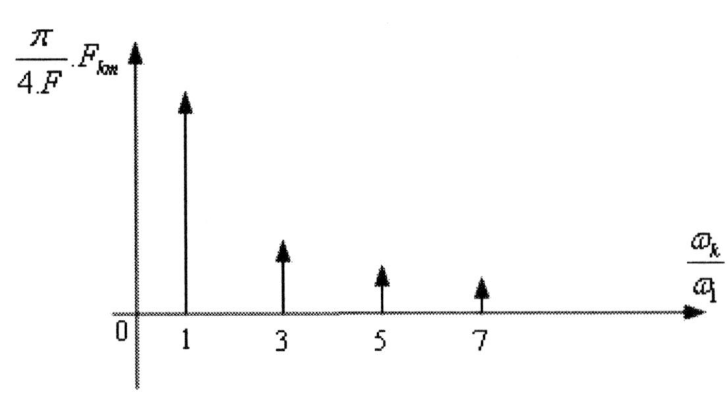

Equation (A.10) shows that here again the relationship between the values of the higher harmonics and this of the first harmonic is valid found as (A.7).

The equations found may be used to find the harmonic coefficient K_H (see (4.7), which is also called "total harmonic distortion (THD)", for the examined typical functions.

At first, this is made for the function shown in Fig.A.2.b. For this purpose, the effective value of the function is found as:

$$F_{RMS} = \sqrt{\frac{1}{\pi}\int_0^{\pi-\alpha} F^2.d\vartheta} = F\sqrt{1-\frac{\alpha}{\pi}} \qquad \text{(A.11)}$$

The effective value f the first harmonic using (A.8) is:

$$F_1 = \frac{2.\sqrt{2}}{\pi}.F.\sin\frac{\pi-\alpha}{2} \qquad \text{(A.12)}$$

The distortion coefficient for the current waveform is found form (A.11) and (A.12):

$$\nu = \frac{F_1}{F_{RMS}} = \frac{2.\sqrt{2}}{\sqrt{\pi.(\pi-\alpha)}}.\sin\frac{\pi-\alpha}{2} \qquad \text{(A.13)}$$

Using the equation (4.8), K_H is derived as:

$$K_H \equiv THD = \frac{1}{\nu}.\sqrt{1-\nu^2} = 100.\sqrt{\frac{\pi.(\pi-\alpha)}{8.\sin^2\left(\frac{\pi-\alpha}{2}\right)}-1}, \quad \% \qquad \text{(A.14)}$$

An analysis of function (A.14) for an extremium may be performed and the minimum value of K_H is found to be 29% at $\alpha = 46.44^\circ$.

Using (A.14), the value of K_H for the function shown in Fig.A.2.a at $\alpha = 0$ is found to be:

Figure A.4. Distribution of harmonics of the function shown in Figure A.2.c

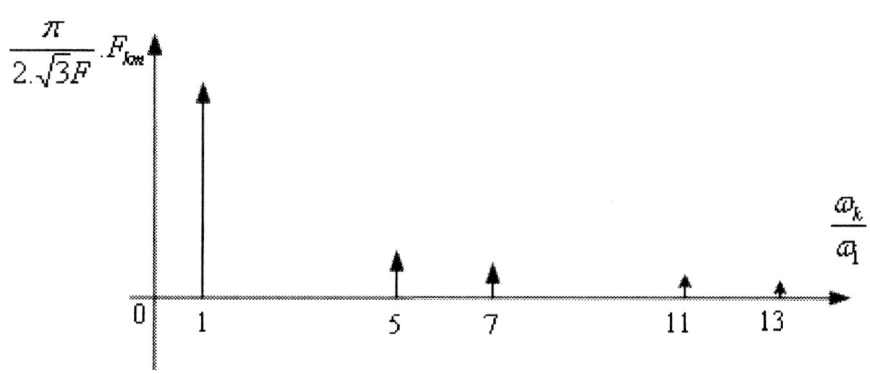

$$K_H = 100.\sqrt{\frac{\pi^2}{8} - 1} = 48.34\% \quad \text{(A.15)}$$

For the function shown in Fig.A.2.c $\alpha = \frac{\pi}{3}$, and again using (A.14), it is found:

$$K_H = 100.\sqrt{\frac{\pi^2}{9} - 1} = 31.08\% \qquad \text{(A.16)}$$

Thus, equations for Fourier series and for harmonic coefficient have been found for the most typical waveforms of the power electronic converters.

In the cases of some more complicated waveforms of certain variables (for example, sinusoidal pulse-width modulation, selective elimination of harmonics see Chapter 7, Figures 9, 11, 13, 22), as well as in the case of several leveled modulation (see Chapter 7, Figure 16), a particular approach is applied in solving the above-stated topics.

Only an example will be examined here – sinusoidal pulse-width modulation shown in Fig.A.5. The waveform shown is typical for a single-phase voltage inverter (see Chapter 5, Figure 10) and also for the source current of the single-phase rectifier(see Chapter 4, Figure 41.b).

The duration of the i-th conductivity interval is defined from the equation:

$$\Delta x_i = \frac{\pi}{N+1} \cdot M \cdot \sin\left(i \cdot \frac{\pi}{N+1}\right); \quad i = 1, 2, \ldots N, \qquad \text{(A.17)}$$

where in $0 \le M \le 1$ is a modulation coefficient.

The turn on and off angles of the switches are:

$$\begin{aligned} \alpha_i &= i \cdot \frac{\pi}{N+1} - \frac{1}{2} \cdot \frac{\pi}{N+1} \cdot M \cdot \sin\left(i \cdot \frac{\pi}{N+1}\right); \quad i = 1, 2, \ldots N \\ \beta_i &= i \cdot \frac{\pi}{N+1} + \frac{1}{2} \cdot \frac{\pi}{N+1} \cdot M \cdot \sin\left(i \cdot \frac{\pi}{N+1}\right); \quad i = 1, 2, \ldots N \end{aligned} \qquad \text{(A.18)}$$

Figure A.5. Sinusoidal pulse-width modulation

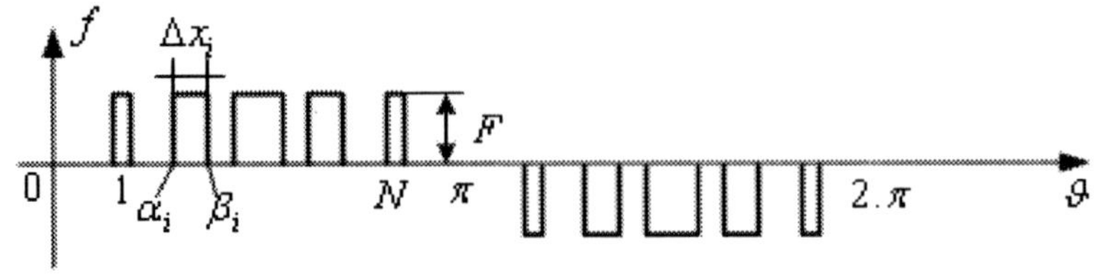

Using Figure A.5, the maximum value of the k-th harmonic may be determined:

$$F_{km} = \frac{2}{\pi} = \sum_{i=1}^{N} \int_{\alpha_i}^{\beta_i} F.sin(n\vartheta).d\,\vartheta \qquad \text{(A.19)}$$

After solving (A.19), the following equation is found:

$$\frac{F_{km}}{F} = \frac{2}{n\pi} \sum_{i=1}^{N} (\cos n.\alpha_i - \cos n.\beta_i) \qquad \text{(A.20)}$$

After substitution using (A.18), an equation for the relative maximum value of the k-th harmonic is found as:

$$\frac{\pi.F_{km}}{4.F} = \frac{1}{n} \sum_{i=1}^{N} \sin\left(i\frac{n\pi}{N+1}\right).\sin\left[\frac{n\pi}{2.(N+1)}.M.\sin\left(i\frac{\pi}{N+1}\right)\right] \qquad \text{(A.21)}$$

At linear pulse-width modulation (see Chapter 4, Figure 41.a), the relationships already found may be uses, having in mind that:

$$\Delta x_i = \Delta x = \frac{\pi}{N+1} \cdot M; \quad i = 1,2,...N \qquad \text{(A.22)}$$

Therefore:

$$\alpha_i = \frac{\pi}{N+1} \cdot \left(i - \frac{M}{2}\right); \quad i = 1,2,...N$$

$$\beta_i = \frac{\pi}{N+1} \cdot \left(i + \frac{M}{2}\right); \quad i = 1,2,...N \qquad \text{(A.23)}$$

After substituting in (A.20), it is found:

$$\frac{\pi.F_{km}}{4.F} = \frac{1}{n} \sum_{i=1}^{N} \sin\left(i\frac{n\pi}{N+1}\right).\sin\left(\frac{n\pi}{2.(N+1)}.M\right) \qquad \text{(A.24)}$$

Figure A.6 depicts the distribution of the harmonics at sinusoidal pulse-width modulation corresponding to the equation (A.21), while Fig.A.7 – at linear pulse-width modulation corresponding to the equation (A.24). The graphics are obtained using *MATLAB*[1] (see Chapter 3). The relationships shown are made for $N = 20$, $M = 0.9$.

The following example illustrates the possibility to made harmonic analysis using *ORCAD*[2] simulation software. The operation of a single-phase voltage inverter controlled according to the method shown in Chapter 7, Figure 11 is simulated. The schematic for the computer simulation is shown in Fig.A.8. The effective value gained of the first harmonic of the output voltage is 230V at active-inductive load of 10Ω, 10mH. The frequency of the saw voltage is 1 kHz (see Chapter 7, Figures 11 and 12). The ratio of the amplitude of the reference sinusoidal waveforms to the amplitude of the saw voltage is 0.9. These

Figure A.6. Distribution of the harmonics at sinusoidal pulse-width modulation

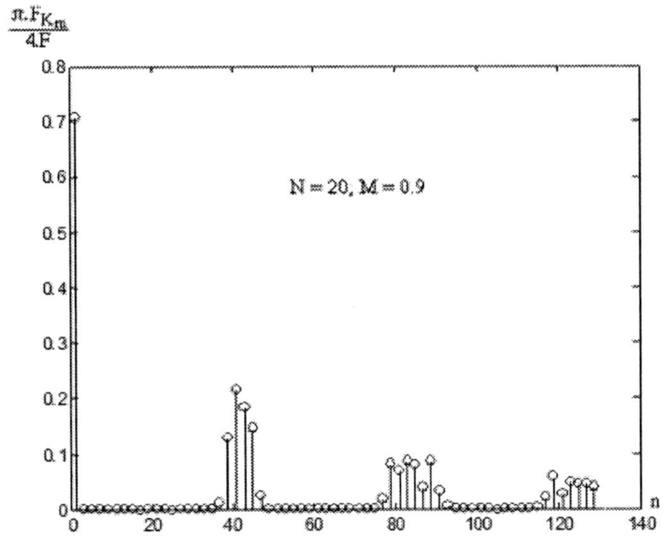

data correspond to the output data used to gain the graphical relationship shown in Figure A.6 using *MATLAB*[1].

The waveforms of the inverter output voltage, as well as the sinusoidal voltage across the load, are shown in Figure A.9.

Figure A.10 depicts the harmonic spectrum of the inverter output voltage.

Figure A.7. Distribution of the harmonics at linear pulse-width modulation

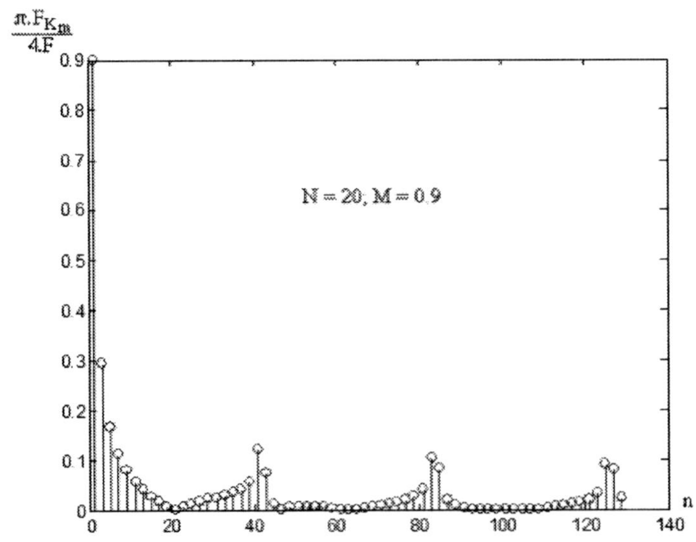

Figure A.8. Computer simulation schema

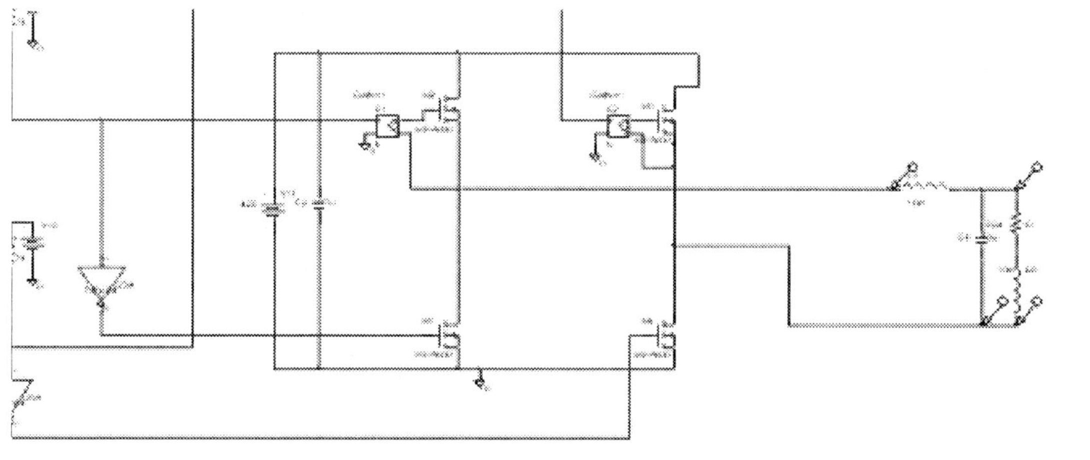

ENDNOTES

[1] MATLAB is registered trademark of the MathWorks Inc.

[2] ORCAD, CAPTURE, PSPICE and PROBE are copyright by the Cadence Design Systems, Inc.

Figure A.9. The inverter output voltage and the load voltage

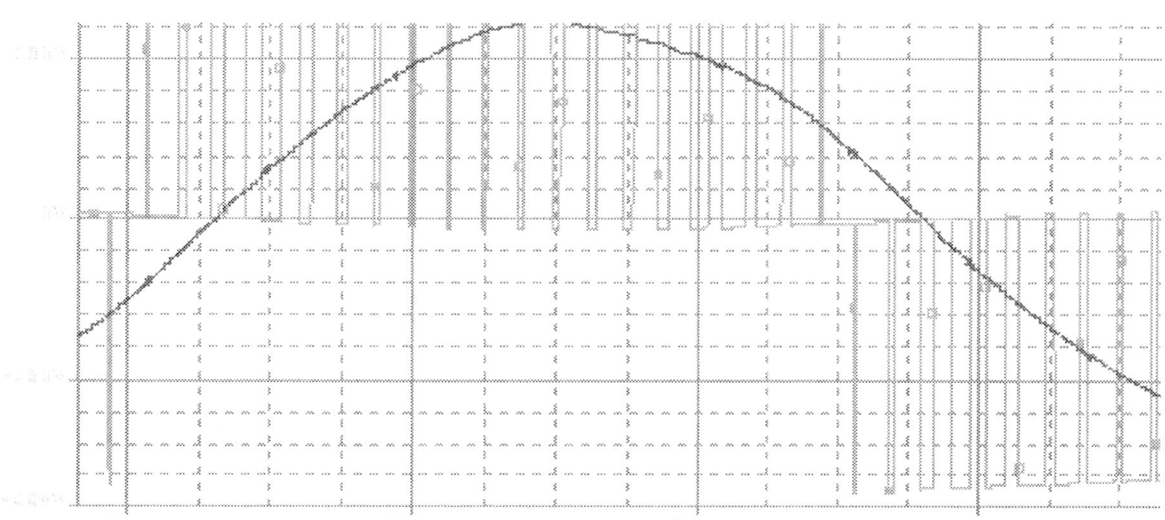

Figure A.10. The harmonic spectrum of the inverter output voltage

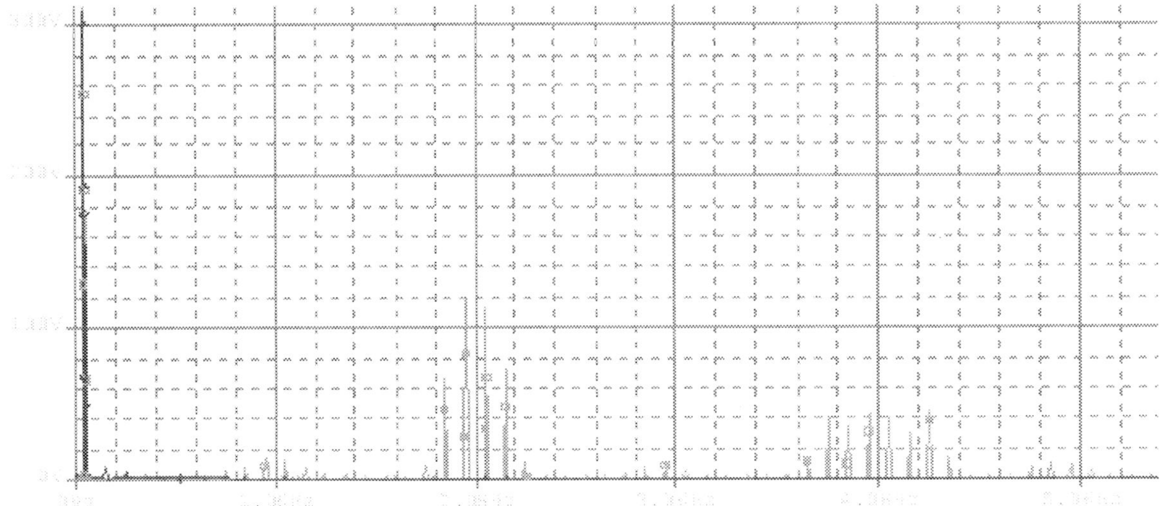

About the Author

Mihail Antchev received BS and MS degrees in electronic engineering with highest distinction from Technical University, Sofia, Bulgaria. He also received his PhD degree in power electronics from Technical University, Sofia, Bulgaria. His PhD thesis was focused on power factor improvement in controlled rectifiers. Dr. Antchev is a professor of power electronics in the Technical University, Sofia, Bulgaria. He has published 70 journal and conference papers, 5 books and also he is a reviewer of IEEE journal and conference papers. He was listed in the *Who's Who in the World* 24th Edition. His areas of interest include design and control of power electronic converters and systems, active power filters and UPS systems.

Index

F

G

H

I

L

M

N

O

P

R

S